Recent Progress in Controlling Chaos

SERIES ON STABILITY, VIBRATION AND CONTROL OF SYSTEMS

Founder and Editor: Ardéshir Guran
Co-Editors: M. Cloud & W. B. Zimmerman

About the Series

Rapid developments in system dynamics and control, areas related to many other topics in applied mathematics, call for comprehensive presentations of current topics. This series contains graduate level textbooks, monographs, and a collection of thematically organized research or pedagogical articles addressing key topics in applied dynamics.

The material is ideal for a general scientific and engineering readership, and is also mathematically precise enough to be a useful reference for research specialists in mechanics and control, nonlinear dynamics, and in applied mathematics and physics.

Reporting on academic/industrial research from institutions around the world, the SVCS series reflects technological advances in mechanics and control. Particular emphasis is laid on emerging areas such as modeling of complex systems, bioengineering, mechatronics, structronics, fluidics, optoelectronic sensors, micromachining techniques, and intelligent system design.

Selected Volumes in Series A

Vol. 12 The Calculus of Variations and Functional Analysis: With Optimal Control and Applications in Mechanics
Authors: L. P. Lebedev and M. J. Cloud

Vol. 13 Multiparameter Stability Theory with Mechanical Applications
Authors: A. P. Seyranian and A. A. Mailybaev

Vol. 14 Stability of Stationary Sets in Control Systems with Discontinuous Nonlinearities
Authors: V. A. Yakubovich, G. A. Leonov and A. Kh. Gelig

Vol. 15 Process Modelling and Simulation with Finite Element Methods
Author: W. B. J. Zimmerman

Vol. 16 Design of Nonlinear Control Systems with the Highest Derivative in Feedback
Author: V. D. Yurkevich

Vol. 17 The Quantum World of Nuclear Physics
Author: Yu. A. Berezhnoy

Selected Volume in Series B

Vol. 14 Impact and Friction of Solids, Structures and Intelligent Machines
Editor: A. Guran

Vol. 15 Advances in Mechanics of Solids: In Memory of Prof. E. M. Haseganu
Editors: A. Guran, A. L. Smirnov, D. J. Steigmann and R. Vaillancourt

SERIES ON STABILITY, VIBRATION AND CONTROL OF SYSTEMS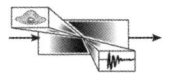

Series B   **Volume 16**

Founder & Editor: **Ardéshir Guran**

Co-Editors: **M. Cloud & W. B. Zimmerman**

Recent Progress in Controlling Chaos

Miguel A F Sanjuán
Universidad Rey Juan Carlos, Spain

Celso Grebogi
University of Aberdeen, UK

World Scientific

NEW JERSEY · LONDON · SINGAPORE · BEIJING · SHANGHAI · HONG KONG · TAIPEI · CHENNAI

Published by

World Scientific Publishing Co. Pte. Ltd.
5 Toh Tuck Link, Singapore 596224
USA office: 27 Warren Street, Suite 401-402, Hackensack, NJ 07601
UK office: 57 Shelton Street, Covent Garden, London WC2H 9HE

British Library Cataloguing-in-Publication Data
A catalogue record for this book is available from the British Library.

Series on Stability, Vibration and Control of Systems, Series B
RECENT PROGRESS IN CONTROLLING CHAOS — Vol. 16

Copyright © 2010 by World Scientific Publishing Co. Pte. Ltd.

All rights reserved. This book, or parts thereof, may not be reproduced in any form or by any means, electronic or mechanical, including photocopying, recording or any information storage and retrieval system now known or to be invented, without written permission from the Publisher.

For photocopying of material in this volume, please pay a copying fee through the Copyright Clearance Center, Inc., 222 Rosewood Drive, Danvers, MA 01923, USA. In this case permission to photocopy is not required from the publisher.

ISBN-13 978-981-4291-69-9
ISBN-10 981-4291-69-2

Printed in Singapore by Mainland Press Pte Ltd.

Preface

The field of Nonlinear Dynamics and Complex Systems is a very active research field in which many scientists from many different fields are contributing to its growth and development.

Among the many different and interesting problems covered in this field, we mention those in chaotic dynamics, which manifest themselves in many different scenarios in the natural and technical sciences. And the attempts of scientists to control this chaotic dynamics has contributed to a research line known as chaos control. Research on chaos control was strongly influenced by the seminal paper of Edward Ott, Celso Grebogi and James Yorke. This paper was chosen by the American Physical Society as a milestone in the last fifty years. It opened up a whole new area of research, changing philosophically our way of thinking about chaos. Previously, it was strongly believed by scientists that chaos was intrinsically uncontrollable. We understand now that chaos not only can be controlled and manipulated but that chaos can give a great inherent flexibility in choosing a large number of controllable states by applying tiny perturbations to the system under consideration. Since that publication, a lot of progress has been done in this topic, through the implementation of different techniques and algorithms with a lot of applications as well.

This present book, *Recent Progress in Controlling Chaos*, contains 15 contributions of world-renowned scientists and experts in the field of controlling chaos, and the editors of this volume are well known scientists in the field of Nonlinear Dynamics, Chaos and Complex Systems, in particular, Chaos Control. Among the topics covered are techniques used for the reduction of chaotic transport in magnetized plasmas, methods of control applied to chaotic neural networks, different methods of adaptive feedback, pulsive and delayed feedback control techniques, phase control technique, applications of control techniques to traffic, mechanical structures and neuron networks, the method of partial control of chaos useful for controlling transient chaos in presence of noise and certain applications to celestial mechanics.

Our main goal is to offer readers a broad view of the more recent progress in the field. Besides the progress in key techniques and concepts in the field of control, some of the contributions are also very useful for those who look for applications in Science and Technology. The contributors have been carefully selected and include researchers from different areas and background. Some authors come from a more fundamental and basic background while others, a more applied and engineering background.

The book will certainly be of much use for graduate students and scientists working in Nonlinear Dynamics, Chaos and Complex Systems, in general. It provides the attractiveness for the study of the complex nonlinear systems and it brings new results and ideas to students and researchers.

Miguel A. F. Sanjuán and Celso Grebogi
Editors
Móstoles, Madrid and Aberdeen, 5 May 2009

Contents

Preface   v

1. Reduction of the Chaotic Transport of Impurities in Turbulent Magnetized Plasmas   1

   C. Chandre, G. Ciraolo and M. Vittot

2. Controlling Chaos in a Chaotic Neural Network   21

   G. He, P. Zhu, J. Kuroiwa and K. Aihara

3. Adaptive Feedback Control of Periodic Orbits in Chaotic Systems   45

   H. Ando, S. Boccaletti and K. Aihara

4. Feedback Anti-control of Chaos   73

   Y. Shi and G. Chen

5. Delayed Feedback Control Techniques   103

   K. Pyragas and V. Pyragas

6. Phase Control in Nonlinear Systems   147

   S. Zambrano, J. M. Seoane, I. P. Mariño, M. A. F. Sanjuán and R. Meucci

7. Recent Advances in Control of Complex Dynamics in Mechanical and Structural Systems   189

   G. Rega and S. Lenci

8. Clipping Chaos to Cycles 239
   S. Sinha

9. A Minimal Model of City Traffic: Chaos, Critical
   Behavior and Control 267
   J. A. Valdivia, B. A. Toledo, V. Muñoz and J. Rogan

10. Controlling Chaotic Bursting in Map-based Neuron
    Network Models 291
    R. L. Viana, J. C. A. de Pontes, S. R. Lopes,
    C. A. S. Batista and A. M. Batista

11. Partial Control of Chaotic Systems 315
    S. Zambrano and M. A. F. Sanjuán

12. Continuous and Pulsive Feedback Control of Chaos 337
    G. Litak, L. M. Saha and M. Ali

13. Chaos Control 371
    L. F. R. Turci and E. E. N. Macau

14. Chaos Stabilization in the Three Body Problem 395
    A. R. Dzhanoev, A. Loskutov, J. E. Howard and
    M. A. F. Sanjuán

15. Controlling the Chaos using Fuzzy Estimation in a
    Gyrostat Satellite 409
    A. Guran

Index 427

Chapter 1

REDUCTION OF THE CHAOTIC TRANSPORT OF IMPURITIES IN TURBULENT MAGNETIZED PLASMAS

C. Chandre[1,*], G. Ciraolo[2], M. Vittot[1]

[1] *Centre de Physique Théorique, CNRS — Aix-Marseille Universités, Luminy, Case 907, F-13288 Marseille cedex 9, France*
[2] *M2P2, IMT La Jetée, Technopôle de Château Gombert, Marseille cedex 20, F-13451, France*
[*] *chandre@cpt.univ-mrs.fr*

1. Introduction

The control of transport in magnetically confined plasmas is of major importance in the long way to achieve controlled thermonuclear fusion. Two major mechanisms have been proposed for such a turbulent transport: transport governed by the fluctuations of the magnetic field and transport governed by fluctuations of the electric field. There is presently a general consensus to consider, at low plasma pressure, that the latter mechanism agrees with experimental evidence [1]. In the area of transport of trace impurities, i.e. that are sufficiently diluted so as not to modify the electric field pattern, the $\mathbf{E} \times \mathbf{B}$ drift motion of test particles should be the exact transport model. The possibility of reducing and even suppressing chaos (combined with the empirically found states of improved confinement in tokamaks) suggest to investigate the possibility to devise a strategy of control of chaotic transport through some smart perturbations acting at the microscopic level of charged particle motions.

Chaotic transport of particles advected by a turbulent electric field with a strong magnetic field is associated with Hamiltonian dynamical systems under the approximation of the guiding center motion due to $\mathbf{E} \times \mathbf{B}$ drift velocity. For an appropriate choice of turbulent electric field, it has been shown that the resulting diffusive transport is found to agree with the experimental counterpart [2, 3].

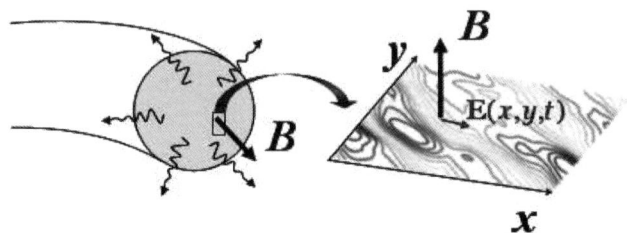

Fig. 1. The interaction between the electric turbulent field **E** and the toroidal magnetic field **B** produces the **E** × **B** motion which causes a drift of particles from the center of the plasma toward the edge. The guiding center motion of charged test particles on a transversal plane to the toroidal direction can be described by a Hamiltonian formalism with the spatial coordinates (x, y) which represents canonical conjugate variables and with the electrostatic potential V, $\mathbf{E} = -\boldsymbol{\nabla} V$, which plays the role of the Hamiltonian, as described by Eq. (1).

Here we address the turbulent transport of particles governed by the interaction of the electric turbulent potential **E** generated by the plasma itself and the strong confining magnetic field **B** produced by external coils. The resulting **E** × **B** drift motion is perpendicular to the confining magnetic field and causes losses of particles and energy from the inner region to the border of the device, with a consequent decrease of plasma temperature (see Fig. 1). The equations of this drift motion for charged test particle, in the guiding center approximation, have a Hamiltonian structure and are given by

$$\dot{\mathbf{x}} = \frac{d}{dt}\begin{pmatrix} x \\ y \end{pmatrix} = \frac{c}{B^2}\mathbf{E}(\mathbf{x},t) \times \mathbf{B} = \frac{c}{B}\begin{pmatrix} -\partial_y V(x,y,t) \\ \partial_x V(x,y,t) \end{pmatrix}, \qquad (1)$$

where $\mathbf{x} = (x, y)$ represents the spatial coordinates of the transversal section to the confining toroidal magnetic field, B the norm of the magnetic field **B**, c the velocity of light and V is the electric turbulent potential, that is $\mathbf{E} = -\boldsymbol{\nabla} V$. We notice that the resulting dynamics is of Hamiltonian nature with a pair of canonically conjugate variables (x, y) which consists of the position of the guiding center. Since the potential which plays the role of the Hamiltonian is time-dependent, it is expected that the dynamics is chaotic (with one and a half degrees of freedom).

The sensitivity of chaotic systems to small perturbations triggered a strong interdisciplinary effort to control chaos [4–11]. After the seminal work on optimal control by Pontryagin [12], new and efficient methods were proposed for controlling chaotic systems by nudging targeted trajectories

[13–18]. However, for many body experiments like the magnetic confinement of a plasma or the control of turbulent flows, such methods are hopeless due to the high number of trajectories to deal with simultaneously.

Here the control is performed with the addition of a control term f to the electric potential. It consists as a small and apt modification of the electric potential with relevant effects on the reduction of chaotic particle diffusion. More generally, controlling Hamiltonian systems here means to achieve the goal of suppressing chaos when it is harmful as well as to enhance chaos when it is useful, by driving the dynamical evolution of a system toward a "target behavior" by means of small and apt perturbations. For magnetic confinement devices for controlled thermonuclear fusion, the two possibilities of increasing chaos and suppressing it are of interest. In fact, in the core of a magnetic confinement device the aim is to have a stable and regular dynamics in order to induce as many fusion reactions as possible, that is a high rate of energy production. At the same time, in order to spread heat over a large area, which is fundamental both for a good recycling of energy and for the preservation of plasma facing components, at the edge of the confinement device the problem of draining highly energetic particles requires the capacity of increasing the degree of chaoticity of the dynamics.

For reducing chaos, KAM theory gives a path to integrability and the possibility of controlling Hamiltonian systems by modifying it with an apt and small perturbation which preserves the Hamiltonian structure. In fact, if on one hand it is evident that perturbing an integrable system with a generic perturbation gives rise to a loss of stability and to the break-up of KAM tori, on the other hand the structure of the Hamiltonian system changes on the set of KAM tori with a continuity C^∞ [19] with respect to the amplitude of the perturbation. The idea is that a well selected perturbation instead of producing chaos makes the system integrable or more regular, recovering structures as the KAM tori present in the integrable case.

Starting from these ideas we have addressed the problem of control in Hamiltonian systems. For a wide class of chaotic Hamiltonians, expressed as perturbations of integrable Hamiltonians, that is $H_0 + \varepsilon V$, the aim is to design a control term f such that the dynamics of the controlled Hamiltonian $H_0 + \varepsilon V + f$ has more regular trajectories (e.g. on invariant tori) or less chaotic diffusion than the uncontrolled one. Obviously $f = -\varepsilon V$ is a solution since the resulting Hamiltonian is integrable. However, it is a useless solution since the control is of the same magnitude of the perturbation while, for energetic purposes, the desired control term should be small with respect to the perturbation εV. For example the control term should be of order ε^2 or higher. In many physical situations, the control is only conceivable or interesting in some specific regions of phase space where the Hamiltonian can be known and/or an implementation is possible. Moreover, it is desirable to control the transport properties without significantly altering the original setup of the system under investigation nor its overall chaotic structure. The possibility of recreating specific regular structures in phase space such as barriers to diffusion, can be used to bound the motion of particles without changing the phase space on both sides of the barrier.

In addition, the construction of the control term has to be robust. In fact, one expects from a KAM theorem also a given robustness with respect to modifications of the well selected and regularizing perturbation. This fact still comes from the previous results that we mentioned above about the transition from integrability to chaos that Hamiltonian systems generically show. In other words, if an apt modification of a chaotic Hamiltonian gives rise to a regularization of the dynamics, with for example the creation of KAM tori, one expects that also some approximations of this apt modification are still able to perform a relevant regularization of the system. This is a fundamental requirement for any kind of control scheme in order to guarantee the stability of the results and in view of possible experimental implementations.

2. Control Term for Hamiltonian Flows

In what follows, we explain a method to compute control terms of order ε^2 where ε is the amplitude of the perturbation, which aims at restoring the stable structures which were present in the case where $\varepsilon = 0$. We notice that such stable structures are present for Hamiltonian systems $H_0 + \varepsilon' V$ where $\varepsilon' < \varepsilon_c < \varepsilon$ (where ε_c is the threshold of break-up of the selected invariant torus), even though they are deformed (by an order ε').

For a function H, let $\{H\}$ be the linear operator such that

$$\{H\}H' = \{H, H'\},$$

for any function H', where $\{\cdot,\cdot\}$ is the Poisson bracket. The time-evolution of a function V following the flow of H is given by

$$\frac{dV}{dt} = \{H\}V,$$

which is formally solved as

$$V(t) = e^{t\{H\}}V(0),$$

if H is time independent. Let us now consider a given Hamiltonian H_0. The operator $\{H_0\}$ is not invertible since a derivation has always a non-trivial kernel (which is the set of constants of motion). Hence we consider a pseudo-inverse of $\{H_0\}$. We define a linear operator Γ such that

$$\{H_0\}^2\,\Gamma = \{H_0\}, \tag{2}$$

i.e.

$$\forall V, \quad \{H_0, \{H_0, \Gamma V\}\} = \{H_0, V\}.$$

The operator Γ is not unique (see the remark at the end of this section).

We define the *non-resonant* operator $\mathcal{N}$ and the *resonant* operator $\mathcal{R}$ as

$$\mathcal{N} = \{H_0\}\Gamma,$$
$$\mathcal{R} = 1 - \mathcal{N},$$

where the operator 1 is the identity. We notice that the range $\mathrm{Rg}\,\mathcal{R}$ of the operator $\mathcal{R}$ is included in $\mathrm{Ker}\{H_0\}$. A consequence is that $\mathcal{R}V$ is constant under the flow of H_0.

Let us now assume that H_0 is integrable with action-angle variables $(\mathbf{A}, \boldsymbol{\varphi}) \in \mathbb{B} \times \mathbb{T}^L$ where $\mathbb{B}$ is an open set of $\mathbb{R}^L$ and $\mathbb{T}^L$ is the L-dimensional torus. Thus $H_0 = H_0(\mathbf{A})$ and the Poisson bracket $\{H, H'\}$ between two elements H and H' of $\mathcal{A}$ is

$$\{H, H'\} = \frac{\partial H}{\partial \mathbf{A}} \cdot \frac{\partial H'}{\partial \boldsymbol{\varphi}} - \frac{\partial H}{\partial \boldsymbol{\varphi}} \cdot \frac{\partial H'}{\partial \mathbf{A}}.$$

The operator $\{H_0\}$ acts on V expanded as follows

$$V = \sum_{\mathbf{k} \in \mathbb{Z}^L} V_{\mathbf{k}}(\mathbf{A}) e^{i\mathbf{k}\cdot\boldsymbol{\varphi}},$$

as

$$\{H_0\}V(\mathbf{A}, \boldsymbol{\varphi}) = \sum_{\mathbf{k}} i\boldsymbol{\omega}(\mathbf{A}) \cdot \mathbf{k}\, V_{\mathbf{k}}(\mathbf{A}) e^{i\mathbf{k}\cdot\boldsymbol{\varphi}},$$

where

$$\boldsymbol{\omega}(\mathbf{A}) = \frac{\partial H_0}{\partial \mathbf{A}}.$$

A possible choice of Γ is

$$\Gamma V(\mathbf{A}, \boldsymbol{\varphi}) = \sum_{\substack{\mathbf{k} \in \mathbb{Z}^L \\ \boldsymbol{\omega}(\mathbf{A})\cdot\mathbf{k} \neq 0}} \frac{V_{\mathbf{k}}(\mathbf{A})}{i\boldsymbol{\omega}(\mathbf{A}) \cdot \mathbf{k}}\, e^{i\mathbf{k}\cdot\boldsymbol{\varphi}}. \tag{3}$$

We notice that this choice of Γ commutes with $\{H_0\}$. For a given $V \in \mathcal{A}$, $\mathcal{R}V$ is the resonant part of V and $\mathcal{N}V$ is the non-resonant part:

$$\mathcal{R}V = \sum_{\omega(\mathbf{A})\cdot \mathbf{k}=0} V_\mathbf{k}(\mathbf{A})e^{i\mathbf{k}\cdot\varphi}, \tag{4}$$

$$\mathcal{N}V = \sum_{\omega(\mathbf{A})\cdot \mathbf{k}\neq 0} V_\mathbf{k}(\mathbf{A})e^{i\mathbf{k}\cdot\varphi}. \tag{5}$$

From these operators defined from the integrable part H_0, we construct a control term for the perturbed Hamiltonian $H_0 + V$ where $V \in \mathcal{A}$, i.e. f is constructed such that $H_0 + V + f$ is canonically conjugate to $H_0 + \mathcal{R}V$. We have the following equation

$$e^{\{\Gamma V\}}(H_0 + V + f) = H_0 + \mathcal{R}V, \tag{6}$$

where

$$f = e^{-\{\Gamma V\}}\mathcal{R}V + \frac{1 - e^{-\{\Gamma V\}}}{\{\Gamma V\}}\mathcal{N}V - V. \tag{7}$$

We notice that the operator $(1 - e^{-\{\Gamma V\}})/\{\Gamma V\}$ is defined by the expansion

$$\frac{1 - e^{-\{\Gamma V\}}}{\{\Gamma V\}} = \sum_{n=0}^{\infty} \frac{(-1)^n}{(n+1)!}\{\Gamma V\}^n.$$

The control term can be expanded in power series as

$$f = \sum_{n=1}^{\infty} \frac{(-1)^n}{(n+1)!}\{\Gamma V\}^n(n\mathcal{R} + 1)V. \tag{8}$$

We notice that if V is of order ϵ, f is of order ϵ^2. Therefore the addition of a well chosen control term f makes the Hamiltonian canonically conjugate to $H_0 + \mathcal{R}V$.

If H_0 is non-resonant then with the addition of a control term f, the Hamiltonian $H_0 + V + f$ is canonically conjugate to the integrable Hamiltonian $H_0 + \mathcal{R}V$ since $\mathcal{R}V$ is only a function of the actions [see Eq. (4)]. If H_0 is resonant and $\mathcal{R}V = 0$, the controlled Hamiltonian $H = H_0 + V + f$ is conjugate to H_0. In the case $\mathcal{R}V = 0$, the series (8) which gives the expansion of the control term f, can be written as

$$f = \sum_{s=2}^{\infty} f_s, \tag{9}$$

where f_s is of order ε^s, where ε is the size of the perturbation V, and is given by the recursion formula

$$f_s = -\frac{1}{s}\{\Gamma V, f_{s-1}\}, \qquad (10)$$

where $f_1 = V$.

Remark: Non-unicity of Γ– The operator Γ is not unique. Any other choice Γ' satisfies that the range $\mathrm{Rg}(\Gamma'-\Gamma)$ is included into the kernel $\mathrm{Ker}(\{H_0\}^2)$. It is straightforward to see that adding a constant to Γ does not change the expression of the control term. However, different control terms are obtained by adding a more complex linear operator to Γ. For instance, we choose

$$\Gamma' = \Gamma + \mathcal{R}\mathcal{B},$$

where $\mathcal{B}$ is any linear operator. The operator Γ' satisfies $\{H_0\}^2\Gamma' = \{H_0\}$ since $\{H_0\}\mathcal{R} = 0$. The control term we obtain is given by Eq. (8) where Γ is replaced by Γ'. We notice that the new resonant operator defined as $\mathcal{R}' = 1 - \{H_0\}\Gamma'$ is equal to $\mathcal{R}$ since $\{H_0\}\mathcal{R} = 0$. The new control term is in general different from f. For instance, its leading order is (still of order ε^2)

$$f_2' = f_2 - \frac{1}{2}\{\mathcal{R}\mathcal{B}V, (\mathcal{R}+1)V\}.$$

This freedom in choosing the operator Γ can be used to simplify the control term (for withdrawing some Fourier coefficients) or more generally, to satisfy some additional constraints.

Remark: Higher-order control terms– It is possible to construct higher order control terms, i.e. control terms of order ε^n with $n > 2$, and there are many ways to do so. In this paragraph, we add ε in the control term for bookkeeping purposes. For instance, we notice that $H_0^{(c)} = H + \varepsilon V + \varepsilon^2 f$ is integrable. The perturbed Hamiltonian can be written as

$$H = H_0^{(c)} - \varepsilon^2 f.$$

For simplicity, we assume a non-resonant condition on V, i.e. $\mathcal{R}V = 0$. We define a new operator $\tilde{\Gamma}$ as

$$\tilde{\Gamma} = e^{-\varepsilon\{\Gamma V\}}\Gamma e^{\varepsilon\{\Gamma V\}},$$

which is ε-close to Γ. It is straightforward to check that $\tilde{\Gamma}$ satifies

$$\{H_0^{(c)}\}^2\tilde{\Gamma} = \{H_0^{(c)}\},$$

since we have $\{H_0^{(c)}\} = e^{-\varepsilon\{\Gamma V\}}\{H_0\}e^{\varepsilon\{\Gamma V\}}$ from Eq. (6). By applying again Eq. (6) and replacing V by $-\varepsilon^2 f$ and H_0 by $H_0^{(c)}$ (and consequently ε by $-\varepsilon^2$), we have the existence of a control term $\varepsilon^4 g$ which satisfies :

$$H_0 + \varepsilon V + \varepsilon^4 g = e^{\varepsilon^2\{\tilde{\Gamma} f\}}\left(H_0^{(c)} - \varepsilon^2 \tilde{\mathcal{R}} f\right),$$

where $\tilde{\mathcal{R}}$ is the resonant operator associated with $\tilde{\Gamma}$ defined as $\tilde{\mathcal{R}} = 1 - \{H_0^{(c)}\}\tilde{\Gamma}$ (and is ε-close to $\mathcal{R}$). It follows from the commutation of $\{\tilde{\mathcal{R}} V\}$ and $\{H_0^{(c)}\}$ that the controlled Hamiltonian $H_0 + \varepsilon V + \varepsilon^4 g$ is integrable. An expansion in ε of the control term g gives its leading order

$$g_4 = -\frac{\varepsilon^4}{8}\{\Gamma\{\Gamma V, V\}, \{\Gamma V, V\}\}.$$

However, we notice that $\Gamma\{\Gamma V, V\}$ introduces additional small denominators. It is not obvious that, given a value of ε, the control term $\varepsilon^4 g$ is smaller than $\varepsilon^2 f$ (using a standard norm of functions).

3. Reduction of Chaotic Transport for E × B Drift in Magnetized Plasmas

3.1. *A paradigmatic model of electric potential*

We consider the following model of electrostatic potential [20]

$$V(\mathbf{x}, t) = \sum_{\mathbf{k}\in\mathbb{Z}^2} V_{\mathbf{k}} \sin\left[\frac{2\pi}{L}\mathbf{k}\cdot\mathbf{x} + \varphi_{\mathbf{k}} - \omega(\mathbf{k})t\right], \quad (11)$$

where $\varphi_{\mathbf{k}}$ are random phases (uniformly distributed) and $V_{\mathbf{k}}$ decrease as a given function of $\|\mathbf{k}\|$, in agreement with experimental data [21], and L is the typical size of the elementary cell as represented in Fig. 1. In principle one should use for $\omega(\mathbf{k})$ the dispersion relation for electrostatic drift waves (which are thought to be responsible for the observed turbulence) with a frequency broadening for each $\mathbf{k}$ in order to model the experimentally observed spectrum $S(\mathbf{k}, \omega)$.

Here we consider a quasiperiodic approximation of the turbulent electric potential with a finite number K of frequencies. We assume that $\omega_k \neq 0$ (otherwise, see remark at the end of Sec. 3.2). The phases $\varphi_{\mathbf{k}}$ are chosen at random in order to mimic a turbulent field with the reasonable hope that the properties of the realization thus obtained are not significantly different from their average. In addition we take for $\|V_{\mathbf{k}}\|$ a power law in $\|\mathbf{k}\|$ to reproduce the spatial spectral characteristics of the experimental

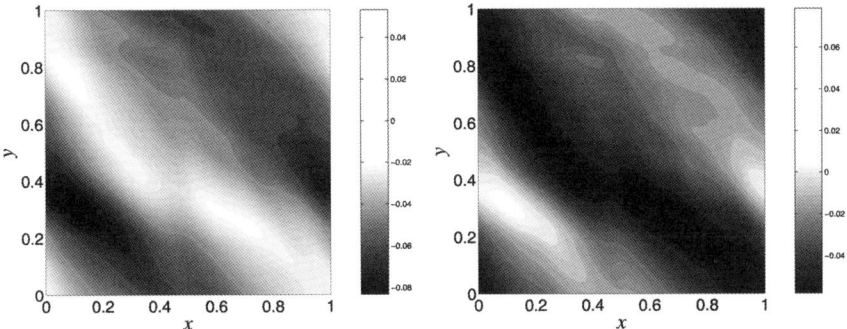

Fig. 2. Contour plots of $V(x, y, t)$ given by Eq. (13) for $a = 1$, $N = 25$, $t = 0$ (left panel) and $t = 1/2$ (right panel).

$S(\mathbf{k})$, see Ref. [21]. Thus we consider the following explicit form of the electric potential:

$$V(x,y,t) = \sum_{k=1}^{K} \sum_{m,n=1}^{N} \frac{a_k}{2\pi(n^2+m^2)^{3/2}} \sin[2\pi(nx+my) + \varphi_{kmn} - 2\pi\omega_k t], \quad (12)$$

where φ_{kmn} are random phases.

For example, Fig. 2 shows contour plots of the potential (12) with only one frequency $\omega_1 = 1$, that is a periodic potential with period 1 in x and y, given by

$$V(x,y,t) = \frac{a}{2\pi} \sum_{\substack{m,n=1 \\ n^2+m^2 \leq N^2}}^{N} \frac{1}{(n^2+m^2)^{3/2}} \sin\left[2\pi(nx+my) + \varphi_{nm} - 2\pi t\right], \quad (13)$$

with $N = 25$.

Since the model is fluctuating in time, the eddies of Fig. 2 are rapidly modified in time and where a vortex was initially present, an open line appears, and so on.

Two particular properties of the model, anisotropy and propagation have been observed : each image of the potential field shows an elongated structure of the eddies and superposing images obtained at different times a slight propagation in the $y = x$ direction is found. However, this propagation can easily be proved not to disturb the diffusive motion of the guiding centers. The property of propagation can be easily understood analytically. In fact, restricting ourselves to the most simplified case of an electric potential given only by a dominant mode ($n = m = 1$) it is immediately evident that at any given time the maxima and minima of the sine are located on the lines $y = -x +$ constant. As the amplitudes are decreasing functions of n and m, this structure is essentially preserved also in the case of many waves. The property of anisotropy is an effect of the random phases in producing eddies that are irregular in space.

We notice that there are two typical time scales in the equations of motion: the drift characteristic time τ_d, inversely proportional to the parameter a, and the period of oscillation τ_ω of all the waves that enter the potential. The competition between these two time scales determines what kind of diffusive behaviour is observed [2, 3]. In what follows we consider the case of weak or intermediate chaotic dynamics (coexistence of ordered and chaotic trajectories) which corresponds to the quasi-linear diffusion regime (see Sec. 3.5.1). Whereas in the case of fully developed chaos, that corresponds to the so-called Bohm diffusion regime, one has to introduce a slightly more complicated approach (see remark at the end of Sec. 3.2).

3.2. Control term for a potential which varies rapidly in time

We consider an electric potential of the form $V(\mathbf{x}, \mathbf{y}, t/\epsilon)$, i.e. such that the time scale of its variation is of order ε.

We consider a time-dependent Hamiltonian system described by the function $V(\mathbf{x}, \mathbf{y}, t/\epsilon)$, with $(\mathbf{x}, \mathbf{y}) \in \mathbb{R}^{2L}$ canonically conjugate variables and t the time. This Hamiltonian system has $L + 1/2$ degrees of freedom.

We map this Hamiltonian system with $L + 1/2$ degrees of freedom to an autonomous Hamiltonian with $L + 1$ degrees of freedom by extending the phase space from $(\mathbf{x}, \mathbf{y})$ to $(\mathbf{x}, \mathbf{y}, E, \tau)$ where the new dynamical variable τ evolves as $\tau(t) = t + \tau(0)$ and E is its canonically conjugate variable. The autonomous Hamiltonian is given by

$$H(\mathbf{x}, \mathbf{y}, \tau, E) = E + V(\mathbf{x}, \mathbf{y}, \tau/\epsilon).$$

Rescaling τ by a canonical change of variable,

$$\hat{\tau} = \tau/\epsilon, \quad \hat{E} = \epsilon E,$$

one obtains

$$H(\mathbf{x}, \mathbf{y}, \tau, E) = E + \epsilon V(\mathbf{x}, \mathbf{y}, \tau), \tag{14}$$

where we have renamed $\hat{E} = E$ and $\hat{\tau} = \tau$. In the case of rapidly time-varying potentials, Hamiltonian (14) is in the form $H = H_0 + \epsilon V$, that is an integrable Hamiltonian H_0 plus a small perturbation ϵV.

In our case $H_0 = E$, i.e. independent of $\mathbf{x}, \mathbf{y}, \tau$, therefore we have

$$\{H_0\} = \frac{\partial}{\partial \tau}.$$

If we consider a potential $V(\mathbf{x}, \mathbf{y}, \tau)$, in the periodic case we can write

$$V(\mathbf{x}, \mathbf{y}, \tau) = \sum_k V_k(\mathbf{x}, \mathbf{y}) e^{ik\tau},$$

and the action of Γ, $\mathcal{R}$ and $\mathcal{N}$ on V is given by

$$\Gamma V = \sum_{k \neq 0} \frac{V_k(\mathbf{x}, \mathbf{y})}{ik} e^{ik\tau}, \tag{15}$$

$$\mathcal{R} V = V_0(\mathbf{x}, \mathbf{y}), \tag{16}$$

$$\mathcal{N} V = V(\mathbf{x}, \mathbf{y}, \tau) - V_0(\mathbf{x}, \mathbf{y}). \tag{17}$$

Otherwise, in the more general case of a non periodic potential, one can write, under suitable hypotheses,

$$V(\mathbf{x}, \mathbf{y}, \tau) = \int_{-\infty}^{+\infty} \hat{V}(\mathbf{x}, \mathbf{y}, k) e^{ik\tau} dk,$$

and the action of Γ, $\mathcal{R}$ and $\mathcal{N}$ operators on V is given by

$$\Gamma V = PV \int_{-\infty}^{+\infty} \frac{\hat{V}(\mathbf{x}, \mathbf{y}, k)}{ik} e^{ik\tau} dk, \tag{18}$$

$$\mathcal{R} V = \hat{V}(\mathbf{x}, \mathbf{y}, 0), \tag{19}$$

and

$$\mathcal{N} V = V(\mathbf{x}, \mathbf{y}, \tau) - \hat{V}(\mathbf{x}, \mathbf{y}, 0). \tag{20}$$

The computation of the control term is now a straightforward application of Eq. (8).

Following the previous section, first we map the Hamiltonian system with $1+1/2$ degrees of freedom given by Eq. (12) to an autonomous Hamiltonian with two degrees of freedom. This is obtained by extending the phase space from (x, y) to (x, y, E, τ) that is considering E the variable conjugate to the new dynamical variable τ. This autonomous Hamiltonian is

$$H(x, y, E, \tau) = E + \sum_{m,n,k} \frac{a_k}{2\pi(n^2 + m^2)^{3/2}} \sin[2\pi(nx + my) + \varphi_{kmn} - 2\pi\omega_k \tau]. \tag{21}$$

The integrable part of the Hamiltonian from which the operators Γ, $\mathcal{R}$ and $\mathcal{N}$ are constructed is isochronous

$$H_0 = E.$$

We notice that H_0 is resonant (since it does not depend on the action variable x). From the action of Γ and $\mathcal{R}$ operators computed using Eqs. (18)–(19), we obtain the control term using Eq. (8). For instance, the expression of f_2 is

$$f_2(x,y,t) = \frac{1}{8\pi} \sum_{\substack{k,m,n \\ k',n',m'}} \frac{a_k a_{k'}(n'm - nm')}{\omega_k(n^2+m^2)^{3/2}(n'^2+m'^2)^{3/2}}$$

$$\times \{\sin[2\pi((n+n')x + (m+m')y) + \varphi_{kmn} + \varphi_{k'm'n'} - 2\pi(\omega_k + \omega_{k'})t]$$
$$+ \sin[2\pi((n-n')x + (m-m')y) + \varphi_{kmn} - \varphi_{k'm'n'} - 2\pi(\omega_k - \omega_{k'})t]\}. \tag{22}$$

Remark: Similar calculations can be done in the case where there is a zero frequency, e.g., $\omega_0 = 0$ and $\omega_k \neq 0$ for $k \neq 0$. The first term of the control term is

$$f_2 = -\frac{1}{2}\{\Gamma V, (\mathcal{R}+1)V\},$$

where

$$\mathcal{R}V = \sum_{m,n} \frac{a_0}{2\pi(n^2+m^2)^{3/2}} \sin[2\pi(nx+my) + \varphi_{0mn}].$$

If we add the exact expression of the control term to $H_0 + V$, the effect on the flow is the confinement of the motion, i.e. the fluctuations of the trajectories of the particles, around their initial positions, are uniformly bounded for any time [22].

In Sec. 3.5, we show that truncations of the exact control term f, like for instance f_2 or $f_2 + f_3$, are able to regularize the dynamics and to slow down the diffusion.

We notice that for the particular model (13) the partial control term f_2 is independent of time and is given by

$$f_2(x,y,\tau) = \frac{a^2}{8\pi} \sum_{\substack{n_1,m_1 \\ n_2,m_2}} \frac{n_1 m_2 - n_2 m_1}{(n_1^2+m_1^2)^{3/2}(n_2^2+m_2^2)^{3/2}}$$

$$\times \sin\left[2\pi\left[(n_1-n_2)x + (m_1-m_2)y\right] + \varphi_{n_1 m_1} - \varphi_{n_2 m_2}\right]. \tag{23}$$

Figure 3 depicts a contour plot of f_2 given by Eq. (23). Figure 4 depicts a contour plot of $V + f_2$ for $a = 0.8$ at $t = 0$. We notice that the controlled

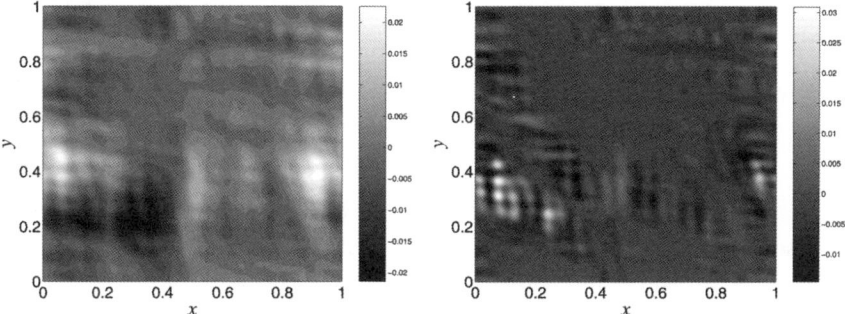

Fig. 3. Contour plot of f_2 (left panel) given by Eq. (23) and f_3 at $t = 0$ for $a = 1$ and $N = 25$.

potential is a small modification of the potential V since this contour plot looks very similar to the one depicted in Fig. 2.

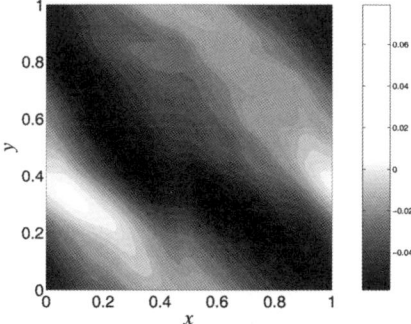

Fig. 4. Contour plot of $V + f_2$ for $a = 0.8$ and $N = 25$ at $t = 1/2$.

The computation of the other terms of the series (see Eq. (9)) can be done recursively by using Eq. (10). Again, for $a < 1$, $V + f_2 + f_3$ is close to V, meaning that the control in this regime is based on a small modification of the potential.

3.3. *Properties of the control term*

In this section, we state that for a sufficiently small, the exact control term exists and is regular. Then we give estimates of the partial control terms in order to compare the relative sizes of the different terms with respect to the perturbation. The computations are performed for the model (see Eq. (13)) but can be easily generalized to the more general case (12).

Concerning the existence of the control term, we have the following proposition [20]:

Proposition 1.1. – *If the amplitude a of the potential is sufficiently small, there exists a control term f given by the series (see Eq. (9)) such that $E + V + f$ is canonically conjugate to E, where V is given by Eq. (13).*

We have shown [20] that the exact control term exists for $a \lesssim 7 \times 10^{-3}$. As usual, such estimates are very conservative with respect to realistic values of a. In the numerical study, we consider values of a of order 1.

Concerning the regularity of the control term of the potential (see Eq. (13)), we notice that each term f_s in the series (see Eq. (9)) is a trigonometric polynomial with an increasing degree with s. The resulting control term is not smooth but its Fourier coefficients exhibit the same power law mode dependence as V:

Proposition 1.2. – *All the Fourier coefficients $f^{(s)}_{nmk}$ of the functions f_s of the series (see Eq. (9)) satisfy:*

$$|f^{(s)}_{nmk}| \leq \frac{a^s C^s}{(n^2 + m^2)^{3/2}}, \qquad (24)$$

for $(n, m) \neq (0, 0)$. Consequently, for a sufficiently small, the Fourier coefficients of the control term f given by Eq. (9) satisfy:

$$|f_{nmk}| \leq \frac{C_\infty}{(n^2 + m^2)^{3/2}},$$

for $(n, m) \neq (0, 0)$ and for some constant $C_\infty > 0$.

3.4. Magnitude of the control term

A measure of the relative sizes of the control terms is defined by the electric energy density associated with each electric field V, f_2 and f_3. From the potential we get the electric field and hence the motion of the particles. We define an average energy density $\mathcal{E}$ as

$$\mathcal{E} = \frac{1}{8\pi} \langle \|\mathbf{E}\|^2 \rangle,$$

where $\mathbf{E}(x, y, t) = -\nabla V$. In terms of the particles, it corresponds to the mean value of the kinetic energy $\langle \dot{x}^2 + \dot{y}^2 \rangle$ (up to a multiplicative constant). For $V(x, y, t)$ given by Eq. (13),

$$\mathcal{E} = \frac{a^2}{8\pi} \sum_{\substack{n,m=1 \\ n^2+m^2 \leq N^2}}^{N} \frac{1}{(n^2 + m^2)^2}. \qquad (25)$$

We define the contribution of f_2 and f_3 to the energy density by

$$e_2 = \frac{1}{8\pi}\langle \|\nabla f_2\|^2 \rangle. \tag{26}$$

For $N = 25$, these contributions satisfy:

$$\frac{e_2}{\mathcal{E}} \approx 0.1 \times a^2.$$

It means that the control terms f_2 can be considered as small perturbative terms with respect to V when $a < 1$.

Remark on the number of modes in V: In this section, all the computations have been performed with a fixed number of modes $N = 25$ in the potential V given by Eq. (13). The question we address in this remark is how the results are modified as we increase N. First we notice that the potential and its electric energy density are bounded with N since

$$|V(x,y,t)| \le a \sum_{n,m=1}^{\infty} \frac{1}{(n^2+m^2)^{3/2}} < \infty,$$

$$\mathcal{E} \le \frac{a^2}{8\pi} \sum_{n,m=1}^{\infty} \frac{1}{(n^2+m^2)^2} < \infty.$$

Concerning the partial control term f_2, we see that it is in general unbounded with N. From its explicit form it grows like $N \log N$. Less is known on the control term since it is given by a series whose terms are defined by recursion. However, from Proposition 1.2, we can show that the value a of existence of the control term decreases like $1/(2N \log N)$. This divergence of the control term comes from the fact that the Fourier coefficients of the potential V are weakly decreasing with the amplitude of the wavenumber.

Therefore, the exact control term might not exist if we increase N by keeping a constant. However for practical purposes the Fourier series of the control term can be truncated to its first terms (the Fourier modes with highest amplitudes). Furthermore in the example we consider as well as for any realistic situation the value of N is bounded by the resolution of the potential. In the case of electrostatic turbulence in plasmas $k\rho_i \sim 1$ determines an upper bound for k, where k is the transverse wave vector related to the indices n,m and ρ_i the ion Larmor radius. The physics corresponds to the averaging effect introduced by the Larmor rotation.

3.5. Efficiency and robustness

With the aid of numerical simulations (see Refs. [2, 3] for more details on the numerics), we check the effectiveness of the control term by comparing the dynamics of particles obtained from the uncontrolled Hamiltonian and from the same Hamiltonian with the control term f_2 and with a more refined control term $f_2 + f_3$.

3.5.1. Diffusion of test particles

The effect of the control terms can first be seen from a few randomly chosen trajectories. We have plotted Poincaré sections (stroboscopic plots of the trajectories of V) on Fig. 5 of two trajectories issued from generic initial conditions computed without and with the control term f_2 respectively. Similar pictures are obtained for many other randomly chosen initial conditions. The stabilizing effect of the control term (see Eq. (23)) is illustrated by such trajectories. The motion remains diffusive but the extension of the phase space explored by the trajectory is reduced.

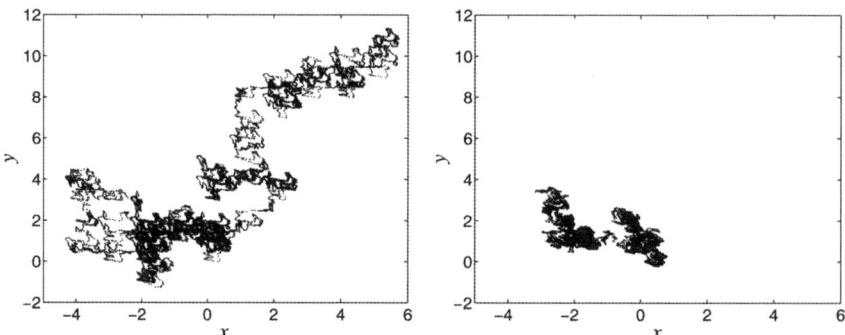

Fig. 5. Poincaré sections of a trajectory obtained using a generic initial condition for Hamiltonian (see Eq. (13)) with $a = 0.8$, without control term (upper left panel) and with control term (see Eq. (23)) (upper right panel).

The dynamics is more clearly seen on a Poincaré section on the $[0, 1]^2$ torus (i.e. by taking x and y modulo 1). Such Poincaré sections are depicted in Fig. 6 for V (upper left panel), $V + f_2$ (upper right panel) and $V + f_2 + f_3$ (lower panel). These figures shows that the Poincaré sections are composed of two types of trajectories : ones which are trapped around resonant islands, and diffusive (chaotic) ones which lead to global transport properties. We notice that the number of resonant islands has been drastically increased by the control term. The mechanism of reduction of chaos seems to be that a significant number of periodic orbits has been stabilized by the addition of the control term.

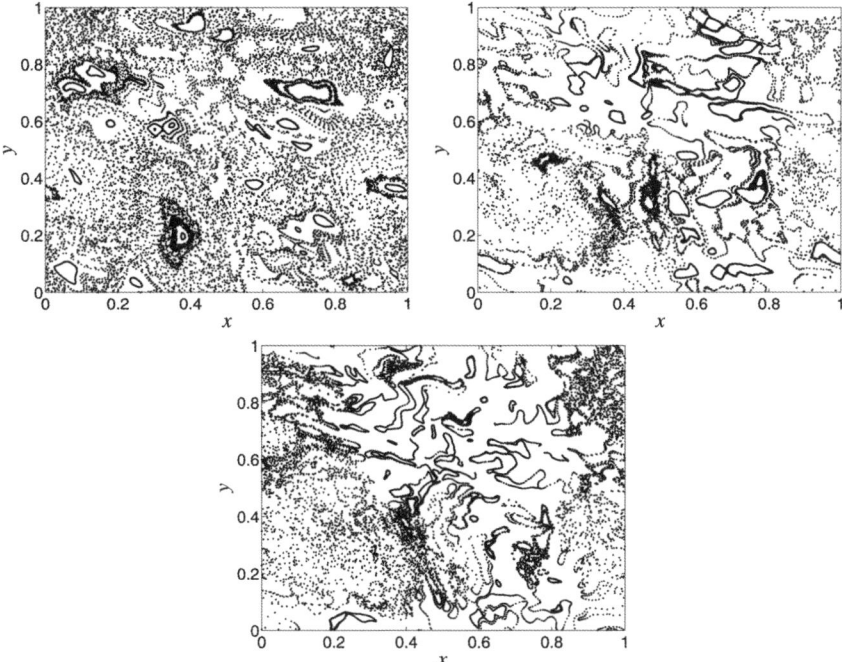

Fig. 6. Poincaré sections on the $[0,1]^2$ torus for V (upper left panel), $V + f_2$ (upper right panel) and $V + f_2 + f_3$ (lower panel) with $a = 0.6$.

A clear evidence is found for a relevant reduction of the diffusion in presence of the control term (22). In order to study the diffusion properties of the system, we have considered a set of $\mathcal{M}$ particles (of order 1000) uniformly distributed at random in the domain $0 \leq x, y \leq 1$ at $t = 0$. We have computed the mean square displacement $\langle r^2(t) \rangle$ as a function of time

$$\langle r^2(t) \rangle = \frac{1}{\mathcal{M}} \sum_{i=1}^{\mathcal{M}} \|\mathbf{x}_i(t) - \mathbf{x}_i(0)\|^2$$

where $\mathbf{x}_i(t) = (x_i(t), y_i(t))$ is the position of the i-th particle at time t obtained by integrating Hamilton's equations with initial conditions $\mathbf{x}_i(0)$. Figure 7 (left panels) shows $\langle r^2(t) \rangle$ for three different values of a. For the range of parameters we consider, the behavior of $\langle r^2(t) \rangle$ is always found to be linear in time for t large enough. The corresponding diffusion coefficient is defined as

$$D = \lim_{t \to \infty} \frac{\langle r^2(t) \rangle}{t}.$$

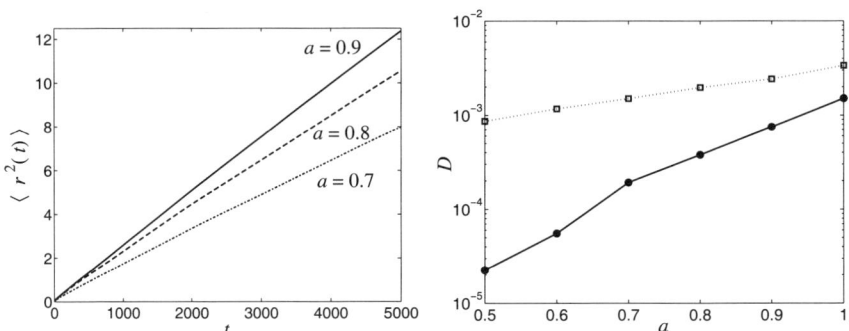

Fig. 7. On the left panel, mean square displacement $\langle r^2(t) \rangle$ versus time t obtained for Hamiltonian (see Eq. (13)) with three different values of $a = 0.7$, $a = 0.8$, $a = 0.9$. In the right panel, diffusion coefficient D versus a obtained for Hamiltonian (see Eq. (13)) (open squares) and Hamiltonian (see Eq. (13)) plus control term (see Eq. (23)) (full circles).

Figure 7 (right panel) shows the values of D as a function of a with and without control term. It clearly shows a significant decrease of the diffusion coefficient when the control term is added. As expected, the action of the control term gets weaker as a is increased towards the strongly chaotic phase. We notice that the diffusion coefficient is plotted on a log-scale. For $a = 0.7$, the control reduces the diffusion coefficient by a factor approximately equal to 10.

3.5.2. Robustness of the control

In the previous sections, we have seen that a truncation of the series defining the control term by considering the first f_2 or the two first terms $f_2 + f_3$ in the perturbation series in the small parameter a, gives a very efficient control on the chaotic dynamics of the system. This reflects the robustness of the method.

In this section we show that it is possible to use other types of approximations of the control term and still get an efficient control of the dynamics.

We check the robustness of the control by increasing or reducing the amplitude of the control term [20]. We replace f_2 by $\delta \cdot f_2$ and we vary the parameter δ away from its reference value $\delta = 1$. Figure 8 shows that both the increase and the reduction of the magnitude of the control term (which is proportional to $\delta \cdot a^2$) result in a loss of efficiency in reducing the diffusion coefficient. The fact that a larger perturbation term – with respect to the computed one – does not work better, also means that the control is not a "brute force" effect.

The interesting result is that one can significantly reduce the amplitude of the control ($\delta < 1$) and still get a reduction of the chaotic diffusion. We notice that the average energy density $e_2(\delta)$ associated with a control term $\delta \cdot f_2$ is equal to $e_2(\delta) = \delta^2 e_2$, where e_2 is given by Eq. (26). Therefore,

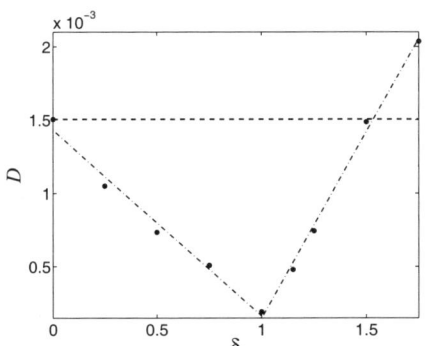

Fig. 8. Diffusion coefficient D versus the magnitude of the control term δf_2 where f_2 is given by Eq. (23) for $a = 0.7$. The horizontal dashed line corresponds to the value of D without control term ($\delta = 0$). The dash-dotted line is a piecewise linear interpolation.

for $\delta = 0.5$ where the energy necessary for the control is one fourth of the optimal control, the diffusion coefficient is significantly smaller than in the uncontrolled case $\delta = 0$ (by nearly a factor 3).

References

[1] Scott, B.D., 2003, Phys. Plasmas, 10, 963.
[2] Pettini, M., Vulpiani, A., Misguich, J.H., De Leener, M., Orban, J., and Balescu, R., 1988, *Phys. Rev. A*, 38, 344.
[3] Pettini, M., 1989, Non-linear dynamics, G. Turchetti (Ed.), World Scientific, Singapore.
[4] Lichtenberg, A.J. and Lieberman, M.A., 1983, Regular and Stochastic Motion. Applied Mathematical Sciences. Springer-Verlag, New York, 1983.
[5] L.P. Kadanoff. *From Order to Chaos: Essays: Critical, Chaotic and Otherwise*. World Scientific, Singapore, 1998.

[6] G. Chen and X. Dong. *From Chaos to Order.* World Scientific, Singapore, 1998.

[7] D. J. Gauthier. Resource Letter: CC-1: Controlling chaos. *Am. J. Phys.*, 71:750–759, 2003.

[8] W.L. Ditto, S.N. Rauseo, and M.L. Spano. Experimental control of chaos. *Phys. Rev. Lett*, 65:3211–3215, 1990.

[9] V. Petrov, V. Gaspar, J. Masere, and K. Showalter. Controlling chaos in the Belousov-Zhabotinsky reaction. *Nature*, 361:240–243, 1993.

[10] S.J. Schiff, K. Jerger, D.H. Duong, T. Chang, M.L. Spano, and W.L. Ditto. Controlling chaos in the brain. *Nature*, 370:615–620, 1994.

[11] Y. Braiman, J.F. Lindner, and W.L. Ditto. Taming spatiotemporal chaos with disorder. *Nature*, 378:465–467, 1995.

[12] L.S. Pontryagin, V.G. Boltyanskii, R.V. Gamkrelidze, and E.F. Mishchenko. *The Mathematical Theory of Optimal Processes.* Wiley, New York, 1961.

[13] E. Ott, C. Grebogi, and J.A Yorke. Controlling chaos. *Phys. Rev. Lett.*, 64:1196–1199, 1990.

[14] R. Lima and M. Pettini. Suppression of chaos by resonant parametric perturbations. *Phys. Rev. A*, 41:726–733, 1990.

[15] L. Fronzoni, M. Giocondo, and M. Pettini. Experimental evidence of suppression of chaos by resonant parametric perturbations. *Phys. Rev. A*, 43:6483–6487, 1991.

[16] T. Shinbrot, C. Grebogi, E. Ott, and J.A. Yorke. Using small perturbations to control chaos. *Nature*, 363:411–417, 1993.

[17] E. Ott and M. Spano. Controlling chaos. *Phys. Today*, 48:34–40, 1995.

[18] R. Lima and M. Pettini. Parametric resonant control of chaos. *Int. J. Bif. Chaos*, 8:1675–1684, 1998.

[19] J. Poschel. Integrability of Hamiltonian systems on Cantor sets. *Comm. Pure Appl. Math.*, 25:653–695, 1982.

[20] G. Ciraolo, F. Briolle, C. Chandre, E. Floriani, R. Lima, M. Vittot, M. Pettini, C. Figarella, and Ph. Ghendrih. Control of Hamiltonian chaos as a possible tool to control anomalous transport in fusion plasmas. *Phys. Rev. E*, 69:056213, 2004.

[21] A.J. Wootton, H. Matsumoto, K. McGuire, W.A. Peebles, Ch.P. Ritz, P.W. Terry, and S.J. Zweben. Fluctuations and anomalous transport in tokamaks. *Phys. Fluids B*, 2:2879, 1990.

[22] M. Vittot. Perturbation theory and control in classical or quantum mechanics by an inversion formula. *J. Phys. A: Math. Gen.*, 37:6337–6357, 2004.

Chapter 2

CONTROLLING CHAOS IN A CHAOTIC NEURAL NETWORK

Guoguang He[1,2,3,*], Ping Zhu[2], Jousuke Kuroiwa[4] and Kazuyuki Aihara[1,3]

[1] Aihara Complexity Modelling Project, ERATO, JST,
Uehara 3-23-5-201, Shibuya-ku, Tokyo 151-0064, Japan
[2] Department of Physics, College of Science, Zhejiang University,
Hangzhou 310027, China
[3] Institute of Industrial Science, the University of Tokyo,
Komaba 4-6-1, Meguro-ku, Tokyo 153-8505, Japan
[4] Department of Human and Artificial Intelligent Systems,
Faculty of Engineering, University of Fukui,
Bunkyo 3-9-1, Fukui 910-8507, Japan
*gghe@zju.edu.cn

1. Introduction

Chaotic phenomena are ubiquitous in the real world including biological neurons and neural networks. For example, chaotic behavior has been observed in nerve membranes by electrophysiological experiments on squid giant axons [1–3] and in measurements of brain electroencephalograms (EEG) [4, 5]. Therefore, various neural network models with chaotic dynamics have been investigated in order to understand brain functions, explore mechanisms of information processing, and subsequently realize a flexible intelligent information processing that mimics the brain [5–13]. Aihara et al. [6] have shown that even a simple neuron model can exhibit chaotic dynamics. Tsuda [7] has presented a nonequilibrium neural network model for a dynamic linking of memory and has suggested that the cortical chaos may serve for dynamically linking true memory as well as for memory search. Nara and Davis [8] have shown that a network comprising simple binary neurons can exhibit chaotic wandering among cycle memories because of the emergence of complex dynamics in systems with finite but large degrees of freedom. Kuroiwa et al. [9] have studied the dynamic response properties of a single chaotic neuron to stochastic inputs and have indicated that the response of a chaotic neuron is sensitive to its inputs and robust to noise. Nozawa [10] and Chen and Aihara [11, 12] have demonstrated that chaotic neural networks have a remarkable capability to search for global optimal solutions in combination optimization problems. Adachi and Aihara [13] have investigated chaotic dynamics in a chaotic neural network and have shown that the chaotic neural network can generate spatio-temporal chaotic dynamics with associative memory over several parameter regions. Shiino and Fukai [14] and Nakagawa [15] have investigated the memory storage capacity of neural networks and concluded that chaotic neural networks have a larger memory storage capacity than the conventional neural networks. These characteristics demonstrate that chaotic neural networks can be promising techniques in information processing. However, the outputs of chaotic neural networks wander around all the learned patterns (or stored patterns) and cannot be stabilized to one of the stored patterns or a periodic orbit due to the presence of chaos. It is difficult to judge when to terminate chaotic dynamics, thereby hampering applications of chaotic neural networks to information processing.

On the other hand, Freeman's studies on the olfactory bulb of a rabbit have shown that the dynamics of the neural system is chaotic in the basic state; however, when a familiar scent is presented, the system rapidly simplifies its behavior, and its dynamics becomes more ordered and almost periodic [5]. This suggests an attractive model for the recognition and learning process in biological neural systems. In other words, self-induced chaos control in biological neural systems may be important in the recognition and learning process of biological neural systems. The control of chaotic behavior in chaotic neural networks is an important problem to apply them to information processing; such studies would also be useful in understanding the mechanism of the recognition and learning process of the real brain.

2. Chaotic Neural Network Model

There exist some chaotic neural networks models, for example, Freeman's KIII model [5], Aihara model [6], Tsuda's nonequilibrium neural network model [7], Nara's recurrent neural network model [8], and Hayashi's excitatory-inhibitory model [16]. Here, we focus on the Aihara model which was proposed based on electrophysiological experiments on squid giant axons [1–3], and theoretical study of the Hodgkin-Huxley equation [17] as well as the Nagumo-Sato model [18]; moreover the model is simple but the

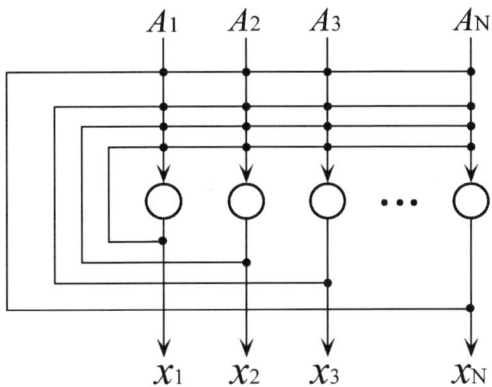

Fig. 1. The structure of a chaotic neural network.

dynamics is very complex. Aihara's chaotic neural network (CNN) model comprises chaotic neurons by considering the spatiotemporal summation of both external and feedback inputs from other chaotic neurons [6]. As shown in Fig. 1, the structure of the CNN is similar to that of a Hopfield network [19]. The output x_i of the ith chaotic neuron in the network is determined by the external stimulus A_i, feedback inputs from the other constituent neurons in the CNN, and refractoriness of the chaotic neuron. When A_i is temporally constant, it can be included in the threshold term. The dynamics of the ith chaotic neuron in the CNN at time t can be described as follows [6, 13]:

$$x_i(t+1) = f(\eta_i(t+1) + \zeta_i(t+1)), \tag{1}$$

$$\eta_i(t+1) = k_f \eta_i(t) + \sum_{j}^{N} w_{ij} x_j(t), \tag{2}$$

$$\zeta_i(t+1) = k_r \zeta_i(t) - \alpha g(x_i(t)) + a_i, \tag{3}$$

where t denotes a discrete time ($t = 0, 1, 2, \cdots$); $x_i(t)$, the output of the ith chaotic neuron; $\eta_i(t)$ and $\zeta_i(t)$, the internal state variables of the feedback inputs from other constituent neurons in the network and refractoriness of the chaotic neuron, respectively; N, the number of neurons in the CNN; parameter α, the refractory scaling of the neuron; parameter a_i, the threshold of the ith neuron; k_f and k_r, the decay parameters of the feedback inputs and refractoriness, respectively; and the function $g(\cdot)$, the refractory function of the neuron. For the sake of simplicity, we assume $g(x) \equiv x$. For the output function of a neuron $f(\cdot)$, we employ the following sigmoidal function with the steepness parameter ε:

$$f(x) = \frac{1}{1 + exp(-x/\varepsilon)}. \tag{4}$$

Here, w_{ij} in Eq. (2) denotes the synaptic weight on the ith constituent neuron from the jth constituent neuron. The weight is defined according to the following symmetric auto-associative matrix of n binary patterns:

$$w_{ij} = \frac{1}{n} \sum_{p=1}^{n} (2x_i^p - 1)(2x_j^p - 1), \tag{5}$$

where x_i^p denotes the ith component of the pth binary pattern with a discrete value of zero or one; and n, the total number of stored memory patterns. A neuron does not receive the synaptic connection from itself, i.e., $w_{ii} = 0$, and the binary patterns can be stored as basal memory patterns in the CNN.

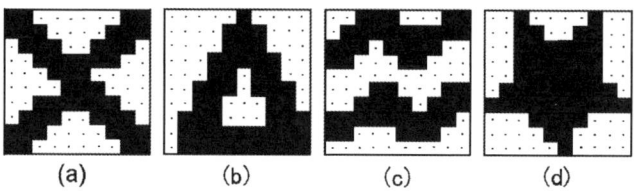

Fig. 2. Four stored patterns.

Figure 2 shows four patterns [13], which are used as the stored patterns to be embedded in the CNN according to the synaptic weights learning rule provided in Eq. (5). Each pattern comprises 10 by 10 binary pixels. Correspondingly, the CNN comprises 100 neurons, that is, $N = 100$. Each neuronal state is represented by a block (■) when its output, x_i, is equal to one; this indicates that the neuron is "exciting." When its output equals to zero, which indicates that the neuron is "resting," the neuronal state is denoted by a dot (·). The dynamics of the CNN is dependent on the network parameters. When the network parameters are fixed at: $k_r = 0.95, k_f = 0.20, \alpha = 10.0, a_i = 2.0 (i = 1, 2, \cdots, 100)$, and $\varepsilon = 0.02$, the CNN exhibits associative memory dynamics with all the four stored patterns and their

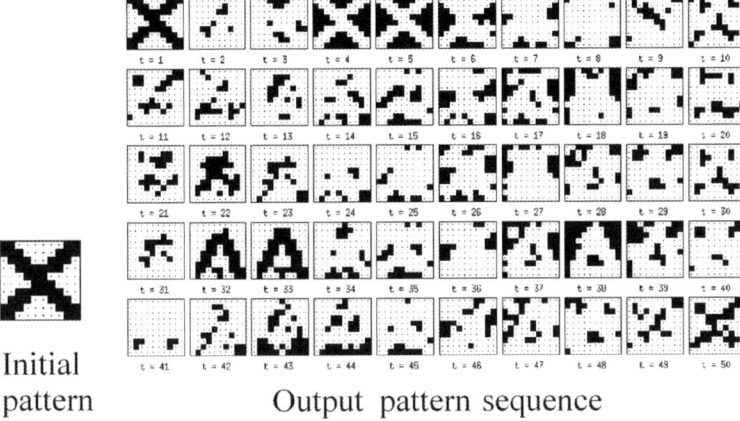

Fig. 3. The first 50 patterns in the output sequence of the CNN.

reverse patterns appearing in its output sequence [6, 13]. In Fig. 3, the first 50 patterns in the output sequence of the CNN are shown, where the initial pattern is the stored pattern shown in Fig. 2(a). Evidently, the CNN has associative memory dynamics, and its output sequence is nonperiodic. The largest Lyapunov exponent of the CNN is calculated to be equal to 0.0308, which clearly indicates that the CNN is chaotic when the abovementioned parameters are used.

3. Chaos Control Methods in the CNN

The first chaos control method was proposed by Ott, Grebogi, and Yorke (the OGY method) [20]. Since this pioneering work of OGY, various methods such as the occasional proportional feedback (OPF) method [21], continuous feedback control [22], and pinning method [23] have been proposed for chaos control. The chaos control has been applied to a wide variety of systems, for example, electric circuits [21], mechanical systems [24], chemical reactions [25, 26], lasers [27], cardiac tissue [28], and the brain [29].

There are also many studies on controlling chaos in chaotic neurons and chaotic neural networks. Gong *et al.* employed an open-loop (or entrainment) method to change the dynamics of a neuron. By specifying a target-dynamics for the neuron, the neuron can be controlled from chaos to order, and also from order to chaos, chaos to chaos, and order to order [30]. Sinha employed the OPF method to control Hayashi's chaotic neural network model from chaos to periodic [31]. Bondarenko used an external sinusoidal force to change the dynamics in high-dimensional chaotic neural network and obtained "chaos–order," "order–chaos," and "chaos–chaos" transitions [32]. A time delay feedback control for chaos in chaotic neural networks was investigated by Tsui *et al.* [33] and Bondarenko [34]. Cortes *et al.* used a synaptic noise to change the chaotic behavior [35].

The above investigations indicate that the dynamics of the chaotic neuron or chaotic neural networks can be changed from chaotic behavior to a periodic state or a fixed point. However, these works did not refer to the associative memory of the controlled chaotic neural networks or the relation between the stable output of a controlled chaotic neural network and its initial state. From the viewpoint of the application of chaotic neural networks, the associative memory is important with possible information processing, such as memory retrieval and pattern recognition. In general, the essential features of memory retrieval and pattern recognition are similar for networks with associative memory. When an initial pattern is input to a network with associative memory, the network can retrieve a template pattern or a periodic orbit that is related to one of its stored patterns. If the output has a relation with the initial pattern, it is said that the stored pattern has been retrieved by memory recall and that the pattern has been identified by pattern recognition. Therefore, it is an important problem to consider the associative memory dynamics and the relation between the stable output and the initial state when chaos control is performed in chaotic neural networks. Adachi *et al.* calculated the eigenvalues of the Jacobian Matrix of the CNN and concluded that by using simple chaos control methods, the controlled CNN cannot be stabilized to a stored pattern, and a new control strategy must be developed for the CNN [36]. For the associative memory dynamics, Nakamura *et al.* [37], Kushibe *et al.* [38], He *et al.* [39]–[45], and Shrimali *et al.* [46]–[47] have proposed some chaos control methods for chaotic neural networks and investigated the controlled dynamics. In this chapter, we focus on several chaos control methods concerned with associative memory and the relation between the stable output and the initial state in the CNN.

3.1. *Pinning method*

The Pinning method was first proposed by Hu and Qu to control spatial-temporal chaos in coupled map lattice systems [23]. He *et al.* adapted a Pinning method to the CNN [40]. The Pinning method in the CNN is described as follows:

$$x_i(t+1) = f(\eta_i(t+1) + \zeta_i(t+1)), \quad (6)$$

$$\eta_i(t+1) = k_f \eta_i(t) + \sum_j^N w_{ij} \left\{ x_j(t) + \sum_l^{N/I} \delta(j - Il - 1) K u_j(t) \right\}, \quad (7)$$

$$u_j(t) = x_j(t) - (1 - \tilde{x}_j), \quad (8)$$

$$\zeta_i(t+1) = k_r \zeta_i(t) - \alpha g(x_i(t)) + a_i, \quad (9)$$

where $u_j(t)$ is the control signal, which is constructed by comparing the last output $x_j(t)$ and the desired output (or the target output) $\tilde{x}_j$ for each pinned neuron. $\delta(\cdot)$ is the delta function, which is used to distinguish the pinned neurons from the unpinned ones. The jth neuron is pinned if $j - Il - 1 = 0$. I is the pinning distance, the distance between the two nearest pinned neurons, and K is the control strength.

The Pinning method is based on the fact that each neuron in the CNN is coupled with others and evolves in a nonlinear manner. If a part of the neurons in the network are controlled, the controlling effect can be extended to the entire neural network as time elapses because of its spatial coupling and delay feedback of the internal state. Thus, by means of the Pinning method, the chaos in the CNN is controlled. The output of the controlled CNN is stabilized in the target stored pattern when a suitable pinning distance and control strength are considered.

3.2. *Phase space constraint control*

The CNN can be stabilized to a stored pattern by the Pinning method, but a control target pattern must be specified. It is well known that chaos appears when the orbit of a nonlinear system is divergent in one or more directions. Therefore, one type of chaos control method was proposed by limiting the phase space of a chaotic system [39, 48]. An essential difference between the CNN described in Eqs. (1)–(3) and the Hopfield networks [19] is that the refractoriness term is introduced into the CNN [6]. In the phase space constraint control method, thresholds of the refractoriness term, ζ_{max} and ζ_{min}, are set to control the chaos in the CNN [39] as follows.

$$x_i(t+1) = f(\eta_i(t+1) + \hat{\zeta}_i(t+1)), \quad (10)$$

$$\eta_i(t+1) = k_f \eta_i(t) + \sum_j^N w_{ij} x_j(t), \tag{11}$$

$$\hat{\zeta}_i(t+1) = \begin{cases} \zeta_{max} & \zeta_i(t+1) \geq \zeta_{max} \\ \zeta_i(t+1) & \zeta_{min} < \zeta_i(t+1) < \zeta_{max} \\ \zeta_{min} & \zeta_i(t+1) \leq \zeta_{min} \end{cases}, \tag{12}$$

$$\zeta_i(t+1) = k_r \hat{\zeta}_i(t) - \alpha g(x_i(t)) + a_i. \tag{13}$$

In the method, the values of the refractoriness ζ are limited to be between ζ_{min} and ζ_{max}. The phase space of the CNN is therefore constrained. By means of this method, the chaos in the CNN is controlled, and the output pattern of the controlled CNN is one of the stored patterns or its reverse pattern with the smallest Hamming distance from the initial state of the network, which provides a desired controlled state of the neural network. In this control method, the control target need not be specified, and any knowledge of the local dynamics of the attractor around unstable periodic orbits is not necessary.

The phase space can be limited at a certain time kt_c (t_c is a variable time interval and can be taken as natural numbers $1, 2, 3, \cdots$, and $k = 1, 2, 3, ...$). Only at a certain time kt_c, the values of $\zeta(kt_c)$ are checked. Then, Eq. (12) is replaced by the following equation [46]:

$$\hat{\zeta}_i(kt_c) = \begin{cases} \zeta_{max} & \zeta_i(kt_c) \geq \zeta_{max} \\ \zeta_i(kt_c) & \zeta_{min} < \zeta_i(kt_c) < \zeta_{max} \\ \zeta_{min} & \zeta_i(kt_c) \leq \zeta_{min} \end{cases}, \tag{14}$$

The controlled CNN can then be stabilized to one of the stored patterns and its reverse pattern related to the initial pattern with a temporal period when ζ_{min}, ζ_{max}, and t_c are suitably chosen.

3.3. Threshold coupling method

In the phase space constraint control method, the chaos in the CNN is controlled by setting thresholds of the refractoriness to limit the phase space of the CNN. The excess values of ζ are simply removed from the corresponding neuron. A threshold coupling method is extended from the idea of the phase space constraint method to reuse the excess values

[43, 45, 47] as follows.

$$\eta_i(t+1) = k_f \eta_i(t) + \sum_j^N w_{ij} x_j(t), \quad (15)$$

$$\zeta_i(t+1) = k_r \zeta_i(t) - \alpha g(x_i(t)) + a_i; \quad (16)$$

if $|\zeta_i(t+1)| > \zeta^*$ then,

$$\zeta_i(t+1) \to \pm \zeta^*, \quad (17)$$

$$\delta = \pm(|\zeta_i(t+1)| - \zeta^*)/M, \quad (18)$$

$$\zeta_j(t+1) \to \zeta_j(t+1) \pm \delta; \quad (19)$$

$$x_i(t+1) = f(\eta_i(t+1) + \zeta_i(t+1)), \quad (20)$$

where ζ^* is a critical positive value. δ is the excess value, which is transported to another neuron/neurons. The value M and the transporting method of the excess value determine the coupling method. When M is taken as 1, if δ is transported to a neighboring neuron j ($j = i+1$), the method is known as unidirectional local threshold coupling. If δ is transported to a neuron j chosen randomly in the network, the method is known as unidirectional random threshold coupling. By both these coupling methods, the CNN can be controlled to the desired patterns with the shortest Hamming distance from the initial state. When $M = 2$, the excess values δ are transported to two neurons, and the threshold coupling is bidirectional. If the control is triggered at a certain time kt_c, the excess is transported equally to both the neighbors of the supercritical neuron, and the CNN is then stabilized to a periodic state in which one of the stored patterns or/and its reverse pattern related to the initial state are obtained with a temporal period. The threshold coupling method is robust against noise in the initial pattern.

3.4. *Parameter modulated method*

Generally speaking, chaos control methods can be divided into two categories. In one category, the chaos is controlled by limiting its state spaces [39, 43, 45, 48], while in the other, the chaos is controlled by perturbing its parameters based on their sensitive effects on the chaotic dynamics [37, 38, 49]. In the methods proposed by Nakamura *et al.* [37] and

Kushibe et al. [38], the chaos is controlled by changing the parameters of the chaotic neural networks. The chaotic neural networks, therefore, reduce to a kind of the Hopfield network, in which the outputs of the controlled chaotic neural networks are fixed points; thus, pattern recognition and memory retrieval are achieved. However, mapping of the initial pattern to a stored pattern should be assigned *a priori*. Moreover, a fixed output from the controlled chaotic neural networks makes chaotic neural networks lose their advantage, namely dynamics. This limits applications of the chaotic neural networks in information processing. It has been known that the dynamics of the CNN are dependent on the refractoriness parameter α [13, 44]. The dynamics in the CNN changes if the value of α in the chaotic neurons is perturbed by a control signal. A parameter modulated method was therefore proposed [44].

$$x_i(t+1) = f(\eta_i(t+1) + \zeta_i(t+1)), \tag{21}$$

$$\eta_i(t+1) = k_f \eta_i(t) + \sum_j^N w_{ij} x_j(t), \tag{22}$$

$$\zeta_i(t+1) = k_r \zeta_i(t) - \alpha \beta^{k_c u(t)} x_i(t) + a_i, \tag{23}$$

$$u(t) = \sum_i^N |x_i(t) - x_i(t-\tau)|, \tag{24}$$

where $u(t)$ is the control signal, and $\beta^{k_c u(t)}$ is the perturbation term to the refractory scaling parameter α. β is a control parameter, which takes a positive value less than 1.0, and k_c is the control strength. The control signal $u(t)$ is determined by the delay feedback signals in the network. When the output of the CNN is chaotic, $u(t)$ is not zero. The term $\alpha \beta^{k_c u(t)}$ in Eq. (23) will be less than α. The $\alpha \beta^{k_c u(t)}$ term falls in a non-chaotic region if a suitable value of parameter β is chosen. Therefore, the controlled CNN escapes from the chaotic state. On the other hand, when the controlled CNN is at a fixed point or in a periodic state with period τ, $u(t)$ is zero, and $\alpha \beta^{k_c u(t)}$ returns to α. Therefore, the controlled CNN returns to the chaotic state. The CNN repeats this process, and the controlled CNN is at the edge of chaos. Therefore, the chaos in the CNN can be controlled in this manner. The output patterns of the controlled CNN are a periodic sequence in which only one of the stored patterns and its reverse pattern are included. The controlled CNN has a superior ability to distinguish between initial conditions with a small difference when compared with the Hopfield model, that is, the controlled CNN utilizes the advantage that the chaotic dynamics is sensitively dependent on the initial conditions.

In real neural networks, some neurons may not transfer signals to some of the others. Therefore, it is impossible for all the neurons to participate in the control process. Inspired by the property of the real neural networks and the pinning control method [23, 40], an improved parameter modulated method was proposed [42]. Compared with the parameter modulated method, the difference between the two method is on the control signals. In the improved parameter modulated method, the control signal is derived from some neurons (control neurons) chosen randomly from all the neurons in the network, while the control signal is derived from all the neurons in the parameter modulated method. In the improved parameter modulated method, the control signal is given by

$$u(t) = \sum_{j}^{n_c} |x_j(t) - x_j(t-\tau)|, \qquad (25)$$

where j shows the index of the control neurons; n_c is the total number of control neurons between 1 and N. With a different n_c chosen, different periodic spatio-temporal behavior of the CNN is obtained. The improved parameter modulated method retains some of the advantages of the parameter modulated method, and may be considered to be a natural method of a self–induced chaos controlling mechanism in the neural networks.

3.5. Delay feedback method

Now, we consider a delay feedback control method in the CNN. The key points of the method are as follows. First, the delay feedback control method does not specify a control target, but simply specifies the delay time and the control strength. Second, the delay feedback control method does not rely on *a priori* knowledge of the local dynamics around unstable periodic orbits that are to be stabilized. Third, delays are actually inherent in biological neural networks, such as propagation delays of nerve impulses along axons and synaptic delays. Moreover, neurons require a certain amount of time to receive their inputs and produce a response. Finally, the feedback control method is a well-known control method, and has been successfully implemented in some chaotic systems [21, 22, 50–52]. The delay feedback control method for the CNN is described as follows:

$$x_i(t+1) = f(\eta_i(t+1) + \zeta_i(t+1)), \qquad (26)$$

$$\eta_i(t+1) = k_f \eta_i(t) + \sum_j^N w_{ij} x_j(t), \qquad (27)$$

$$\zeta_i(t+1) = k_r \zeta_i(t) - \alpha g(x_i(t) + u_i(t)) + a_i, \qquad (28)$$

$$u_i(t) = K(x_i(t-\tau) - x_i(t)), \qquad (29)$$

where $u_i(t)$ denotes the control signal; K, the control strength; and τ, the delay time.

The chaos in the CNN is controlled by using a control signal to weaken or counteract the refractoriness of the chaotic neurons since the refractoriness of the chaotic neurons is the essential difference from conventional neural networks such as the Hopfield network. We should note that in the delay feedback control method, $u_i(t)$ is determined by two outputs x_i with delay time τ.

4. Illustrative Example

We show an example of controlling the chaos by the delay feedback method in the CNN. We perform the task of controlling the chaos in the CNN as described by Eqs. (26)–(29). Here, K and τ are set to 2.0 and 4, respectively. The stored pattern shown in Fig. 2(a) is taken as an initial state. The initial state is injected into the CNN by representing it as $x_i(0)$ in the internal states η_i and ζ_i. The other initial internal states are assigned random values.

4.1. *Controlled dynamics*

Figure 4(a) shows the output sequences of the controlled CNN. For comparison, the output sequences of an uncontrolled CNN are shown in Fig. 4(b). Apart from the fact that control signals were injected for the simulation shown in Fig. 4(a), all the other conditions are the same for both the simulations. In Fig. 4, it is observed that the output sequences of the controlled CNN are ordered, while those of the uncontrolled CNN are chaotic; this indicates that the dynamics of the CNN changes from a chaotic state to an ordered state when the control signals are injected into the system.

In general, stability of a controlled system can be studied by calculating the Lyapunov spectrum and the autocorrelation function [53, 54]. For the sake of simplicity, we calculate the largest Lyapunov exponent instead of the entire Lyapunov spectrum. The largest Lyapunov exponents of the controlled CNN is calculated to be -0.3762, which implies that the dynamics

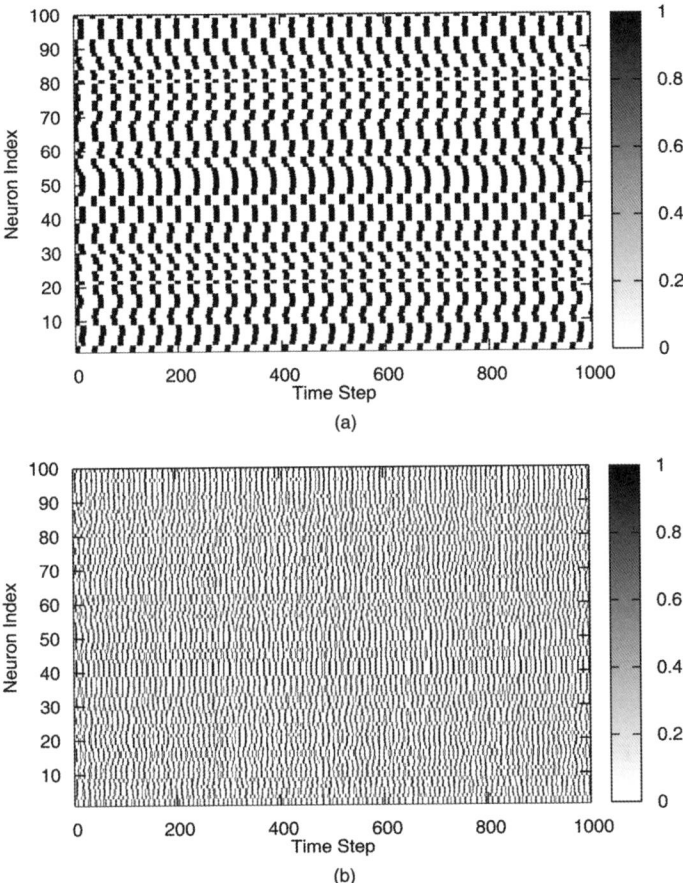

Fig. 4. The output sequences of the neurons in the CNN: (a) the controlled CNN and (b) the uncontrolled CNN.

of the controlled CNN is not chaotic. The autocorrelation function not only determines whether a system is stable or not but also determines the period of a periodic system. The autocorrelation function is defined as follows:

$$C(\gamma) = \frac{1}{N} \sum_{i=1}^{N} \lim_{T_m \to \infty} \frac{1}{T_m} \sum_{t=0}^{T_m} x_i(t) x_i(t+\gamma), \qquad (30)$$

where $x_i(t)$ denotes the output of neuron i at time t; γ, the time difference; N, the number of neurons in the CNN. By this definition, the autocorrelation function is periodic if the identified system is periodic. The dependence

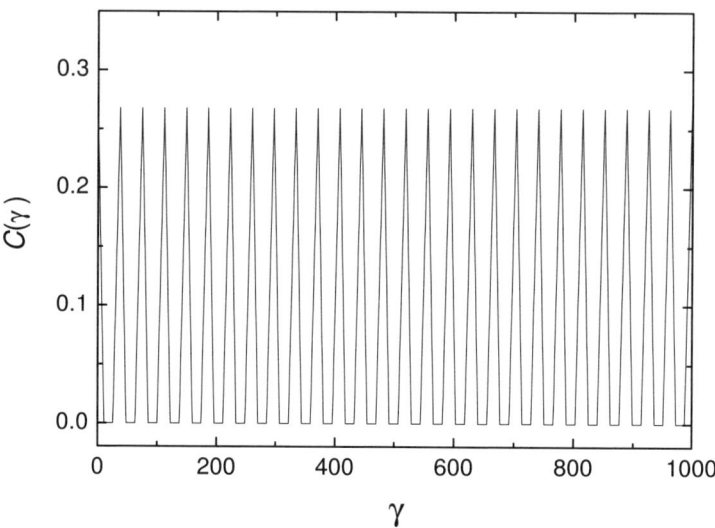

Fig. 5. The autocorrelation function of the controlled CNN. K and τ are set to 2.0 and 4, respectively.

of $C(\gamma)$ on γ for the controlled CNN is shown in Fig. 5. The periodicity of $C(\gamma)$ clearly confirms that the controlled CNN is stable and periodic.

The dynamics of the controlled CNN depends on the control parameters K and τ. We investigate the correlation between the dynamics of the controlled CNN and the control parameters. First, we set K to 2.0 and calculate the autocorrelation function of the controlled CNN at different values of τ. The results are shown in Fig. 6, which shows that the outputs of the CNN are periodic or mostly periodic except when $\tau = 3$, and the period is dependent on the value of τ; the period increases with τ. Furthermore, we find that the outputs of the CNN are accurately periodic when τ is even. However, when τ is odd, the outputs are mostly periodic except when $\tau = 5$. In the mostly periodic state, the outputs of the neurons in the network are periodic except for those of one or two neurons. The mostly periodic state is therefore different from the states with a long-term period or no-period.

Next, we investigate the relation between the period T and the controls parameter K. We fix τ at 4 in the following simulations for which the output of the controlled CNN are periodic. In fact, the simulations are the same when τ is set to other values such as 5 or 6. We obtain T for different K by calculating the autocorrelation function of the output sequences. The

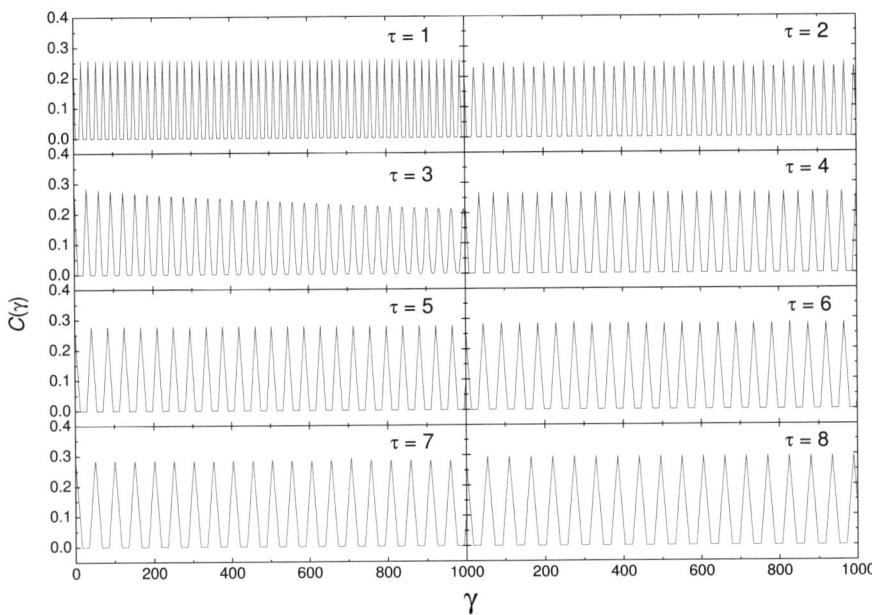

Fig. 6. The autocorrelations of the output of the controlled CNN for different delay times τ. The control strength K is set to 2.0.

relation between T and K is shown in Fig. 7. T increases almost linearly with K. This phenomenon also exists in some feedback systems [22].

4.2. Associative memory dynamics

The associative memory dynamics is the essential feature of memory retrieval and pattern recognition. Now, we investigate the associative memory dynamics of the controlled CNN. As demonstrated in Subsect. 4.1, the dynamics of the controlled CNN depends on the control parameters. In the following simulations, the values of K and τ are set to 2.0 and 4, respectively. The controlled CNN with these parameter values is periodic. Evidently, the results would be the same if τ and K are fixed at other parameter values where the controlled CNN is also periodic. The four stored patterns and four other patterns with small differences from the stored patterns, i.e., noisy stored patterns, are provided as the initial patterns. Figure 8 shows the initial patterns (on the left) and periodic sequences of the corresponding output patterns generated by the controlled CNN (on the right). It should be noted that there is a transient state in the controlled CNN.

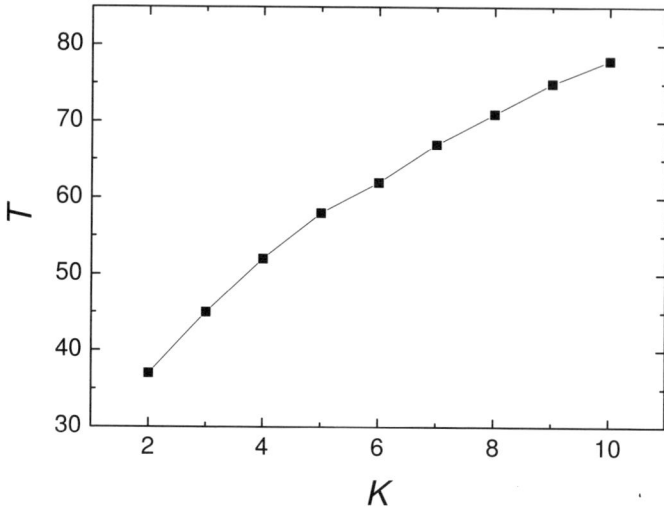

Fig. 7. The period T of the controlled CNN vs. the control strength K for the delay time $\tau = 4$.

Since we focus on the stable sates of the controlled CNN, we did not plot the transient states of the controlled CNN in Fig. 8 and provided the stable periodic states of the controlled CNN only. Figure 8 shows that the output sequences are different for different initial patterns; in other words, the output sequence is related to the initial pattern. In addition, most of the patterns appearing in the output sequences are the stored patterns nearest to the initial patterns or their reverse patterns. Inevitably, several blank patterns as well as patterns similar to the corresponding stored patterns are also included in the output sequences; however, other stored patterns never appear in the output sequences.

Figure 8 also shows that there are differences between the two output sequences generated by the controlled CNN when the initial pattern is a stored pattern or a noisy stored pattern. When the initial pattern is a noisy stored pattern, the output sequences generated by the controlled CNN are similar to those when the initial state corresponds to the exact stored pattern; however, the output sequences and the frequencies of the stored pattern or its reverse pattern appearing in the output sequences are different, even when the periods are the same. We then measure the frequencies of recalling the stored patterns and their reverse patterns in the output sequence. The retrieval rate of a given stored pattern in the output sequence of the CNN is calculated as the fraction of times when the stored

Controlling Chaos in a Chaotic Neural Network 37

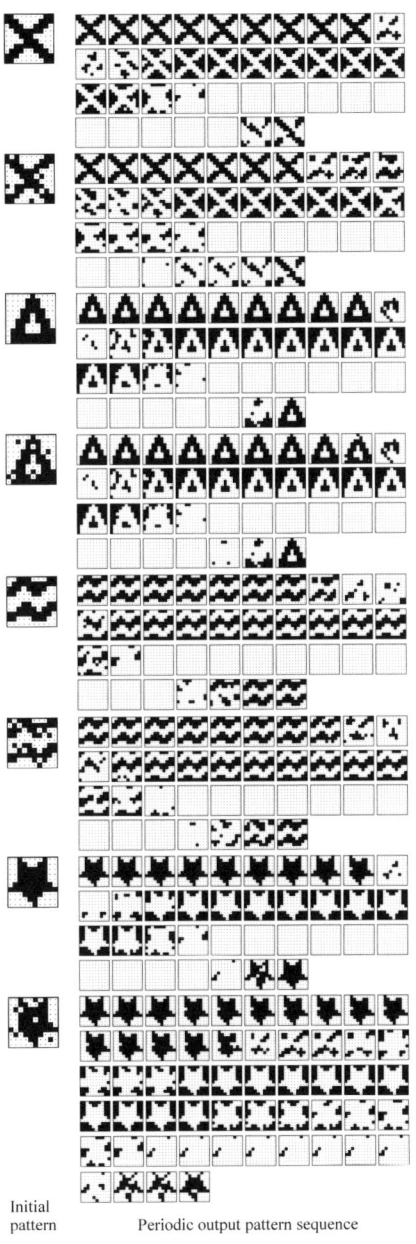

Initial pattern   Periodic output pattern sequence

Fig. 8. The initial patterns and corresponding periodic output pattern sequences generated by the controlled CNN. The figures on the left are the initial patterns. The figures on the right are the corresponding periodic output pattern sequences.

Table 1. The rates of recalling the patterns from different initial patterns.

Initial pattern	Stored pattern (a)	Noisy pattern of (a)	Stored pattern (b)	Noisy pattern of (b)	Stored pattern (c)	Noisy pattern of (c)	Stored pattern (d)	Noisy pattern of (d)
Stored pattern	0.2432	0.1622	0.2703	0.2432	0.2432	0.2162	0.2432	0.0556
Reverse pattern	0.2162	0.1622	0.2162	0.2162	0.2432	0.2162	0.2162	0.2037

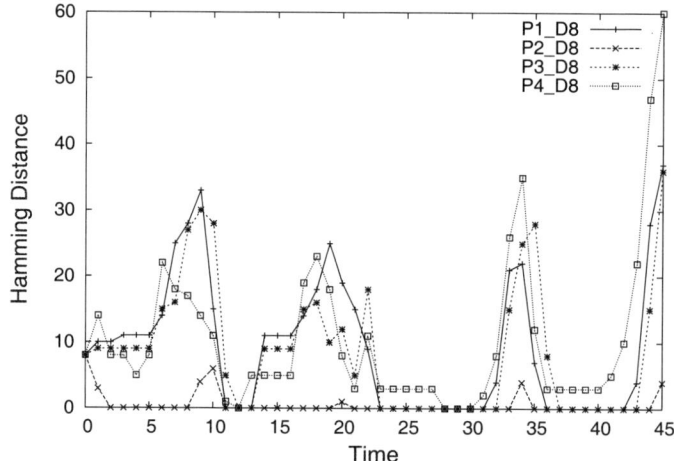

Fig. 9. The evolution of the Hamming distance between two output patterns with different initial patterns in Fig. 8. P1_D8, P2_D8, P3_D8, and P4_D8 indicate the cases where the initial patterns are the stored patterns (a), (b), (c), and (d) in Fig. 2, respectively, and their noisy patterns with the Hamming distance of 8.

pattern appears in the output sequence. In Table 1, we summarize the rates of recalling a stored pattern and its reverse pattern in the output sequences. We find that the rates are higher when the stored patterns are taken as the initial patterns than when the noisy stored patterns are chosen.

To further confirm whether the controlled CNN can distinguish between two initial patterns with small differences, we investigated the orbits of the controlled CNN by measuring the Hamming distance evolution between two output pattern sequences of the controlled CNN in which the initial states have small difference. Figure 9 shows the evolution of the Hamming distance between two output pattern sequences with two different initial patterns in Fig. 8. The curves "P1_D8," "P2_D8," "P3_D8," and "P4_D8"

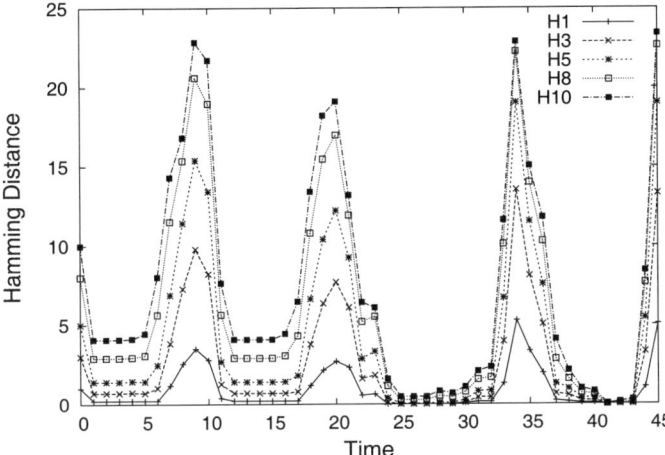

Fig. 10. The evolution of the average Hamming distance between two output patterns of different initial patterns. H1, H3, H5, H8, and H10 means that one initial pattern is stored pattern (a) in Fig. 2, and the other initial patterns are noisy patterns the Hamming distances of which with the stored pattern (a) are 1, 3, 5, 8, and 10, respectively.

are the cases with the initial states being stored patterns (a), (b), (c), (d), respectively, and their noisy patterns with the Hamming distance of 8. Figure 9 shows that the Hamming distance changes as the outputs evolve. Meanwhile, we measured the evolution of the average Hamming distances between two output pattern sequences with different initial patterns. We fixed the stored pattern (a) as one initial state and a randomly chosen noisy pattern with a fixed Hamming distance to the stored pattern (a) as another initial state. The average is taken over noisy initial patterns with the same Hamming distance. Figure 10 shows the average Hamming distances with the noises 1%, 3%, 5%, 8%, and 10%, which are represented by the curves "H1," "H3," "H5," "H8," and "H10," respectively. Figures 9 and 10 show that the orbits of the controlled CNN are different when the initial states are not the same, even if the Hamming distance between the two initial states is just 1.

Furthermore, we measured the average rates of recalling the stored patterns and their reverse patterns. The initial states are noisy patterns of the stored patterns, which are generated randomly with a fixed amount of noise. In other words, the Hamming distance between a stored pattern and its noisy patterns is fixed. Figure 11 shows the results where the curve

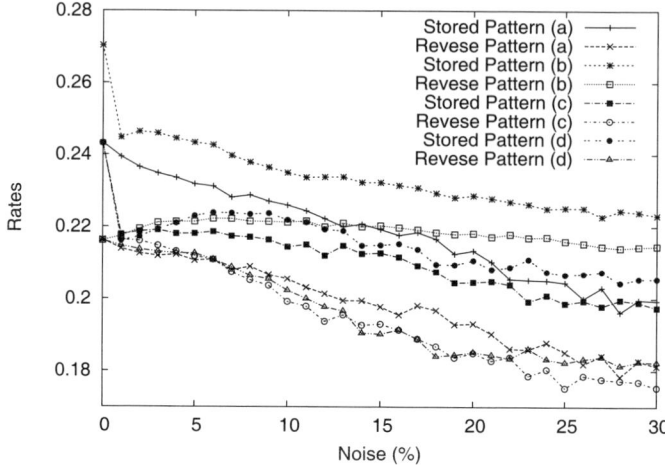

Fig. 11. The average rates of recalling the stored patterns and their reverse patterns from noisy initial patterns.

labeled by "Stored Pattern (a)" or "Reserve Pattern (a)" represents the average rates of recalling stored pattern (a) or its reverse pattern for the initial noisy patterns of the stored pattern (a) or its reverse pattern. Figure 11 shows that the average rates of recalling a stored pattern and its reserve pattern are dependent on the noises in the initial states, and the average recalling rates decrease as the noises increase. The above investigations further confirm that the controlled CNN can distinguish between two initial states with a small difference.

5. Conclusions

In this chapter, several chaos control methods concerned with the associative memory in the CNN have been reviewed. As an example, the delay feedback control method and its dynamics have been illustrated in details. The results show that the controlled CNN can been stabilized to a periodic sequence in which only one of the stored patterns and its reverse pattern related with the initial state are included. This indicates that the controlled CNN can be feasibly applied to information processing after controlling the chaos. The dynamics of the controlled CNN is usually periodic, and the period of the output sequence depends on the initial state. These characteristics are similar to those of the olfactory bulb in which the dynamics becomes ordered and nearly periodic when the brain identifies a familiar scent [5]. Therefore, the characteristics of the chaos control in the CNN may contribute to understanding the mechanism of the recognition in the real brain.

Figure 8 and Table 1 demonstrate that the stable output sequences of the controlled CNN are sensitive to the initial patterns. From the viewpoint of the control theory, the controlled CNN is characterized by "nonrobustness." However, for application to information processing, this nonrobustness can be an advantage. It is this nonrobustness that enables the controlled CNN to distinguish between two initial patterns with a small difference. The controlled CNN retains its characteristics that the dynamics is sensitive to the initial conditions. The conventional artificial neural networks such as the Hopfield network have associative memory functions and have been applied to pattern recognition and memory retrieval. In such networks, the outputs usually converge to the same fixed point from similar initial patterns with small differences. In other words, the network cannot distinguish between patterns with small differences. As described above, the controlled CNN can distinguish between different initial patterns, even if the two patterns have only a small difference. Moreover, the dynamics of the controlled CNN is more complex than that of the conventional neural networks. These properties can be useful and advantageous when the controlled CNN is applied to information processing.

References

[1] K. Aihara, and G. Matsumoto, "Chaotic oscillations and bifurcations in squid giant axons," In A. V. Holden (Ed.), Chaos, Princeton: Princeton University Press, 1986, pp.257–269.

[2] G. Matsumoto, K. Aihara, Y. Hanyu, N. Takahashi, S. Yoshizawa, and J. Nagumo, "Chaos and phase locking in normal squid axons," Phys. Lett. A, vol. 123, pp. 162–166, 1987.

[3] H. Degn, A. V. Holden, and L. F. Olsen (Eds.), Chaos in biological systems. New York: Plenum, 1987.

[4] A. Babloyantz, and A. Destexhe "Low-dimensional chaos in an instance of epilepsy," Proc. Nat. Acad. Sci. UAS, vol. 83, pp. 3513–3517, 1986.

[5] W. J. Freeman, "Simulation of chaotic EEG patterns with a dynamic model of the olfactory system," Biological Cybernetics, vol. 56, pp. 139–150, 1987.

[6] K. Aihara, T. Takabe, and M. Toyoda, "Chaotic neural networks," Phys. Lett. A, vol. 144, pp. 333–340, Mar. 1990.

[7] I. Tsuda, "Dynamic link of memory-chaotic memory map in nonequilibrium neural networks," Neural Networks, vol. 5, pp. 313–326, 1992.

[8] S. Nara, and P. Davis, "Learning feature constraints in a chaotic neural memory," Phys. Rev. E, vol. 55, pp. 826–830, Jan. 1997.

[9] J. Kuroiwa, S. Nara, and K. Aihara, "Response properties of a single chaotic neuron to stochastic inputs," Int. J. Bifurcation and Chaos, vol. 5, pp. 1447–1460, 2001.

[10] H. Nozawa, "A neural network model as a globally coupled map and applications based on chaos," Chaos, vol. 2, pp. 377–387, 1992.

[11] L. Chen, and K. Aihara, "Chaotic simulated annealing by a neural network model with transient chaos," Neural Networks, vol. 8, pp. 915–930, 1995.

[12] L. Chen, and K. Aihara, "Global searching ability of chaotic neural networks," IEEE Trans. Circuits and Systems I, vol. 46, pp. 974–993, 1999.

[13] M. Adachi, and K. Aihara, "Associative dynamics in a chaotic neural network," Neural Networks, vol. 10, pp. 83–98, 1997.

[14] M. Shiino, T. Fukai, "Self-consistent signal-to-noise analysis of the statistical behavior of analog neural networks and enhancement of the storage capacity," Phys. Rev. E, vol. 48, pp. 867–897, 1993.

[15] M. Nakagawa, "An artificial neuron model with a periodic activation function," J, Phys. Soc. Jpn, vol. 64, pp. 1023–1031, 1995.

[16] Y. Hayashi, "Oscillatory neural network and learning of continuously transformed patterns," Neural Networks, vol. 7, pp. 219–231, 1994.

[17] A. L. Hodgkin, and A. F. Huxley, "A quantitative description of membrane current and its application to conduction and excitation in nerve," J. Physiol. vol. 117, pp. 500–544, 1952.

[18] J. Nagumo, and S. Sato, "On a response characteristic of a mathematical neuron model," Kybernetik, vol. 10, pp. 155–164, 1972.

[19] J. J. Hopfield, "Neural networks and physical systems with emergent collective computation abilities," Proc. Natl. Acad. Sci. USA, vol. 79, pp. 2554–2558, 1982.

[20] E. Ott, C. Grebogi, and J. A. Yorke, "Controlling chaos," Phys. Rev. Lett., vol. 64, pp. 1196–1199, 1990.

[21] E. R. Hunt, "Stabilizing high-period orbits in a chaotic system: The diode resonator," Phys. Rev. Lett., vol. 67, pp. 1953–1955, Oct. 1991.

[22] K. Pyragas, "Continuous control of chaos by self-controlling feedback," Phys. Lett. A, vol. 170, pp. 421–428, 1992.

[23] G. Hu, and Z. Qu, "Controlling spatiotemporal chaos in coupled map lattice systems," Phys. Rev. Lett., vol. 72, pp. 68–77, 1994.

[24] W. L. Ditto, S. N. Rauseo, and M. L. Spano, "Experimental control of chaos," Phys. Rev. Lett., vol. 65, pp. 3211–3214, 1990.

[25] V. Petrov, V. Gaspar, J. Masere, and K. Showwalter, "Controlling chaos in the Belousov–Zhabotinsky reaction," Nature, vol. 361, pp. 240–243, 1993.

[26] R. W. Rollins, P. Parmanada, and P. Sherard, "Controlling chaos in highly dissipative system–a simple recursive algorithm," Phys. Rev. E, vol. 47, pp. 780–783, 1993.

[27] R. Roy, T. W. Murphy, T. D. Maier, and Z. Gills, "Dynamical control of a chaotic laser–experimental stabilization of globally coupled system," Phys. Rev. Lett., vol. 68, pp. 1259–1262, 1992.

[28] A. Garfinkel, M. Spano, W. L. Ditton, and J. Weiss, "Controlling cardiac chaos," Science, vol. 257, pp. 1230–1235, 1992.
[29] S. J. Schiff, K. Jerger, D. H. Duong, T. Chang, M. L. Spano, and W. L. Ditto, "Controlling chaos in the brain," Nature, vol. 370, pp. 615–620, 1994.
[30] X. Gong, F. Li, and H. Zhang, "On the control of chaotic neurons," Chaos, Solitons and Fractals, vol. 7, pp. 1397–1409, 1996.
[31] S. Sinha, "Controlled transition from chaos to periodic oscillations in a neural network model," Physical A, vol. 224, pp. 433–446, 1996.
[32] V. E. Bondarenko, "Control and 'anticontrol' of chaos in an analog neural network with time dely," Chaos, Solitons and Fractals, vol. 13, pp. 139–154, 2002.
[33] A. P. M. Tsui, and A. J. Jones, "Periodic respones to external stimulation of a chaotic neural network with delayed feedback," Int. J. Bifurcation and Chaos, vol. 9, pp. 713–722, 1999.
[34] V. E. Bondarenko, "High-dimensional chaotic neural network under external sinusoidal force," Phys. Lett. A, vol. 236, pp. 513–519, 1997.
[35] J. M. Cortes, J. J. Torres, and J. Marro, "Control of neural chaos by synaptic noise," BioSystem, vol. 87, pp. 186–190, 2007.
[36] M. Adachi, and K. Aihara, "An analysis on instantaneous stability of an associative chaosc neural network," Int. J. Bifurcation and Chaos, vol. 9, pp. 2157–2163, 1999.
[37] K. Nakamura, and M. Nakagawa, "On the associative model with parameter controlled chaos neurons," J. Phys. Soci. Jpn., vol. 62, pp. 2942–2955, 1993.
[38] M. Kushibe, Y. Liu, and J. Ohtsubo, "Associative memory with spatiotemporal chaos control," Phys. Rev. E, vol. 53, pp. 4502–4508, 1996.
[39] G. He, Z. Cao, H. Chen, and P. Zhu, "Controlling chaos in a chaotic neural network based on phase space constraint," Int. J. Modern Physics B, vol. 17, pp. 4209–4214, 2003.
[40] G. He, Z. Cao, P. Zhu, and H. Ogura, "Controlling chaos in a chaotic neural network," Neural Networks, vol. 16, pp. 1195–1200, 2003.
[41] G. He, J. Kuroiwa, H. Ogura, P. Zhu, Z. Cao, and P. Chen, "A type of delay feedback control of chaotic dynamics in a chaotic neural network," IEICE Trans. Fundamentals, vol. E87-A, pp. 1765–1771, 2004.
[42] G. He, M. D. Shrimali, and K. Aihara, "Partial state feedback control of chaotic neural network and its application," Phys. Lett. A, vol. 371, pp. 228–233, 2007.
[43] G. He, M. D. Shrimali, and K. Aihara, "Chaos control in a neural network with threshold activated coupling," Proceedings of the 2007 International Joint Conference on Neural Networks, pp. 350–354, August 12–17, 2007, Orlando, Florida, USA. (doi: 10.1109/ijcnn.2007.4370981)
[44] G. He, L. Chen, and K. Aihara, "Associative Memory with a Controlled Chaotic Neural Network," Neurocomputing, vol. 71, pp. 2794–2805, 2008. (doi: 10.1016/j.neucom.2007.09.005)
[45] G. He, M. D. Shrimali, and K. Aihara, "Threshold control of chaotic neural network," Neural Networks, vol. 21, pp. 114–121, 2008. (doi: 10.1016/j.neunet.2007.12.004)

[46] M. D. Shrimali, G. He, and K. Aihara, "Targeting spatio-temporal patterns in chaotic neural network," Optimization and Systems Biology, pp. 60–67, 2007.

[47] M. D. Shrimali, G. He, S. Sinha, and K. Aihara, "Control and synchronization of chaotic neurons under threshold activated coupling," Lecture Notes in Computer Science, vol. 4668: Part I, pp. 954–962, 2007.

[48] X. Zhang, and K. Shen, "Controlling spatiotemporal chaos via phase space compression," Phys. Rev. E, vol. 63, 046212, 2001.

[49] S. Mizutani, T. Sano, T. Uchiyama, and N. Sonehara, "Controlling chaos in chaotic neural networks," Electronics and Communication in Japan, Part 3, vol. 81, pp. 73–82, 1998.

[50] J. Singer, Y-Z. Wang, and Haim H. Bau, "Controlling a chaotic system," Phys. Rev. Lett., vol. 66, pp. 1123–1125, 1991.

[51] S. Parthasarathy, and S. Sinha, "Controlling chaos in unidimensional maps using constant feedback," Phys. Rev. E, vol. 51, pp. 6239–6242, 1995.

[52] A. Ahlborn, and U. Parlitz, "Controlling dynamical systems using multiple delay feedback control," Phys. Rev. E, vol. 72, 016206, 2005.

[53] A. J. Lichtenberg, and M. A. Lieberman, Regular and Stochastic Motion, Berlin, Springer-Verlag, 1983.

[54] Z. Tan, and M. K. Ali, "Pattern recognition in a neural network with chaos," Phys. Rev. E, vol. 58, pp. 3649–3652, 1998.

Chapter 3

ADAPTIVE FEEDBACK CONTROL OF PERIODIC ORBITS IN CHAOTIC SYSTEMS

Hiroyasu Ando[3,4,*], S. Boccaletti[1,2] and Kazuyuki Aihara[3,4]

[1] *CNR–Istituto dei Sistemi Complessi, Via Madonna del Piano, 10, 50019 Sesto Fiorentino (FI), Italy*
[2] *Embassy of Italy in Tel Aviv, Trade Tower, 25 Hamered Street, Tel Aviv, Israel*
[3] *Institute of Industrial Science, The University of Tokyo, 4-6-1 Komaba, Meguro-ku, Tokyo 153-8505, Japan*
[4] *Aihara Complexity Modelling Project, ERATO, JST, 3-23-5 Uehara, Shibuya-ku, Tokyo 151-0064, Japan*
[*] *andoh@sat.t.u-tokyo.ac.jp*

We describe an adaptive feedback control technique that is able to properly force the evolution of a chaotic system toward a desired periodic motion. First, we present a method enabling chaotic systems to change their dynamics to a stable periodic orbit, based on an adaptive feedback adjustment of an additional parameter of the system, and we discuss its reliability by applying it to several discrete-time systems, mainly focusing on one-dimensional unimodal maps. Then, we propose a strategy for controlling periodic orbits of desired periods in a chaotic dynamics and tracking them toward the set of unstable periodic orbits embedded within the original chaotic attractor. The proposed strategy does not require information on the system to be controlled nor on any reference states for the targets. Assessments on the method's effectiveness and robustness are given by means of the application of the technique for the stabilization of unstable periodic orbits in both discrete- and continuous-time systems. Additionally, we show how this procedure can be exploited to visualize the bifurcation structures of a chaotic dynamical system.

1. Introduction

During the last few decades, lots of technique on controlling chaos have been proposed despite many intractable properties of chaos, e.g. unpredictability, sensitive dependence on initial conditions, and so on. Generally, control of chaos refers to a process wherein a judiciously chosen perturbation is applied to a chaotic system, in order to realize a desirable (chaotic or periodic) behavior. Since the seminal contribution by Ott, Grebogi, and Yorke in 1990 [Ott et al. (1990)], this concept of harnessing a chaotic behavior has been variously developed and found a huge number of applications [Boccaletti et al. (2000)].

The two most widely used approaches to realize such a process are the OGY-based techniques and the methods based on time-delayed feedback control (DFC) [Pyragas (1992)]. While the OGY-based methods stabilize unstable periodic orbits (UPOs) embedded within the chaotic attractor using small time-dependent perturbations on some accessible control parameters, they in general require a reconstruction of the parametrical variations in the UPOs' stable and unstable manifolds in some section of the flow, which may difficult their real-time experimental applications to fast dynamical systems. On the other hand, DFC-based methods and their extensions [Socolar et al. (1994); Ahlborn and Parlitz (2004, 2006)] use a direct time-delayed feedback on an accessible system's variable. On its turn, DFC-based methods need to tune principal parameters, such as the delay time and the feedback gain, in advance.

We here present a technique which is able to circumvent both such difficulties. The strategy is based on a modification of the so called constant feedback (CF) method [Parthasarathy and Sinha (1995)], that makes use of an additional parameter to create a new dynamics in the original system, and stabilizes periodic orbits (POs) in the new dynamics. The CF method, indeed, shares with OGY and DFC the great advantage that no *a priori* information on the original system is required, and has been widely applied [Parekh et al. (1998); Sole et al. (1999)].

Furthermore, as we will see momentarily, the proposed technique tunes adaptively and automatically the CF's additional parameter, and therefore it has the further advantage of not even requiring the knowledge of the bifurcation analysis of the new system with respect to the additional parameter.

In addition, we show how to exploit the main ideas and concept of the method for the visualization of the bifurcation structure in transient processes. It is, indeed, common to analyze such dynamical systems by means of their stationary states, i.e. after having expired the corresponding transient processes. However, such transient evolutions carry sometimes relevant information about the system as a whole, and on its dynamics [Grebogi et al. (1983, 1986); Bezruchko et al. (2001)]. For example, chaotic transients [Grebogi et al. (1983, 1986)], which can be observed just past a crisis, are characteristic behaviors during transient processes. On the other hands, chaotic transients have been proven to be a useful source of information for the measurement of, e.g. the correlation dimension, the Lyapunov exponents, as well as for the detection of unstable periodic orbits [Dhamala et al. (2000, 2001)]. On its turn, the visualization of the bifurcation structure can also extract valuable information on the location of unstable periodic orbits.

2. Constant Feedback Method

In the CF method, the system under study is described as follows:

$$x_{n+1} = f(x_n, a) + k, \quad n = 0, 1, \ldots, N, \quad (1)$$

where $f(\cdot, a)$ is a one-dimensional unimodal map with a control parameter a. x_n is the value of the state variable at time n, and k represents an additional parameter for constant feedback. The chaotic dynamics can be converted to stable periodic one by tuning k externally [Gueron (1998)].

In what we propose here, we suppose that the dynamical system (see Eq. (1)) possesses a region in the parameter k, where chaotic behaviors are observed with a great number of intermingled periodic windows. In general, this kind of bifurcation structure with many periodic windows is ubiquitous for most smooth chaotic systems and it has been observed in many experimental systems.

3. Feedback Adjustment Stabilization of Periodic Orbits

For feedback adjustment to work in the system (see Eq. (1)), we introduce dynamics to k using feedback from the state variable x_n [Ando and Aihara (2006)]. Since k changes its value with time, it can be denoted by k_n. Furthermore, k_n is changing more slowly than x_n, i.e. k_n is updated every T steps of the time unit of x_n, where T is an integer parameter.

Let us consider the following one-dimensional map with an additional variable:

$$x_{n+1} = f(x_n, a) + k_n, n = 0, 1, 2, \ldots, N, \quad (2)$$
$$k_{n+1} = g_n(x_n, k_n), \quad (3)$$

where f represents the original system, n is the time index, $x_n, k_n \in \mathrm{R}$, and a is a control parameter fixed so as the map (2) would produce a

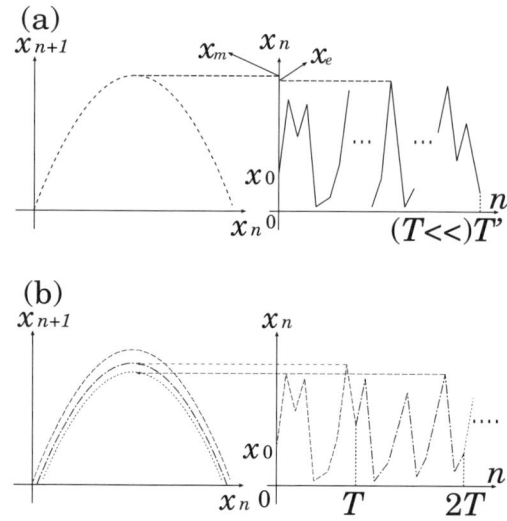

Fig. 1. Schematic illustration of the method. (a) The estimation process of the maximum of the function f in Eq. (2). (b) The mechanism of the feedback function g_n. The first T steps of time series (right) is generated by the map described by the dashed curve (left). Then, the map for the next T steps, which is denoted by the dashed-dotted curve (left), is determined with the largest value of the first T steps, such that the largest value corresponds to the maximum of the dashed-dotted map. The same procedure is repeated every T steps. Reprinted figure with permission from [Ando and Aihara (2006)]. Copyright (2006) American Physical Society.

chaotic orbit without k_n in (2). We assume that function f is upward single-humped. The function g_n is a feedback function depending on x_n, k_n, and n.

We begin to iterate the map (2) from an initial condition for x_0 and k_0. As k_n is changing every T steps according to Eq. (3), the feedback function g_n is defined as follows:

$$g_n(x_n, k_n) = \begin{cases} \hat{x}_i - x_e & \text{if } n = iT, \\ k_n & \text{if } n \neq iT, \end{cases} \quad i = 1, 2, 3, \ldots, \quad (4)$$

$$\hat{x}_i = \max\{x_j\}_{(i-1)T < j \leq iT}, \quad (5)$$

where x_e represents the approximated maximum of the function f, which is derived as the largest value of $T'(\gg T)$-length time series obtained from the iterations of (2) without k_n in advance. Here, T' is a integer parameter which is determined as an appropriate large value so as to conveniently approximate the maximum. The schematic illustration of the function g_n is shown in Fig. 1. The dynamics of k_n around a periodic window is schematic-

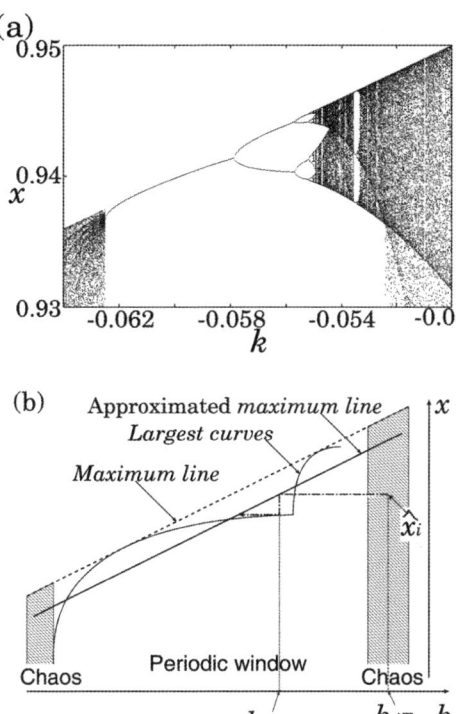

Fig. 2. (a) Upper part of a periodic window in the bifurcation diagram with respect to parameter k of Eq. (1) for the Logistic map. (b) Schematic diagram illustrating convergence mechanism to a stable fixed point of k_n in a periodic window. The dashed-dotted lines show the orbit of k_n. Reprinted figure with permission from [Ando and Aihara (2006)]. Copyright (2006) American Physical Society.

ally shown in Figs. 2(a) and 2(b). When $n = iT$, the feedback function g_n determines $k_{n+1} \equiv k_{iT+1}$ as shown in Fig. 2(b).

The locus of the maximum of $f(x) + k$ in the (k, x)-plane is denoted by $x = h(k)$ and we call the *maximum line*. In practice, using the estimated maximum value x_e, we have approximately the *maximum line* as follows:

$$h(k) = x_e + \epsilon k, \qquad (6)$$

with $\epsilon = 1$. Notice that the value of ϵ is independent on the function f, and therefore, the method is applicable to generic single-humped maps f.

The approximated *maximum line* is less than the real *maximum line*. Therefore, cross points exist between the approximated *maximum line* and the *largest curves* that are composed of the set of the largest values of periodic solutions, as shown in Fig. 2(b). The cross points correspond to the fixed points of the k_n-dynamics. Since there is no cross point in a chaotic region except periodic windows, the attraction of k_n to fixed points is possible only in periodic windows, that is, the orbit of k_n can be attracted by one of the cross points, as shown in Fig. 2(b).

3.1. *Crisis-free modification*

The values of k_{iT}, $i = 0, 1, 2, \ldots$, are almost always decreasing with increasing i. However, some chaotic systems described by Eq. (1) may have a lower limit k_c such that whenever $k < k_c (< 0)$ the orbit of x_n escapes from the chaotic attractor and diverges. Some classes of unimodal maps such as the exponential map have such k_c [Ando and Aihara (2006)].

To prevent such divergence, we keep k_n in a limited interval with the width Δk. We modify the function g_n in Eq. (4) when $n = iT$, $i = 1, 2, 3, \ldots$, such that k_n is kept in the interval $[k_0 - \Delta k, k_0]$, as follows:

$$g_n(x_n, k_n) = k_0 - \tilde{k}_n \quad \text{if } n = iT, \tag{7}$$

$$\tilde{k}_n = |k_0 - (\hat{x}_i - x_e)| \quad (\text{mod } \Delta k). \tag{8}$$

3.2. *Numerical simulations*

The above modification assures the convergence to a periodic window for a number of different initial values and for several systems in numerical simulations. We check 3 different chaotic systems, such as Logistic map with $f(x, a) = ax(1 - x)$, Exponential map with $f(x, a) = x \exp[a(1 - x)]$ [May and Oster (1976)], and the Chaotic neuron map [Aihara *et al.* (1990)] with $f(x, a) = 0.7x - 1/(1 + \exp(-x/0.02)) + a$. The parameter values and the initial setting are as follows: $a = 4, \Delta k = 0.499$ (Logistic map), $a = 2.8, \Delta k = 0.0989$ (Exponential map), $a = 0.35, \Delta k = 0.361$ (Chaotic neuron map), $T = 20$, $T' = 1000$, and $N = 20000$. We can confirm that the time series of k_n converges to periodic windows with respect to more than 95 % of 10000 different initial values for x_0. Figure 3 shows the process of feedback adjustment as time series of k_{iT} for i in the case of Exponential map. Figure 4(a) shows the bifurcation diagram with respect to parameter k of Eq. (1) for the exponential map. Figure 4(b) shows the distribution of the Lyapunov exponents with respect to converged k_n in the case of Exponential map.

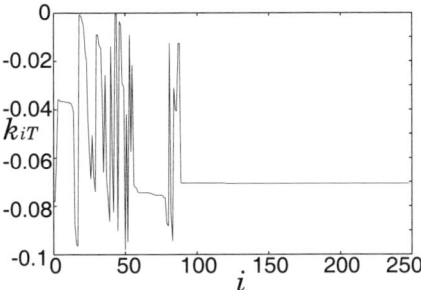

Fig. 3. Time series of k_n (when $n = iT$) showing a process of feedback adjustment to a value at which a periodic state is generated. The time series data are shown till $i = 250$. Reprinted figure with permission from [Ando and Aihara (2006)]. Copyright (2006) American Physical Society.

4. Goal-Oriented Feedback Adjustment Control

In this section, we improve the above method to goal-oriented control of unstable periodic orbits, which has a desired period and is embedded in a chaotic attractor as an ordinary chaos control technique does. Furthermore, we apply the improved method to continuous-time systems with a view to application to experimental systems [Ando, Boccaletti, and Aihara (2007)].

The control process consists of two stages. In the first stage, the CF's parameter is tuned automatically to the value which produces a desired PO. In the second stage, the new dynamics created by the changes in the CF's parameter is tracked back to the original one and therefore the controlled POs are included in the set of UPOs embedded within the original chaotic attractor.

4.1. Feedback adjustment control with period locking

Let us consider again the system described by Eqs. (2) and (3). Here, the function g_n is a variation of the feedback adjustment function in the previous one. The mechanism of the new g_n is illustrated in Fig. 5.

Let L (the targeting period) be integer parameters. Here, one scans the time series of x_n with a constant value for k_n, by T-length time intervals. The locking condition is that the second-largest value and the first-largest value (denoted by $\hat{x}$) in T-length scanning time interval are L-length away from each other in that order. While the locking condition is not satisfied, $k_{n+1} = k_n (\equiv g(x_n, k_n))$. Once the locking condition is satisfied, k_n is adjusted such that $\hat{x}$ is equal to the maximum of $f(x, a) + k_{n+1}$, i.e.

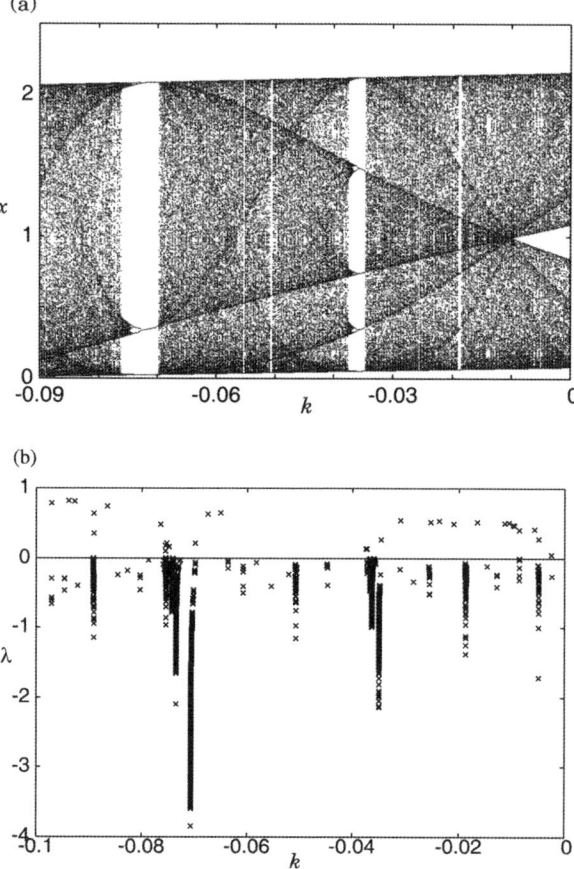

Fig. 4. (a) Bifurcation diagram with respect to parameter k of Eq. (1) for the exponential map. (b) Distribution of Lyapunov exponent with respect to converged values of k_n for 10000 different initial values for x_0 in the case of Exponential map. The horizontal and the vertical axis show k and Lyapunov exponent λ, respectively. Reprinted figure with permission from [Ando and Aihara (2006)]. Copyright (2006) American Physical Society.

$k_{n+1} = \hat{x} - x_e (\equiv g(x_n, k_n))$. We repeat this procedure till $n = N$ for initial conditions x_0 and k_0.

In order to elucidate that the locking process work appropriately to obtain period-L periodic orbits, we do numerical simulations by applying the method to the Logistic map with $a = 4$. Figures 6(a) and 6(b) show the time series x_n and k_n respectively when the locking is applied with $L = 7$.

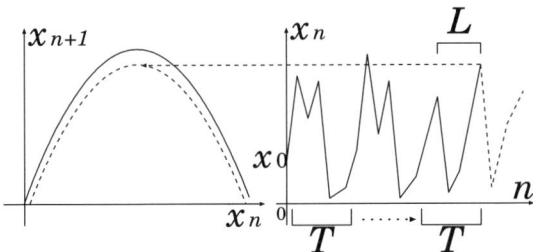

Fig. 5. Schematic illustration of the locking process. The solid map (left) generates the time series (right). The dashed map (left) is adjusted one when the locking condition hold. Reprinted figure with permission from [Ando, Boccaletti, and Aihara (2007)]. Copyright (2007) American Physical Society.

Table 1. Number of POs whose period is L with respect to 128 initial values of x_0. The lowest row shows the number of period-L POs obtained by the original method, i.e. without locking when $T = 20$. $N = 20000$, $T' = 20000$.

L	2	3	5	7	11
T	3	8	16	23	17
Number of POs	109	119	77	56	5
Number of POs (original)	107	49	25	1	0

Table 1 shows the numbers of POs whose period are related to the value of L with respect to 5 cases of L and 128 initial values of x_0. The initial values are chosen in $[0, 0.5]$ and $k_0 = 0$. Here we determine that the period of POs is L by that the period is L or its multiples. The values of T are chosen such that the number of period-L POs becomes maximum in $T < 30$. The lowest row of the table shows the number of period-L POs obtained by the original method proposed in the previous Section for $T = 20$. We can find out that there are significant differences between the original method and the locking method.

The adjustment action towards periodic windows is similar to the original method in the previous Section. In the original method, locally stable attractors of the dynamics for k_n exist in periodic windows (see Fig. 2(b)), and k_n is changing its value till trapped into a periodic window and converging to the attractors.

Moreover, the locking condition limits the adjustment action. Once the orbit is trapped into a periodic window, the order pattern of the time series in every T-length scanning interval is also changing periodically. Hence, the adjustment action is going to be terminated in the trapping periodic

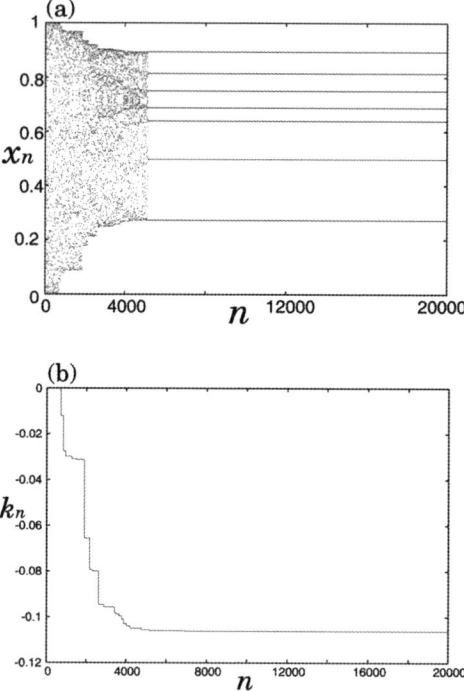

Fig. 6. (a) Time series of x_n when the locking control is applied. (b) Corresponding time series of k_n. $L = 7, T = 24, N = 20000$, and $T' = 20000$. Reprinted figure with permission from [Ando, Boccaletti, and Aihara (2007)]. Copyright (2007) American Physical Society.

window, especially where the period of the window is L. The period of the obtained PO is mL, where m is a positive integer satisfying the condition: $mL < T$.

More precisely, if $1 < m$, then the obtained POs correspond to the locally stable attractors encountered in the original method. In this case, k_n converges to such attractors. On the other hand, if $m = 1$, then the locking condition terminates the adjustment action completely and thus k_n stops changing its value not exactly at the attractors. This is because the largest and the second largest values have coalesced.

The trapping processes to a period-L periodic window are illustrated in Figs. 7(a) to 7(d). The shaded (green) part represents the regions where the second largest points can exist with the locking condition. The upper bound of the shaded regions correspond to period-L UPOs. As shown in Figs. 7(a) and 7(c), the largest and the second-largest points move along the

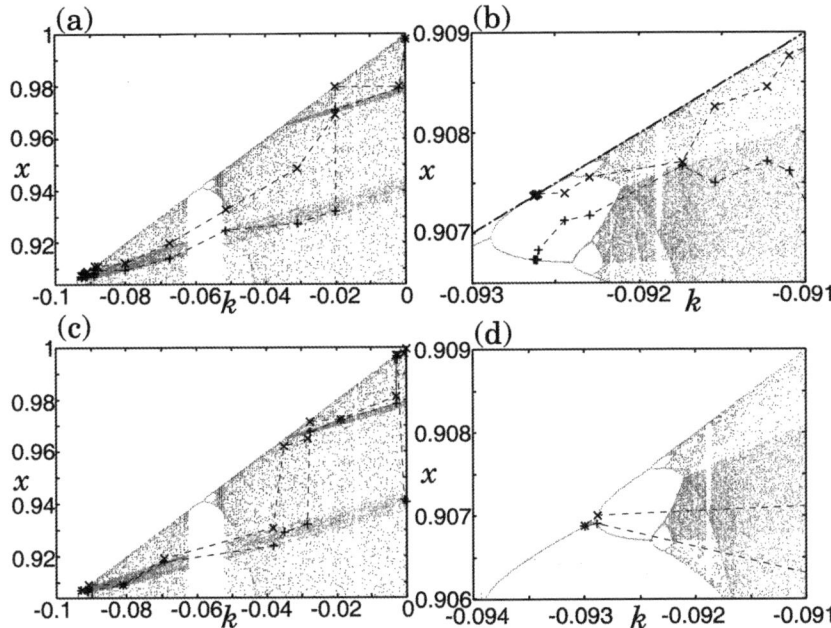

Fig. 7. Adjustment processes in the bifurcation diagram with respect to k. The shaded parts represent the regions where the second largest points can exist when the locking condition holds. The × and + denote the largest and the second largest values (L-length apart) for each k_n. (a)–(b) In the case that the period of the obtained PO is $2L$, (b) The enlargement around the trapping periodic window in (a). The dashed-dotted line shows $x = h(k)$. (c)–(d) In the case that the period is L, (d) The enlargement around the trapping periodic window in (c). $L = 5$, $T = 16$, $N = 20000$, and $T' = 20000$. Reprinted figure with permission from [Ando, Boccaletti, and Aihara (2007)]. Copyright (2007) American Physical Society.

upper bounds. Therefore, the period of the trapping periodic window is L. In Fig. 7(b), the locally stable attractor for k_n is shown as the intersection between the dashed-dotted line and the largest points of POs in the periodic window.

Let us next apply the method to continuous-time systems, such as the Rössler model [Rössler (1976)] and the three-dimensional BZ reaction model [Györgi and Field (1992)]. As these systems are strongly dissipative, i.e. the attractors are stretched in one direction on the Poincaré surface of section, and hence they can be reduced to one-dimensional maps. Thus, we add a control feedback impulsively only into the one system variable on the Poincaré surface of section. Here, the amplitude of the impulsive control

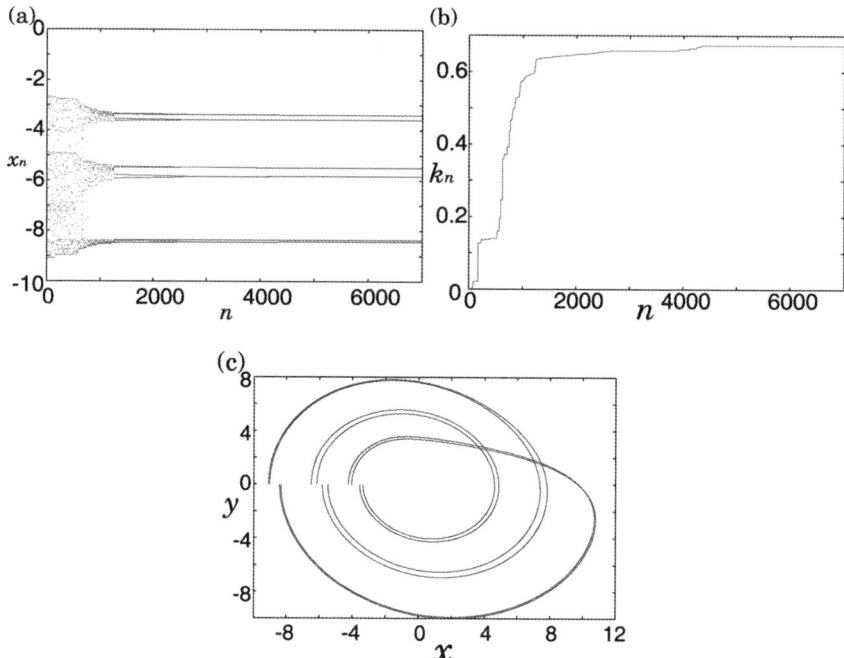

Fig. 8. (a) Time series of x_n on the Poincaré surface of section with respect to minima of x in the Rössler model, (b) Corresponding time series of the feedback k_n, (c) Controlled period-6 orbit. $L = 6, T = 20, N = 7000, T' = 10000$, and $k_0 = 0$.

feedback corresponds to k_n in discrete-time systems. It should be noticed that, in the case of the Rössler model and the BZ reaction model, the minima of the reduced return maps are focused, since the reduced maps are downward single-humped. Hence, in this case, $h(k)$ corresponds to the approximated minimum of the function $f(\cdot) + k$.

Figure 8 shows the results when the locking method with $L = 6$ is applied to the Rössler model by means of the reduced one-dimensional map. Figures 8(a) and 8(b) shows the time series of x_n and k_n. Figure 8(c) shows the controlled period-6 orbit. Figure 9 shows the application to the BZ reaction model with $L = 5$.

4.2. Tracking periodic orbits

As shown in Figs. 8(c) and 9(c), due to the impulsive nature of the feedback considered in the present case, the controlled orbits are discontinuous at

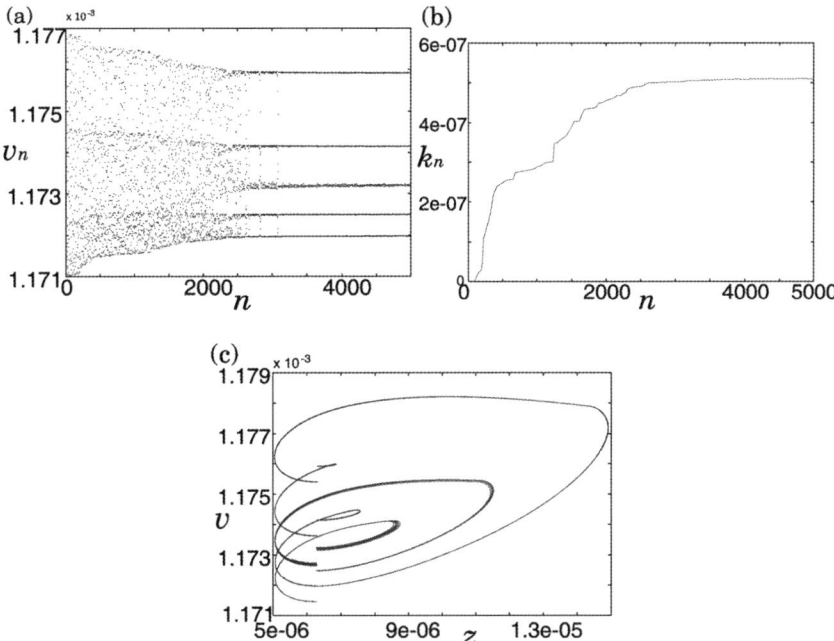

Fig. 9. (a) Time series of v_n on the Poincaré surface of section with respect to minima of v in the BZ reaction model, (b) Corresponding time series of the feedback k_n, (c) Controlled period-5 orbit. $L = 5$, $T = 15$, $N = 5000$, $T' = 10000$, and $k_0 = 0$.

the Poincaé surfaces of section. Thus, this discontinuous periodic orbits are different from the periodic orbits embedded in the original chaotic attractor.

To trace back the controlled POs to the original UPOs, an additional step is introduced, consisting of a tracking method of periodic orbits that changes the value of $k_N (\equiv K)$ slightly in the direction of making it to vanish. First, we present a tracking method in discrete-time systems. Then, the tracking method is applied to continuous-time systems through the one-dimensional reduced map on the Poincaré surface of section.

Let the pre-image of the controlled period-L POs in the reduced map be described as the solution of the following equation:

$$F(x, K) = F^{L+1}(x, K), \qquad (9)$$

where $F(x, k) = f(x, a) + k$ and here f represents the reduced maps when $k_n = 0$, and hence is downward humped. Let us consider the smallest point of the controlled PO. The pre-image of the smallest point is denoted by X_p.

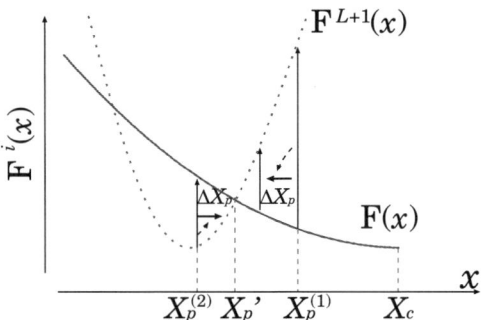

Fig. 10. Schematic illustration for control from X_p to X'_p in the case of a downward-humped map and $X_c > X_p$. If $X_p^{(1)} > X'_p$, then $\Delta X_p < 0$. If $X_p^{(2)} < X'_p$, then $\Delta X_p > 0$. Reprinted figure with permission from [Ando, Boccaletti, and Aihara (2007)]. Copyright (2007) American Physical Society.

We track the pre-image as follows. First, we change the value of K to K' by ΔK. If $|\Delta K|$ is sufficiently small, the pre-image X'_p (the solution of Eq. (9) for K') is close to X_p.

In the next step, thus, it is possible to control X_p to X'_p in the following way. First, we change X_p to $X_p - \Delta X_p$. The ΔX_p is determined as follows:

$$\Delta X_p = \sigma \, \mathrm{sgn}(X_p - X_c)(F(X_p, K') - F^{L+1}(X_p, K')), \qquad (10)$$

where σ is a parameter and $0 < \sigma \ll 1$. X_c denotes the approximated critical point.

The sign of ΔX_p is determined as illustrated in Fig. 10 which shows in case that $X_c > X'_p$. The value of $|F(X_p, K') - F^{L+1}(X_p, K')|$ is decreasing as X_p is approaching X'_p. Therefore, X_p is converging to X'_p.

After convergence to X'_p, then we change K' by ΔK again and repeat the procedure controlling to X'_p till K goes to zero. As a result, we can obtain period-L UPOs embedded in the original chaotic attractor.

We apply the feedback adjustment with the locking condition and the tracking method to the Rössler model and the BZ reaction model by means of the reduced one-dimensional map on the Poincaré surface of section. The tracking processes of $F(X'_p, K')$ with the partial bifurcation diagram are shown in Figs. 11(a) and 12(a). The UPOs embedded in the original chaotic attractor also shown in Figs. 11 and 12(b)–12(d). The discontinuities of the stable POs as shown in Figs. 8(c) and 9(c) disappear as the values of k_n vanish.

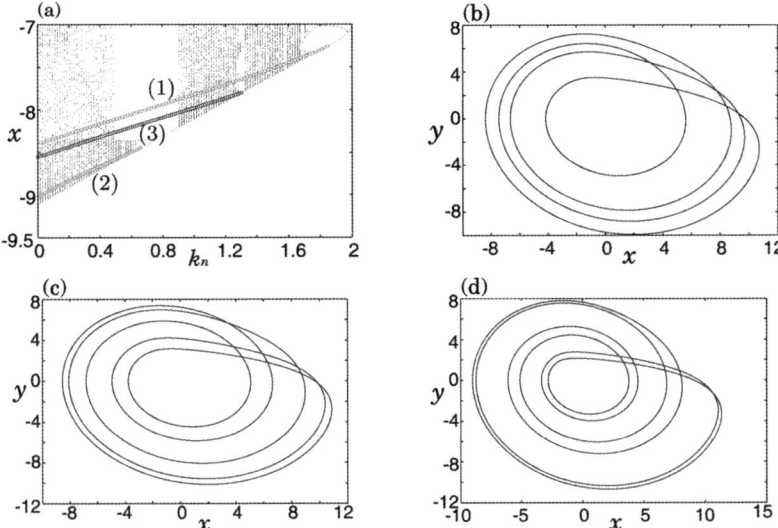

Fig. 11. (a) Tracking process of POs for the Rössler model with the partial bifurcation diagram. The open circles show the loci of $F(X'_p, K')$. (1)–(3) represent the different periods, period-4 (1), period-6 (2), period-5 (3). The UPOs embedded in the original chaotic attractor are shown in (b) period-4 ($T = 12, L = 4, k_0 = 1$), (c) period-5 ($T = 15, L = 5, k_0 = 1$), (d) period-6 ($T = 20, L = 6, k_0 = 0$). $N = 7000, T' = 10000, \sigma = 0.05$, and $\delta k_N = k_N/100$.

4.3. Robustness against noise

Finally, we discuss the robustness of the method against additive noise, as this is a crucial point to be assessed in view of applications to experimental systems. To this purpose, we apply the method to the Rössler system with an additional noise term $D\xi(t)$ in the equation for $\dot{x}$, where D is a amplitude parameter and $\xi(t)$ is a white noise process with zero mean and delta-correlated in time ($<\xi(t)\xi(t')> = \delta(t - t')$).

In Figs. 13(a) and 13(b) we show an example of the behavior of the Rössler system with noise when the method is applied. Namely, Fig. 13(a) shows the results by the time series of x_n and k_n. The controlled and tracked period-3 UPO shows in Fig. 13(b). The amplitude of the additional noise $D = 0.2$ is about one percent of the signal amplitude. Moreover, we also investigate the noise resistance of the method with respect to 100 different initial conditions for the same amplitude of noise, and numerically confirm successes of the method in 66 cases.

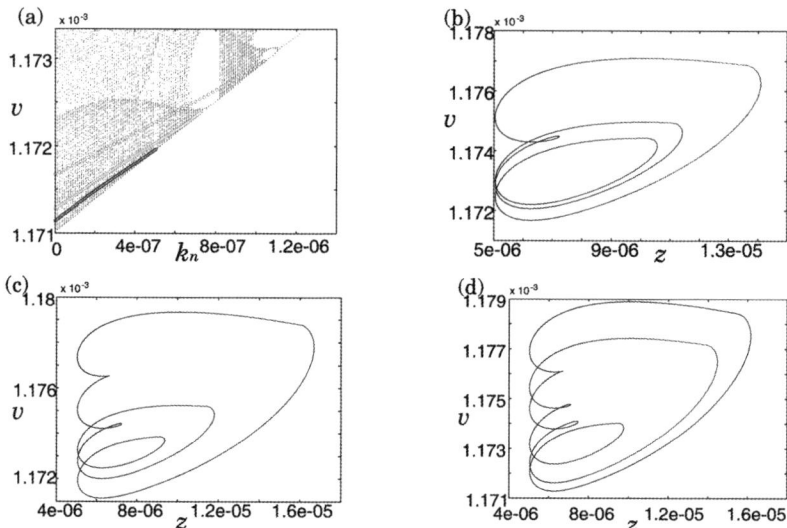

Fig. 12. (a) Tracking process of POs for the BZ reaction model with the partial bifurcation diagram. The open circle and tones are the same as Fig. 11. The UPOs are shown in (b) period-4 ($T = 8, L = 4$), (c) period-5 ($T = 15, L = 5$), (d) period-6 ($T = 15, L = 6$). $N = 5000, T' = 10000, k_0 = 0, \sigma = 0.005. \delta k_N = k_N/50$ for (b) and (c), and $\delta k_N = k_N/100$ for (d).

5. Bifurcation Structures in Transient Processes

In this section, we propose a method related with the above control technique which visualizes the bifurcation structures of dynamical systems by means of their periodic orbits in one-dimensional maps. We focus on the particular kind of transient processes to capture the information of the dynamical systems, which is lost in ordinary analysis using stationary states. In the method, we consider the relation between some characteristic values, such as the largest value and the time steps to the largest value in transient processes, and periodic orbits of the dynamical systems. Generally speaking, transient processes are the behavior of dynamical systems before the systems settle down into their stationary states, e.g. periodic or chaotic attractors. We define transient processes, from a slightly different perspective, as the behavior of the dynamical systems in a fixed length time steps from an initial state. However, this definition of transient processes is included in the general definition, if the fixed length of time steps is not large.

It should be noted that these characteristic values are related with the work known as the extreme value statistics [Balakrishnan et al. (1995)], the

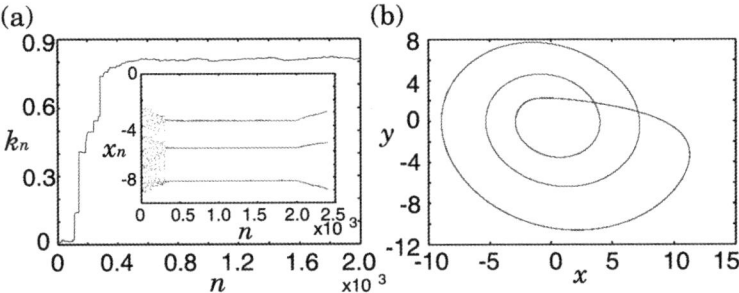

Fig. 13. Control and tracking processes in the Rössler system with noise. $D = 0.2$. (a) Time seriese of k_n (inset: time series of x_n including tracking processes). (b) Controlled and tracked period-3 UPO. $T = 10, L = 3, k_0 = 0, N = 2000, T' = 200, \sigma = 0.05$, and $\Delta K = K/100$. Reprinted figure with permission from [Ando, Boccaletti, and Aihara (2007)]. Copyright (2007) American Physical Society.

order statistics [Valsakumar et al. (1999)], and the turning point analysis [Diakonos and Schmelcher (1997)] in chaotic dynamics. The approaches are similar to our method in that they use the order of the time series. However, the proposed method focuses on visualizing the bifurcation structures.

In the following, we introduce the analysis method using short transient processes in a one-dimensional map and depict the analyzed results by shaded look-up *maps* with varying the control parameter. Then, we compare the results of our analysis with the ordinary bifurcation diagram for a simple one-dimensional map. As a result, we can also observe the period-doubling and tangent bifurcations in the *maps*. Furthermore, we can observe the parameter values where superstable periodic orbits occur. Finally, we will mention the connection between the existing POs for an object system and the *map*.

5.1. *Visualization method of bifurcation structures*

Let $f(x, a) : R \mapsto R, x \in R, a \in R^n$ be a continuous and single-humped function with control parameters a. Although the number of control parameters is not necessarily limited to $n = 1$, we assume $n = 1$ for simplicity. The analysis method is applied to the dynamical system:

$$x_{n+1} = f(x_n, a), \quad n = 0, 1, 2, \ldots. \tag{11}$$

The following is the protocol to capture the bifurcation structure of the system (see Eq. (11)).

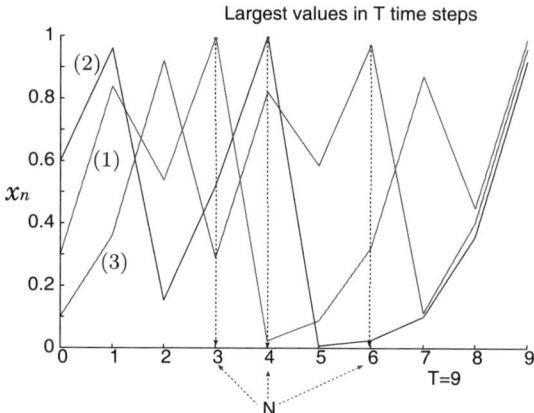

Fig. 14. Schematic illustration for the value N with respect to several initial values x_0. $N = 3, 4$, and 6 for orbits (1), (2) and (3), respectively. $T = 9$.

(1) Determine the time length T and observe T-length transient process from an initial value.
(2) Set initial values with respect to the variable x_0 and parameter a.
(3) Iterate Eq. (11) for T steps from x_0 with a.
(4) Let N be the number of mapping to the largest value of the time series $\{x_i\}_{0<i\leq T}$.
(5) Continue the above procedures for a number of initial conditions x_0 and a in the considering interval for the variable and the parameter.

Figure 14 illustrates the fourth procedure in the above method. The figure shows time series when $T = 9$ for three different initial values x_0. Each x_0 generates the different orbit and N that is the number of time steps to the largest value.

Then, we can obtain a look-up type corresponding *map* of N with respect to initial conditions x_0 and a. We call this *map* BT (*Bifurcation-in-Transience*)-*map*. Although we can take any large value of T, we will consider smaller values of T for simplicity. Figure 15 shows the BT-*maps* for $T = 5, 6, 7$, and 8 when the map $f(\cdot, a)$ is the Logistic map. The horizontal and vertical axes indicate the parameter a and the initial value x_0, respectively. Since the Logistic map has symmetry centering around the critical point 0.5, we show the half domain of the initial conditions for x_0, i.e. $x_0 \in [0, 0.5]$, in Fig. 15.

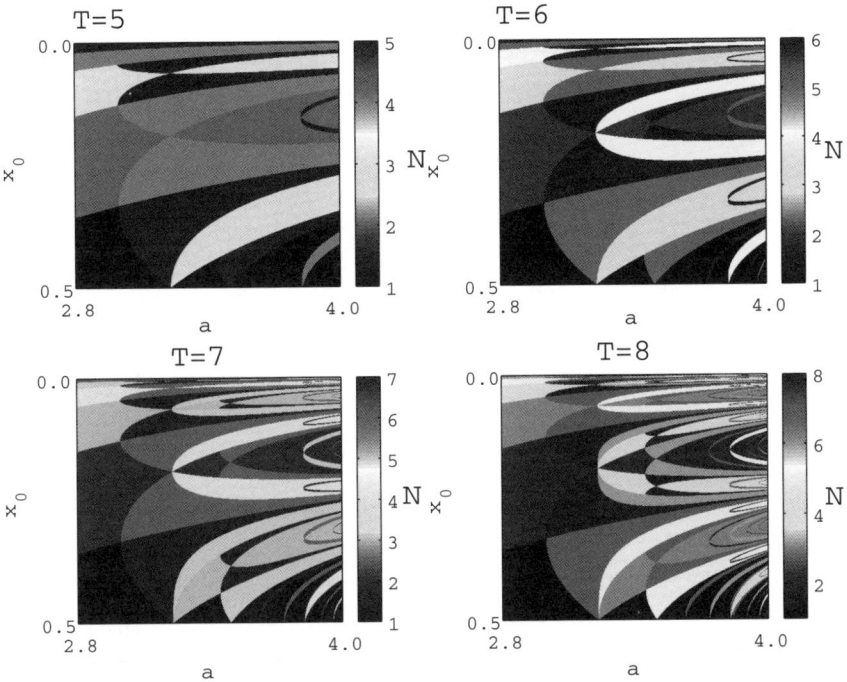

Fig. 15. BT-*map* for $T = 5, 6, 7$, and 8 with respect to the Logistic map. The different tones in the figures denote the values of N following the bar next to each figure. We can find the same structure composed of the boundaries of shaded regions.

We can find the same structure in each *map*, that is, the boundaries of the shaded regions are appearing with increase of T but not disappearing. As discussing at the next Subsection, the boundaries correspond to loci of pre-images of periodic solutions of the considered map (11). Remark that, since the periodic solutions with any periods can co-exist, the loci never disappear even if the length of transient T increases more than the present value.

It should be noted that we can construct BT-*map* without the detailed knowledge of the equation describing the systems except parameter. This fact implies that we can detect unstable periodic orbits embedded in chaotic attractor not only in one-dimensional map but also the continuous-time systems reduced to one-dimensional maps, which cannot be described explicitly, using Poincaré surface of section.

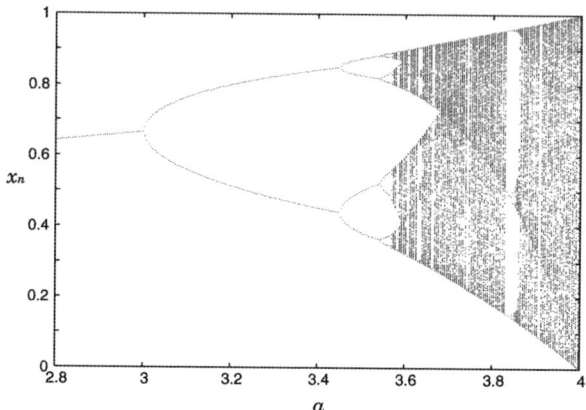

Fig. 16. Bifurcation diagram of the Logistic map. The vertical and horizontal axis denote the variable x_n and a respectively.

5.2. *Boundaries in BT-map*

Before explaining the bifurcation structures in BT-*map*, we consider what the boundaries of shaded regions mean. Let N_1 and N_2 be the numbers which correspond to two regions next to each other, and $N_1 > N_2$ is assumed. We denote N-times map of f as f^N. Since $f^{N_1}(x_0, a)$ and $f^{N_2}(x_0, a)$ are the same largest value at the boundary in T-length time series for the initial condition (x_0, a), the boundary between these two regions are described by the following equation,

$$f^{N_1}(x_0, a) = f^{N_2}(x_0, a). \qquad (12)$$

The solution of Eq. (12) represents the N_2 times pre-image of the largest value of the periodic solution whose period is $N_1 - N_2$ or its divisor. For this reason, it is obvious that the periods of the periodic solutions are less than $T - 1$. Thus, the boundaries in the BT-*map* correspond to the loci of pre-images of the largest values of the periodic solutions of the map (11) with respect to the parameter a.

5.3. *Bifurcation structures in BT-map*

In this Subsection, we compare the bifurcation structures observable in BT-*map* with the ordinary bifurcation diagram using stationary states, regarding unimodal maps such as the Logistic map.

Figure 16 shows the normal bifurcation diagram of the Logistic map for stationary states. In the bifurcation diagram, we can observe two ordinary

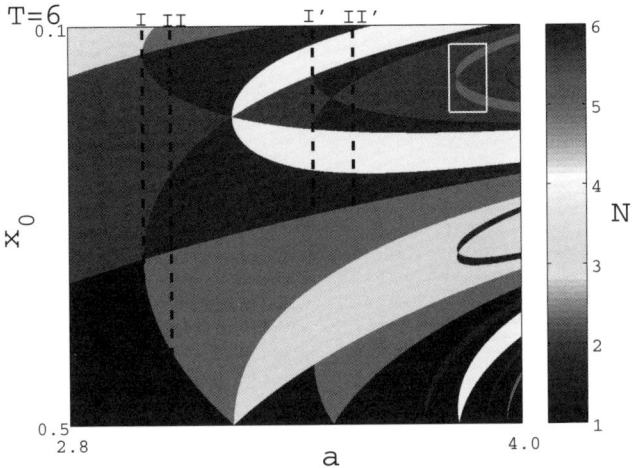

Fig. 17. BT-*map* when $T = 6$. The dashed lines (I) and (I') denote the period-doubling bifurcation points at $a = 3$ and $a \simeq 3.45$ respectively, for the Logistic map. The dashed lines (II) and (II') correspond to the points after period-doubling bifurcation. The detailed mechanism of the bifurcation will be elucidated in the next figure, where we consider the section of the BT-*map* with respect to a fixed value of a. The rectangle at the right upper part will be used to illustrate the saddle-node bifurcation.

types of bifurcations that are well-known as the period-doubling and the saddle-node (tangent) bifurcations. To compare the structures composed of the boundaries in the BT-*map* with the ordinary bifurcation structures found in Fig. 16, we focus on the parameter value around $a = 3$ and $a = 3.45$ for the period-doubling bifurcation, and around $a = 3.83$ for the saddle-node bifurcation.

Figure 17 shows the BT-*map* for $T = 6$ in $x_0 \in [0.1, 0.5], a \in [2.8, 4.0]$. The dashed lines (I) and (I') are the period-doubling points from period-1 to period-2 and from period-2 to period-4, respectively. The dashed lines (II) and (II') are the after the period-doubling bifurcations.

For understanding the period-doubling bifurcation, we consider the section along the dashed lines. In Fig. 18, each figure represents the superimposed map for $(x, f^i(x, a))$, where i is an integer less than T, when the value of a is fixed. The Roman numeral putting above each figure represents that the figure corresponds to the section along the dashed line labeled the same numeral in Fig. 17. The arrows and the numbers beside them mean that the ordinates of the cross points are the largest values of periodic solutions. The periods of the periodic solutions are the difference or its divisor of

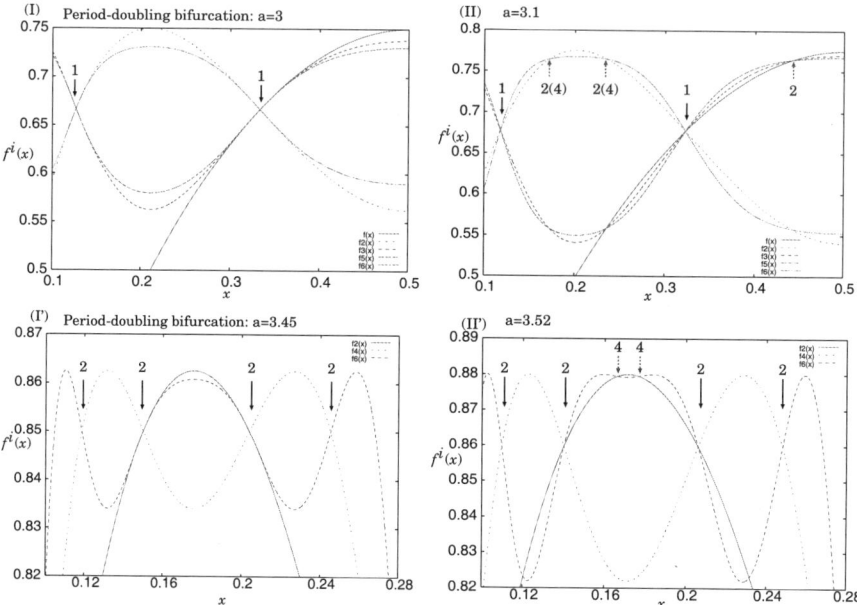

Fig. 18. Sections of the BT-*map* with superimposed maps for $(x, f^i(x, a))$, where a is fixed. $a = 3.0, 3.1, 3.45$, and 3.5 for (I), (II), (I'), and (II'), respectively. i is an integer less than T. The roman numeral putting above each figure corresponds to that in Fig. 17. The arrows and the numbers beside them mean that the ordinates of the cross points are the largest values of periodic solutions with the numbers as periods.

two N's, which are next to each other. The numbers between parentheses are the actual difference of N's, while the numbers before parentheses are the divisor of the difference corresponding to the actual periods. The dotted arrows in (II) and (II') indicate the appeared periodic solutions after the period-doubling bifurcations. We can observe that the periodic points lose their stability and new period-doubled periodic points are appeared through the bifurcation from (I) to (II) and also from (I') to (II').

Figure 19 shows the magnified view of the rectangle region shown in Fig. 17. In this figure, we can observe the saddle-node bifurcation that occurs at the dashed line (III). Then, the light gray region has appeared. The boundaries of the appeared region represent unstable saddles and stable nodes. For capturing the mechanism of the appearances of saddles and nodes, we also consider the section of the BT-map again, along the line (III) and (IV). Furthermore, we consider the section along the line (V) for characterizing the occurrence of superstable periodicity in the BT-*map*.

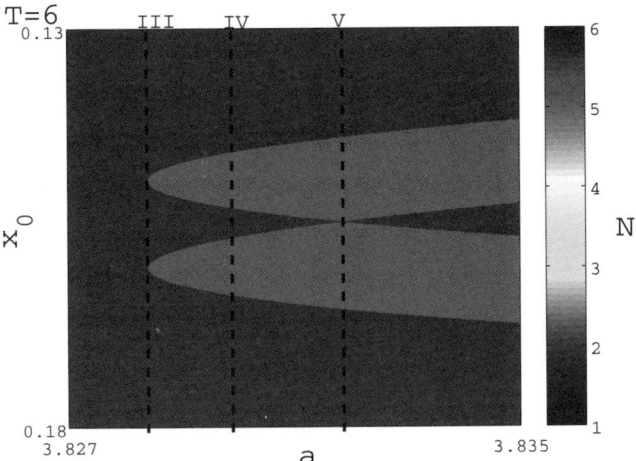

Fig. 19. The magnified view of the part surrounded by the rectangle in Fig. 17. The boundary appearance at the dashed line (III) represents the saddle-node bifurcation, which is also understood by the section along the dashed line (III) and (IV). Additionally, the line (V) corresponds to a superstable point.

Figure 20 shows the superimposed maps of $(x, f^i(x, a))$ representing the section of BT-*map* like Fig. 18. The figure (III) for $a = 3.8283\ldots$ describes the saddle-node bifurcation at the points indicated by the arrows. After the saddle-node bifurcation, unstable periodic points and stable periodic points have appeared, indicated by dashed and dotted arrows respectively in the figure (IV). Further increasing the value of a, we can observe a superstable periodic solution represented by the figure (V), where the two map $f^2(x, a)$ and $f^5(x, a)$ have a tangent point and the derivatives of the both maps at the tangent point are zeros.

Following the above discussion and correspondence between BT-*map* and its section at a fixed value of a, we can extract the pattern changes of describing the period-doubling, saddle-node bifurcations, and superstable points in BT-*map*. It should be noted that such bifurcations and superstable points can be observed at more than two parts at the same value of a in Fig. 18, which can be also observed in Fig. 17. This multiple visualization of the bifurcation structure is a characteristic of BT-*map*. Furthermore, we cannot observe chaotic behavior in the BT-*map*, since we focus on short transient processes.

Finally, let us mention that this visualization method can also apply to detection of UPOs as the boundaries in BT-*maps*. It should be remarked

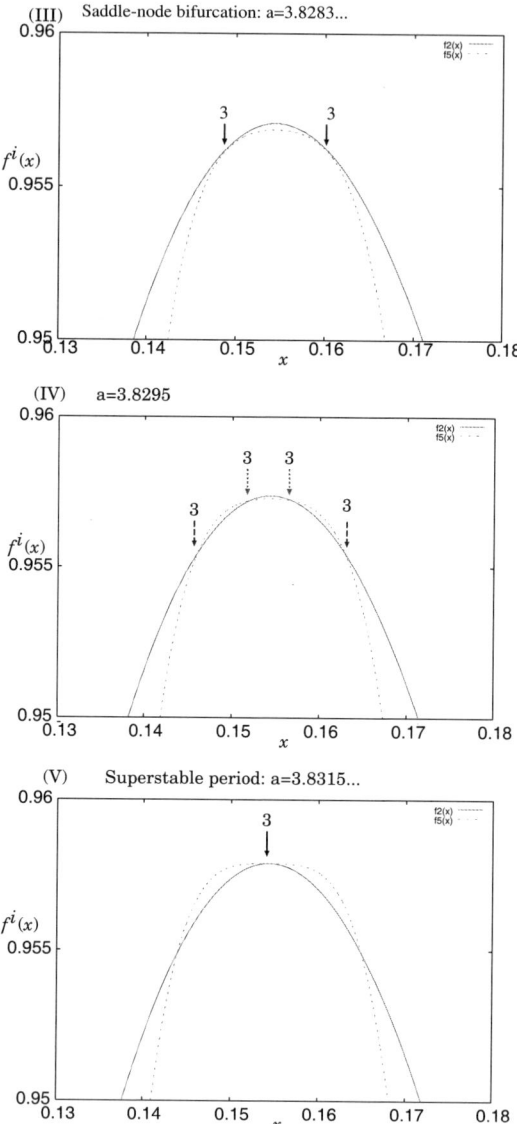

Fig. 20. Sections along the lines shown by Fig. 19 in the same way of Fig. 18. $a = 3.8283\ldots, 3.8295$, and $3.8315\ldots$, respectively. The figure (III) shows the instance of the saddle-node bifurcation with respect to period-3, and the figure (IV) shows after the bifurcation. The figure (IV) represents that the period-3 superstable periodic solution occurs. The arrows and the numbers are as well in Fig. 18. The dotted arrows and the dashed arrows indicate stable and unstable periodic points, respectively.

that since this detection of UPOs uses only transient processes, even if the parameter value is in the period-p stable periodic region, it is possible to detect period-q ($q \neq p$) unstable periodic orbits by the proposed method, where $q < T - 1$. This point is different from ordinary method to detect UPOs embedded in chaotic atractor [Lathrop and Kostelich (1989); So *et al.* (1996); Schmelcher and Diakonos (1997)]. This detection of UPOs by BT-*map* is not so hard that we have succeeded the detection in the Logistic map and Róssler system (not shown here). We do not mention the detailed algorithm here, but it is similar to one with the above tracking process of UPOs.

5.4. *Periodic points in BT-map*

Here let us consider a theorem with respect to the connection between existing POs and the POs in BT-*map*. For explanation, let us define the following notations.

$$F_T(x, a) = \max_{1 \leq i \leq T} \{f^i(x, a)\}, \tag{13}$$

$$N_T(x, a) = m, \tag{14}$$

where m is the number which gives the maximum in Eq. (13) for i. For simplicity, we assume that the map (11) is single-humped and has a control parameter a.

The thick curves in Fig. 21 shows the return plot for x vs. $F_T(x, a)$. Other curves represent $f^i(x, a)$, $1 \leq i \leq T$. The legends are shown at the right-bottom in the figure. The cusp points between the thick curves correspond to the boundaries in the BT-*map*. Therefore, the ordinates of the cusp points are the largest values of periodic orbits. The numbers indicating the cusp points denote the period of the periodic orbits.

It should be noted that the cusp points in Fig. 21 can be connected to whole existing periodic solutions whose periods are less than $T - 1$. (The proof is not shown here.)

6. Conclusion and Discussion

We presented a robust and reliable method able to control a chaotic orbit into a UPO which has the desired period. The method does not require any detailed information of the controlled system such as system parameters, but uses automatically adjusting feedback of the additional parameter.

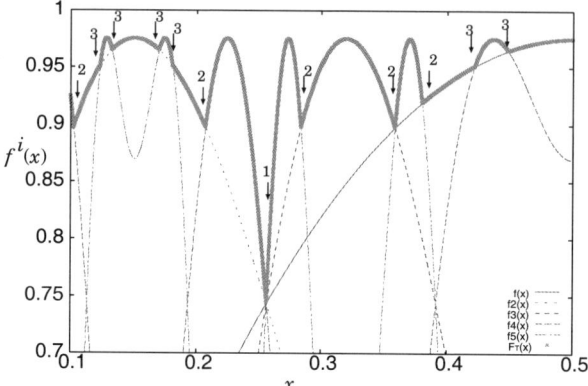

Fig. 21. The set of largest points for $(x, f^i(x, a))$, $1 \leq i \leq T$, $a = 3.9$, $T = 6$. The thick curves are the loci of $F_T(x, a)$ with respect to x. The numbers indicating the cusp points represent the periods of the corresponding periodic solutions.

Moreover, the method also uses a tracking technique by means of the intersections of multiple one-dimensional return maps. The control has been achieved without reference states for targeting POs, given only integers which correspond to the periods. The method does not require a reconstruction of the parametrical variations in the UPOs' stable and unstable manifolds, nor pre-tuning of the principal parameters, and therefore it constitutes as an appropriate strategy to control fast dynamical processes in real time toward any desirable periodic dynamics.

Additionally, we present a method to visualize the bifurcation structure of a one-dimensional general unimodal map through the pre-images of periodic orbits observed in transient processes.

References

E. Ott, C. Grebogi, and J. A. Yorke, Phys. Rev. Lett. **64**, 1196 (1990).
S. Boccaletti, C. Grebogi, Y.-C. Lai, H. Mamcini, and D. Maza, Phys. Rep. **329**, 103 (2000).
K. Pyragas, Phys. Lett. A **170**, 421 (1992).
J. E. Socolar, D. W. Sukow, and D. J. Gauthier, Phys. Rev. E **50**, 3245 (1994).
A. Ahlborn and U. Parlitz, Phys. Rev. Lett **93**, 264101 (2004).
A. Ahlborn and U. Parlitz, Phys. Rev. Lett **96**, 034102 (2006).
S. Parthasarathy and S. Sinha, Phys. Rev. E **51**, 6239 (1995).
N. Parekh, S. Parthasarathy, and S. Sinha, Phys. Rev. Lett. **81**, 1401 (1998).
R. V. Sole, J. Gamarra, M. Ginovart, and D. Lopez, Bull. Math. Biol. **61**, 1187 (1999).

C. Grebogi, E. Ott, and J. A. Yorke, Physica **7D**, 181 (1983).
C. Grebogi, E. Ott, and J. A. Yorke, Phys. Rev. Lett. **57**, 1284 (1986).
B. P. Bezruchko, T. V. Dikanev, and D. A. Smirnov, Phys. Rev. E **64**, 036210 (2001).
M. Dhamala, Y.-C. Lai, and E. J. Kostelich, Phys. Rev. E **61**, 6485 (2000).
M. Dhamala, Y.-C. Lai, and E. J. Kostelich, Phys. Rev. E **64**, 056207 (2001).
S. Gueron, Phys. Rev. E **57**, 3645 (1998).
H. Ando and K. Aihara, Phys. Rev. E **74**, 066205 (2006).
R. M. May and G. F. Oster, Am. Nat. **110**, 573 (1976).
K. Aihara, T. Takabe, and M. Toyoda, Phys. Lett. A. **144**, 333 (1990).
H. Ando, S. Boccaletti, and K. Aihara, Phys. Rev. E **75**, 066211 (2007).
O. E. Rössler, Phys. Lett. A. **57**, 397 (1976).
L. Györgi and R. J. Field, Nature **355**, 808 (1992).
V. Balakrishnan, C. Nicolis, and G. Nicolis, J. Stat. Phys. **80**, 307 (1995).
M. C. Valsakumar, S. V. M. Satyanarayana, and S. Kanmani, J. Phys. A **32**, 6939 (1999).
F. K. Diakonos and P. Schmelcher, Chaos **7**, 239 (1997).
D. P. Lathrop and E. J. Kostelich, Phys. Rev. A **40**, 4028 (1989).
P. So, E. Ott, S. J. Schiff, D. T. Kaplan, T. Sauer, and C. Grebogi, Phys. Rev. Lett. **76**, 4705 (1996).
P. Schmelcher and F. K. Diakonos, Phys. Rev. Lett. **78**, 4733 (1997).

Chapter 4

FEEDBACK ANTI-CONTROL OF CHAOS

Yuming Shi[1,*] and Guanrong Chen[2,†]

[1] *Department of Mathematics, Shandong University,*
Jinan, Shandong 250100, P. R. China
[2] *Department of Electronic Engineering,*
City University of Hong Kong, P. R. China
*ymshi@sdu.edu.cn
† gchen@ee.cityu.edu.hk

This article offers a brief introduction to the notion of feedback anti-control of chaos (namely, chaotification), which means an engineering feedback approach to generate mathematically rigorous chaos from an otherwise non-chaotic dynamical system, or to enhance the existing chaos of a chaotic system. Only discrete systems (maps) are considered, including both finite-dimensional and infinite-dimensional settings, as well as a class of partial difference equations as a case study.

1. Introduction

Chaos control has been rapidly developed to date, mainly in two opposite directions: control of chaos and anti-control of chaos [Chen and Dong (1998); Chen (2003); Chen and Shi (2006)].

A process of stabilizing a chaotic system is referred to as control of chaos. Over the last two decades, research on control of chaos has rapidly evolved and been advanced to become rather mature. This concept of control of chaos is by nature quite classical, regarding chaotic motions harmful. Anti-control of chaos, on the other hand, is a process that makes a non-chaotic system chaotic, or enhances a chaotic system to present stronger or different type of chaos, regarding chaos useful. This process is also called chaotification, aiming to chaotify an otherwise non-chaotic system by means of control. The interest has been continuously motivated by the great potentials of chaos in some non-traditional applications such as those found from electronic, mechanical, optical, and especially biological and medical systems [Chen and Dong (1998); Chen (2003); Chen and Shi (2006)].

In the endeavor of achieving chaotification of discrete maps, the first simple and yet mathematically rigorous algorithm was developed by Chen and Lai in [Chen and Lai (1997), (1998)] from the engineering state-feedback control approach, which yields chaos in the sense of Devaney [Devaney (1989)] when the given system is linear, and is chaotic in the sense of Wiggins [Wiggins (1990)] when the given system is nonlinear. Afterwards, the Chen-Lai algorithm for chaotification was extended to higher-dimensional settings [Wang and Chen (2000)]. Recently, a coupled-expansion theory was developed and applied to show that this feedback anti-control algorithm and its variants can actually be rigorously extended to finite- and also infinite-dimensional systems with chaos in the sense of Devaney and Wiggins as well as Li-Yorke [Shi et al. (2006); Liang et al. (2008)].

It is worthwhile to note that in the pursuit of demonstrating the generated chaos being indeed chaos in a rigorous mathematical sense, the celebrated Li-Yorke theorem (for the one-dimensional case [Li and Yorke (1975)]) and the Marotto theorem (for the n-dimensional case [Marotto (1978)]) were usually employed. The Marotto theorem still is the best-known criterion today for determining and analyzing Li-Yorke chaos for higher-dimensional discrete maps. However, since there was an error in the original Marotto paper [Marotto (1978)], some efforts have been made to correct it towards an elegant result with a perfect proof, leading to the latest complete presentation by Shi and Chen [Shi and Chen 2004b].

It should also be pointed out that although it is possible to use other definitions of chaos, only some most commonly-used definitions are used in the present introductory article. Moreover, for the continuous-time setting, the readers are referred to a brief overview [Wang (2003)] and a small treatise [Chen and Wang (2006)] for more information.

This article is devoted only to discrete chaotic systems (chaotic maps) in the state-space setting. In Sec. 2, some mathematical preliminaries on some basic concepts and criteria of chaos in the sense of Devaney, Wiggins and Li-Yorke are provided. The central problem of chaotification is formulated and described in Sec. 3, where the main results and several anti-control algorithms are summarized. In Sec. 4, the chaotification algorithms established in the preceding section are applied to a class of first-order partial difference equations as a case study. The last section concludes the presentation of the article.

2. Preliminaries

To start with, some basic concepts and several criteria of discrete chaos are introduced.

2.1. *Basic concepts*

This subsection introduces some basic concepts, including periodic points, topological conjugacy, topological transitivity, sensitive dependence on initial conditions, chaos, snap-back repeller, and coupled-expanding map. They will be useful in the following sections of the article, but are also important in their own right.

Consider the following general discrete dynamical system:

$$x_{n+1} = f(x_n), \quad n \geq 0, \tag{2.1}$$

where $f : D \subset X \to X$ is a map and (X, d) is a metric space.

For any $x_0 \in D$, the sequence $\{x_n\}_{n=0}^\infty$ is called the (positive) orbit of system (2.1) starting from x_0, denoted by $O^+(x_0)$, and x_0 is called the initial point of the orbit, where $x_{n+1} = f(x_n)$, $n \geq 0$.

Definition 2.1. A point $x_0 \in D$ is called k-periodic for system (2.1) if $x_{n+k} = x_n$, $n \geq 0$, where $\{x_n\}_{n=0}^\infty$ is the orbit of system (2.1) starting from x_0. Further, k is called the prime period of x_0 if $x_i \neq x_0$, $1 \leq i \leq k-1$. In the special case of $k = 1$, i.e., $f(x_0) = x_0$, x_0 is called a fixed point of system (2.1).

Definition 2.2. Let V be a nonempty subset of D. The set V is called an invariant set of system (2.1) if the orbit of system (2.1), starting from any point in V, always lies in V; that is, $f(V) \subset V$.

Definition 2.3. Let V be a nonempty subset of D. System (2.1) is said to be topologically transitive in V if, for any two nonempty relatively open subsets U and W with respect to V, there exists a positive integer n such that $f^n(U) \cap W \neq \phi$.

Definition 2.4. Let V be a nonempty subset of D. System (2.1) is said to have sensitive dependence on initial conditions in V if there exists a constant $\delta_0 > 0$ such that for any $x_0 \in V$ and any neighborhood U of x_0, there exist $y_0 \in V \cap U$ and a nonnegative integer n such that $d(x_n, y_n) > \delta_0$, where $\{x_i\}_{i=0}^\infty$ and $\{y_i\}_{i=0}^\infty$ are the orbits of system (2.1) starting from x_0 and y_0, respectively. The constant δ_0 is called a sensitivity constant of system (2.1) in V.

Definition 2.5. Let (X, d_1) and (Y, d_2) be two metric spaces, and $f : X \to X$ and $g : Y \to Y$ be two maps. If there exists a homeomorphism $h : X \to Y$ such that $h \circ f = g \circ h$, then f and g are said to be topologically h-conjugate, where "$\circ$" denotes the composition of two maps.

Since Li and Yorke introduced the first definition of chaos [Li and Yorke (1975)], there have been several different definitions of chaos. The three definitions of chaos given by Li and Yorke, Devaney, and Wiggins, respectively, are often used in the literature.

Definition 2.6. A subset S of D is called a scrambled set of f if, for any two different points $x, y \in S$,

(i) $\liminf_{n \to \infty} d(f^n(x), f^n(y)) = 0$;
(ii) $\limsup_{n \to \infty} d(f^n(x), f^n(y)) > 0$.

f is said to be chaotic in the sense of Li-Yorke if there exists an uncountable scrambled set S of f.

Remark 2.1. There are three conditions in the original characterization of chaos in the Li-Yorke theorem [Li and Yorke (1975)]. In addition to the above conditions (i) and (ii), the third one is that for all $x \in S$ and for all periodic points p of f,

$$\limsup_{n \to \infty} d(f^n(x), f^n(p)) > 0.$$

But conditions (i) and (ii) together imply that the scrambled set S contains at most one point x that does not satisfy the above condition. So, the third condition is not essential and therefore may be removed.

Definition 2.7 [Devaney (1989)]. A map $f : V \subset D \to V$ is said to be chaotic on V in the sense of Devaney if

(i) f is topologically transitive in V;
(ii) the periodic points of f are dense in V;
(iii) f has sensitive dependence on initial conditions in V.

By the result of [Banks et al. (1992)], properties (i) and (ii) together imply property (iii) if f is continuous in V. So, property (iii) is redundant in the above definition in this situation. By the same result, if two maps are topological conjugate, then one is chaotic in the sense of Devaney if and only if the other one is so. On the other hand, under some mild conditions, it can be shown that chaos in the sense of Devaney is stronger than that in the sense of Li-Yorke, as shown by the following result [Huang and Ye (2002), Theorem 4.1].

Lemma 2.1. Let V be a compact subset of D, containing infinitely many points. If a map $f : V \to V$ is continuous, surjective, and chaotic in the sense of Devaney on V, then it is chaotic in the sense of Li-Yorke.

Definition 2.8 [Wiggins (1990)]. A map $f : V \subset D \to V$ is said to be chaotic on V in the sense of Wiggins if f satisfies conditions (i) and (iii) in Definition 2.7.

Obviously, chaos in the sense of Devaney is stronger than that in the sense of Wiggins.

To extend the result in [Li and Yorke (1975)] about continuous interval maps to higher-dimensional maps, Marotto in [Marotto (1978)] introduced a concept of snap-back repeller for continuously differentiable maps in $\mathbf{R}^n$ and then showed that a snap-back repeller implies chaos, where the concept is defined in terms of some corresponding eigenvalues and determinants of the Jacobi matrices of the map. Recently, it has been extended to maps in general metric spaces [Shi and Chen (2004a)]. For convenience, its definition is listed below.

Denote by

$$\bar{B}_r(z) = \{x \in X : d(x, z) \leq r\} \text{ and } B_r(z) = \{x \in X : d(x, z) < r\}$$

the closed ball and the open ball of radius $r > 0$ centered at z, respectively.

Definition 2.9 [Shi and Chen (2004a), Definitions 2.1-2.6]. Let (X, d) be a metric space and $f : X \to X$ be a map.

(i) A point $z \in X$ is called an expanding fixed point (or a repeller) of f in $\bar{B}_r(z)$ for some constant $r > 0$, if $f(z) = z$ and f is expanding in distance on $\bar{B}_r(z)$; that is,

$$d(f(x), f(y)) \geq \lambda d(x, y), \quad \forall x, y \in \bar{B}_r(z)$$

for some constant $\lambda > 1$. Furthermore, z is called a regular expanding fixed point of f in $\bar{B}_r(z)$ if z is an interior point of $f(B_r(z))$.

(ii) Assume that z is an expanding fixed point of f in $\bar{B}_r(z)$ for some $r > 0$. Then z is said to be a snap-back repeller of f if there exists a point $x_0 \in B_r(z)$ with $x_0 \neq z$ and $f^m(x_0) = z$ for some positive integer m. Furthermore, z is said to be a non-degenerate snap-back repeller of f if there exist positive constants μ and r_0 such that $B_{r_0}(x_0) \subset B_r(z)$ and

$$d(f^m(x), f^m(y)) \geq \mu \, d(x, y), \quad \forall x, y \in \bar{B}_{r_0}(x_0);$$

z is called a regular snap-back repeller of f if $f(B_r(z))$ is open and there exists a positive constant δ_0 such that $B_{\delta_0}(x_0) \subset B_r(z)$ and z is an interior point of $f^m(B_\delta(x_0))$ for any positive constant $\delta \leq \delta_0$.

Remark 2.2. In [Marotto (1978)], the snap-back repeller is actually a regular and non-degenerate one.

Finally, another useful concept of coupled-expanding map is introduced.

Definition 2.10. Let (X, d) be a metric space and $f : D \subset X \to X$ a map. If there exist $m(\geq 2)$ subsets V_i ($1 \leq i \leq m$) of D with $V_i \cap V_j = \partial_D V_i \cap \partial_D V_j$ for each pair (i, j), $1 \leq i \neq j \leq m$, such that

$$f(V_i) \supset \bigcup_{j=1}^{m} V_j, \ 1 \leq i \leq m, \tag{2.2}$$

then f is said to be coupled-expanding in V_i, $1 \leq i \leq m$, where $\partial_D V_i$ is the relative boundary of V_i with respect to D. Further, the map f is said to be strictly coupled-expanding in V_i, $1 \leq i \leq m$, if $d(V_i, V_j) > 0$ for all $1 \leq i \neq j \leq m$.

Remark 2.3.

(1) The original term of the above concept was a "turbulent map," which was first introduced in [Block and Coppel (1992)] for continuous interval maps. Later on, this concept was extended to maps in general metric spaces [Shi and Yu (2006)], where it was still called turbulent map. Since the term "turbulence" is well-established in fluid mechanics, per referee's suggestion this term was changed to "coupled-expansion" later on [Shi and Chen (2006); Shi and Yu (2008)].
(2) The above concept was further generalized recently by using transition matrices [Shi et al. (2009)].

2.2. Some criteria of chaos

This subsection gives several criteria of chaos that will be used in chaotification to be further discussed later.

Theorem 2.1 [Shi and Chen (2004a), Theorem 3.1]. Let (X, d) be a complete metric space and V_1, V_2 be nonempty, closed, and bounded subsets of X with $d(V_1, V_2) > 0$. If a continuous map $f : V_1 \cup V_2 \to X$ satisfies

(i) f is strictly coupled-expanding in V_1 and V_2;

(ii) there exists a constant $\lambda > 1$ such that
$$d(f(x), f(y)) \geq \lambda\, d(x,y), \quad \forall\, x, y \in V_1,\ i = 1, 2;$$
(iii) there exists a constant $\gamma > 0$ such that
$$d(f(x), f(y)) \leq \gamma\, d(x,y), \quad \forall\, x, y \in V_i,\ i = 1, 2;$$

then there exists a Cantor set $\Lambda \subset V_1 \cup V_2$ such that $f : \Lambda \to \Lambda$ is topologically conjugate to the symbolic dynamical system $\sigma : \Sigma_2^+ \to \Sigma_2^+$. Consequently, system (2.1) is chaotic on Λ in the sense of Devaney as well as in the sense of Li-Yorke.

Remark 2.4.

(1) In Theorem 2.1, the conclusion that system (2.1) is chaotic in the sense of Li-Yorke can be implied by Lemma 2.1.
(2) Condition (iii) can be dropped [Shi et al. (2009), Theorem 5.5].
(3) References [Shi and Yu (2006)], [Shi and Yu (2008)], [Shi and Chen (2006)], and Shi et al. (2009)] provide some other criteria of chaos implied by coupled-expanding maps.

In 1978, motivated by the work [Li and Yorke (1975)], Marotto generalized the result of Li and Yorke for interval maps to higher-dimensional maps, as follows.

Theorem 2.2 [Marotto (1978)]. Let $f : R^n \to R^n$ be continuously differentiable. If f possesses a regular and non-degenerate snap-back repeller, then system (2.1) is chaotic:

(i) there is a positive integer N such that for each integer $p \geq N$, f has a point with period p.
(ii) there is a "scrambled set" of f, namely, an uncountable set S containing no periodic points of f, such that
   (ii$_1$) $f(S) \subset S$,
   (ii$_2$) for every $x, y \in S$ with $x \neq y$,
   $$\limsup_{k \to \infty} \|f^k(x) - f^k(x)\| > 0,$$
   (ii$_3$) for every $x \in S$ and any periodic point y of f,
   $$\limsup_{k \to \infty} \|f^k(x) - f^k(y)\| > 0;$$
(iii) there is an uncountable subset S_0 of S such that for every $x, y \in S_0$,
$$\liminf_{k \to \infty} \|f^k(x) - f^k(y)\| = 0.$$

Remark 2.5. By the above Marotto theorem, a regular and non-degenerate snap-back repeller implies chaos in the sense of Li-Yorke for continuously differentiable maps in $\mathbf{R}^n$. To prove this result, Marotto employed a similar idea to that used in [Li and Yorke (1975)]. Recently, the coupled-expansion theory was employed to show that a regular and non-degenerate snap-back repeller implies chaos in the sense of Devaney as well as in the sense of Li-Yorke for maps in general Banach spaces.

Let $f : X \to X$ be a map and X be a Banach space, and denote the Frechét derivative of f at x by $Df(x)$. For a linear map $L : X \to X$, introduce the following notation:

$$\|L\|^0 := \inf\{\|Lx\| : x \in X \text{ with } \|x\| = 1\}.$$

If a bounded linear map $L : X \to X$ has a bounded linear inverse map, then L is called an invertible linear map.

Theorem 2.3. Let $(X, \|\cdot\|)$ be a Banach space and $f : X \to X$ be a map with a fixed point $z \in X$. Assume that

(i) f is continuously differentiable in $B_{r_0}(z)$ for some $r_0 > 0$ and $Df(z)$ is an invertible linear map satisfying

$$\|Df(z)\|^0 > 1,$$

which is equivalent to that there exists a positive constant $r \leq r_0$ such that z is a regular expanding fixed point of f in $\bar{B}_r(z)$;

(ii) z is a snap-back repeller of f with $f^m(x_0) = z$ for some $x_0 \in B_r(z)$, $x_0 \neq z$, and for some positive integer m. Furthermore, f is continuously differentiable in some neighborhoods of $x_1, \ldots, x_{m-1}$, respectively, satisfying that $Df(x)$ is an invertible linear map for all $x \in B_r(z)$ and for $x = x_j$ ($1 \leq j \leq m - 1$), and $\|Df(x_j)\|^0 > 0$ for $1 \leq j \leq m - 1$, where $x_j = f(x_{j-1})$, $1 \leq j \leq m - 1$.

Then, for any neighborhood U of z, there exist an integer $n > m$ and a Cantor set $\Lambda \subset U$ such that $f^n : \Lambda \to \Lambda$ is topologically conjugate to the symbolic dynamical system $\sigma : \Sigma_2^+ \to \Sigma_2^+$. Consequently, there exists a compact and perfect invariant set $E \subset X$, containing a Cantor set, such that system (2.1) is chaotic on E in the sense of Devaney as well as in the sense of Li-Yorke, and has a dense orbit in E.

Remark 2.6.

(1) Theorem 2.3 is the same as [Shi *et al.* (2006), Theorem 2.1], which is a consequence of [Shi and Chen (2004b), Theorem 3.1].

(2) References [Shi and Chen (2004a)] and [Shi and Yu (2006)] provide some more criteria of chaos implied by snap-back repellers and homoclinic orbits to expanding fixed points in general complete metric spaces.
(3) For finite-dimensional maps, the assumptions in the Marotto theorem (Theorem 2.2) was further weakened so that a regular snap-back repeller implies chaos in the sense of Li-Yorke [Shi and Yu (2008), Theorem 4.1], and that a regular snap-back repeller implies chaos in the sense of Wiggins as well as in the sense of Li-Yorke under some degenerate conditions [Shi and Yu (2008), Theorem 4.2].

3. Chaotification Algorithms

This section introduces several chaotification algorithms by means of feedback control.

Section 3.1 gives a description of the chaotification problem. Sections 3.2–3.4 introduce several chaotification algorithms for systems in a special class of Banach spaces via the mod-operation, sawtooth functions and two general anti-controllers. Section 3.5 introduces two chaotification algorithms by feedback control in general Banach spaces. Section 3.6 provides some illustrative examples with computer simulations.

Let
$$\mathbf{R}^k = \{x = \{x_j\}_{j=1}^k : x_j \in \mathbf{R} \text{ for } 1 \leq j \preceq k\}$$
with $1 \leq k \leq \infty$, and define
$$Y_k = \{x \in \mathbf{R}^k : \|x\|_k < \infty\}$$
with the norm
$$\|x\|_k = \sup\{|x_j| : 1 \leq j \preceq k\},$$
where $j \preceq k$ means "$j \leq k$ for $k < \infty$ and $j < \infty$ for $k = \infty$."

It can be verified that $(Y_k, \|\cdot\|_k)$ is a Banach space. Clearly, in the special case of $k < \infty$, Y_k is the classical k-dimensional real space $\mathbf{R}^k$ and its norm $\|\cdot\|_k$ is the sup-norm, while in the special case of $k = \infty$, $Y_k = l^\infty$ and the norm $\|\cdot\|_k$ is the usual norm of l^∞.

For convenience, denote $I^k := \underbrace{I \times I \times \cdots \times I}_{k}$ for $I \subset \mathbf{R}$. It is evident that $I^k \subset Y_k$ if I is a bounded set of $\mathbf{R}$ and $\bar{B}_r(0) = [-r, r]^k$, $B_r(0) = (-r, r)^k$. Denote
$$\widehat{x} := (x_2, x_3, \cdots, x_k, x_1)^T \qquad (3.1)$$

in the case of $k < \infty$; and
$$\widehat{x} := (x_2, x_3, \cdots)^T \tag{3.2}$$
in the case of $k = \infty$ for $x = (x_1, x_2, \cdots)^T \in Y_k$.

3.1. Problem description

Now, consider the following discrete dynamical system:
$$x_{n+1} = f(x_n), \quad n \geq 0, \tag{3.3}$$
where $f : D \subset X \to X$ is a map, $(X, \|\cdot\|)$ is a Banach space, and system (3.3) may not be chaotic.

The objective is to design a control input sequence, $\{u_n\}$, such that the output of the controlled system
$$x_{n+1} = f(x_n) + u_n, \quad n \geq 0, \tag{3.4}$$
becomes chaotic in the sense of Devaney, Wiggins, or Li-Yorke. The controller to be designed here is in the form of
$$u_n = g(\mu\, x_n) \tag{3.5}$$
or
$$u_n = \mu\, g(x_n), \tag{3.6}$$
where μ is a positive parameter, and the map $g : D' \to X$ is expected to be very simple with D' being a suitable subset of X.

Remark 3.1. In practical design, a meaningful controller should be simple such that the goal of chaotification can be achieved. In other words, a designed controller should be simple, cheap, and easily implementable in engineering applications. The discussions given below are intended to follow this basic engineering principle, trying to design some very simple and implementable anti-controllers.

3.2. Feedback control via mod-operations

The first precise and rigorous anti-control algorithm via feedback control with mod-operation was developed in [Chen and Lai (1997); Chen and Lai (1998)], as follows:
$$x_{n+1} = f(x_n) + u_n \pmod{1} \tag{3.7}$$

with
$$u_n = \mu\, x_n, \qquad (3.8)$$

where $f(0) = 0$, f is continuously differentiable, at least locally in a region containing $x^* = 0$, in the usual k-dimensional real space $\mathbf{R}^k$, and satisfies
$$\|Df(x)\|_1 \leq L$$
for all $x \in \mathbf{R}^k$ or for all x in some region containing $x^* = 0$, and for some constant $L > 0$, with $\|C\|_1$ being the spectral norm of a $k \times k$ matrix $C = (c_{ij})$, $\mu = L + e^c$, and $c > 0$ being a parameter.

It was shown in [Chen and Lai (1998)] that for $c > 0$, the controlled system (3.7)-(3.8) is chaotic in the sense of Devaney in the linear case $f(x) = Ax$, where A is a $k \times k$ real matrix, and is chaotic in the sense of Wiggins in the nonlinear case.

The above chaotification algorithm with the mod-operation is extended to space Y_k lately, as follows.

Theorem 3.1 [Shi et al. (2006), Theorem 4.3]. Consider the chaotification of system (3.1) in space Y_k with $k \leq \infty$. Assume that $f(0) = 0$ and there exists a positive constant r such that f is continuous in $[-r, r]^k$ and furthermore satisfies
$$\|f(x) - f(y)\|_k \leq L\|x - y\|_k, \quad \forall\, x, y \in [-r, r]^k$$
for some constant $L > 0$. Then, for each constant μ satisfying
$$\mu > \mu_0 := \max\{5(1+L), 10L\}, \qquad (3.9)$$
there exists a Cantor set $\Lambda \subset (-\tfrac{5}{2}\mu^{-1}r, \tfrac{5}{2}\mu^{-1}r)^k$ such that $F_\mu(x) := f(x) + \mu x \pmod{r} : \Lambda \to \Lambda$ is topologically conjugate to the symbolic dynamical system $\sigma : \Sigma_2^+ \to \Sigma_2^+$. Consequently, the controlled system
$$x_{n+1} = f(x_n) + \mu x_n \pmod{r},$$
is chaotic on Λ in the sense of Devaney, and also in the sense of both Li-Yorke and Wiggins, where the mod-operation is componentwise.

Remark 3.2.

(1) It was already noticed by the authors that $F_\mu(x) = f(x) + \mu x \pmod{r}$ is discontinuous at some points. So the snap-back repeller theory cannot be applied completely rigorously to this anti-controlled system. Nevertheless, Theorem 3.1 can be proved by employing the coupled-expansion theory as did in [Shi et al. (2006)].

(2) As a consequence of Theorem 3.1, the controlled system (3.7)-(3.8) is chaotic in the sense of Devaney, Wiggins, as well as Li-Yorke.

Theorem 3.2 [Liang et al. (2008), Theorem 2.3]. Consider the following controlled system:

$$x_{n+1} = f(x_n) + \mu \widehat{x}_n \pmod{r} \tag{3.10}$$

in Y_k with $k < \infty$, where $\widehat{x}$ is defined by (3.1). Under all assumptions of Theorem 3.1, for each constant μ satisfying (3.9), there exists a Cantor set $\Lambda \subset (-\frac{5}{2}\mu^{-1}r, \frac{5}{2}\mu^{-1}r)^k$ such that $F_\mu(x) := f(x) + \mu\widehat{x} \pmod{r} : \Lambda \to \Lambda$ is topologically conjugate to the symbolic dynamical system $\sigma : \Sigma_2^+ \to \Sigma_2^+$. Consequently, the controlled system (3.10) is chaotic on Λ in the sense of both Devaney and Li-Yorke.

3.3. Feedback control via sawtooth functions

Introduce the following function in Y_k for each $k \geq 1$:

$$\mathrm{Saw}_r(x) = \{\mathrm{saw}_r(x_j)\}_{j=1}^k,$$

where $\mathrm{saw}_r(x)$ is the classical sawtooth function; that is,

$$\mathrm{saw}_r(x) = (-1)^m(x - 2mr), \ (2m-1)r \leq x < (2m+1)r, \ m \in \mathbf{Z},$$

while $\mathbf{Z}$ denotes the integer set. Clearly, $\mathrm{Saw}_r(x)$ is the sawtooth function $\mathrm{saw}_r(x)$ in $\mathbf{R}$ when $k = 1$, and $\mathrm{Saw}_r(x)$ is the sawtooth function $\mathrm{saw}_r(x)$ in $\mathbf{R}^k$ when $k < \infty$, defined by (11) and (12) in [Wang and Chen (2000)], respectively. So, Saw_r can be regarded as a generalization of the classical sawtooth function.

The following controlled system was considered in [Wang and Chen (2000)]:

$$x_{n+1} = f(x_n) + \mathrm{Saw}_\varepsilon(\mu x_n), \ n \geq 0 \tag{3.11}$$

in $\mathbf{R}^k$ with $k < \infty$, where $\varepsilon > 0$ is a fixed constant.

Theorem 3.3 [Wang and Chen (2000), Theorem 6]. Suppose that f is continuously differentiable in $\mathbf{R}^k$ with $f(0) = 0$, satisfying

$$\|Df(x)\|_\infty < N, \ \forall x \in \bar{B}_{3\mu^{-1}\varepsilon}(0)$$

for some positive constant N and

$$\mu > \max\{3N, 1 + N + (1 + 2N)^{1/2}\},$$

where $\|C\|_\infty = \max\{\sum_{j=1}^k |c_{ij}| : 1 \leq i \leq k\}$ for a $k \times k$ real matrix $C = (c_{ij})$. Then, the controlled system (3.11) is chaotic in the sense of Li-Yorke.

Remark 3.3.

(1) It is evident that $\|\text{saw}_\varepsilon(x)\|_\infty \leq \varepsilon$ for all $x \in \mathbf{R}^k$. Since the constant ε can be chosen very small, the controller can be arbitrarily small in norm (energy). This means that the original system can be forced to become chaotic by using an arbitrarily small-amplitude state-feedback control, which is very desirable for engineering applications.
(2) Theorem 3.3 was proved by using the original Marotto theorem [Marotto (1978)] in [Wang and Chen (2000)].

The above algorithm has been extended to maps in Y_k ($k \leq \infty$) with the following result obtained by employing Theorem 2.1.

Theorem 3.4 [Shi et al. (2006), Theorem 4.1]. Consider the controlled system (3.11) in Y_k with $k \leq \infty$. Assume that $f(0) = 0$ and there exists a positive constant r such that f is continuous in $[-r, r]^k$ and furthermore satisfies

$$\|f(x) - f(y)\|_k \leq L\|x - y\|_k, \quad \forall x, y \in [-r, r]^k \tag{3.12}$$

for some constant $L > 0$. Then, for each constant μ satisfying

$$\mu > \mu_0 := \max\left\{\frac{5}{2}r^{-1}\varepsilon, 5(L+1)\right\}, \tag{3.13}$$

there exists a Cantor set $\Lambda \subset [-\frac{5}{2}\mu^{-1}\varepsilon, \frac{5}{2}\mu^{-1}\varepsilon]^k$ such that $F_\mu(x) = f(x) + \text{Saw}_\varepsilon(\mu x) : \Lambda \to \Lambda$ is topologically conjugate to the symbolic dynamical system $\sigma : \Sigma_2^+ \to \Sigma_2^+$. Consequently, the controlled system (3.11) is chaotic on Λ in the sense of both Devaney and Li-Yorke.

Theorem 3.5 [Liang et al. (2008), Theorem 2.4]. Consider the controlled system

$$x_{n+1} = f(x_n) + \text{Saw}_\varepsilon(\mu \widehat{x}_n), \quad n \geq 0, \tag{3.14}$$

in Y_k with $k < \infty$, where $\widehat{x}$ is defined by (3.1). Let all the assumptions in Theorem 3.4 hold. Then, for each constant μ satisfying (3.13), there exists a Cantor set $\Lambda \subset [-\frac{5}{2}\mu^{-1}\varepsilon, \frac{5}{2}\mu^{-1}\varepsilon]^k$ such that $F_\mu(x) = f(x) + \text{Saw}_\varepsilon(\mu \widehat{x}) : \Lambda \to \Lambda$ is topologically conjugate to the symbolic dynamical system $\sigma : \Sigma_2^+ \to \Sigma_2^+$. Consequently, the controlled system (3.14) is chaotic on Λ in the sense of both Devaney and Li-Yorke.

It is noticed that all the above chaotification algorithms require that the original systems have a fixed point and the algorithms drive the systems to be chaotic near the fixed point. In the case that the original system has no fixed points or is so complex that it is difficult to determine its fixed points, how can one anti-control the system to become chaotic, or how can one make the system chaotic near a point of interest, which is not a fixed point of the system? The following two algorithms provide solutions to these problems. They can be proved by Theorem 2.1.

Without loss of generality, suppose that it is desired to drive the system chaotic near the origin $x^* = 0$, whether or not it is a fixed point of the original system.

Theorem 3.6 [Shi et al. (2006), Theorem 4.2]. Assume that there exist positive constants r and L such that f is continuous in $[-r, r]^k$ with $k \leq \infty$ and satisfies (3.12). Then, for each constant μ satisfying

$$\mu > \mu_0 := \max\{1 + L, 5 + 6(L + \|f(0)\|_k \, r^{-1})\}, \qquad (3.15)$$

there exists a Cantor set $\Lambda \subset (-r, r)^k$ such that $F_\mu(x) := f(x) + \mu \operatorname{Saw}_{\frac{r}{3}}(x) : \Lambda \to \Lambda$ is topologically conjugate to the symbolic dynamical system $\sigma : \Sigma_2^+ \to \Sigma_2^+$. Consequently, the controlled system

$$x_{n+1} = f(x_n) + \mu \operatorname{Saw}_{\frac{r}{3}}(x_n), \quad n \geq 0 \qquad (3.16)$$

is chaotic on Λ in the sense of both Devaney and Li-Yorke.

Theorem 3.7 [Liang et al. (2008), Theorem 2.5]. Consider the controlled system

$$x_{n+1} = f(x_n) + \mu \operatorname{Saw}_{\frac{r}{3}}(\widehat{x}_n), \quad n \geq 0 \qquad (3.17)$$

in Y_k with $k < \infty$, where $\widehat{x}$ is defined by (3.1). Under all the assumptions of Theorem 3.6, for each constant μ satisfying (3.15), there exists a Cantor set $\Lambda \subset (-r, r)^k$ such that $F_\mu(x) = f(x) + \mu \operatorname{Saw}_{\frac{r}{3}}(\widehat{x}) : \Lambda \to \Lambda$ is topologically conjugate to the symbolic dynamical system $\sigma : \Sigma_2^+ \to \Sigma_2^+$. Consequently, the controlled system (3.17) is chaotic on Λ in the sense of both Devaney and Li-Yorke.

3.4. Some general anti-controllers in Y_k

This subsection shows two general anti-controllers for system (3.3) in Y_k ($k \leq \infty$), introduced in [Liang et al. (2008)] and proved by employing Theorem 2.3.

Theorem 3.8. Consider the controlled system (3.4) with controller (3.5) in Y_k ($k \leq \infty$). Assume that

(i) the origin $x^* = 0$ is a fixed point of f and there exist positive constants r and L such that f is continuous in $[-r,r]^k$ and continuously differentiable in $(-r,r)^k$, satisfying

$$\|Df(x)\| \leq L, \quad \forall\, x \in (-r,r)^k;$$

(ii) the map g satisfies the following conditions:

   (iia) g is continuous in $[-r,r]^k \cup [a,b]^k$ and continuously differentiable in $(-r,r)^k \cup (a,b)^k$ with $r < a < b$;

   (iib) $x^* = 0$ is a fixed point of g and there exists a point $\xi \in (a,b)^k$ such that $g(\xi) = 0$;

   (iic) $Dg(x)$ is an invertible linear operator for each $x \in (-r,r)^k \cup (a,b)^k$, and there exists a positive constant N such that

   $$\|g(x) - g(y)\| \geq N\|x - y\|, \quad \forall\, x, y \in [-r,r]^k \text{ and } \forall\, x, y \in [a,b]^k.$$

Then, for any constant μ satisfying

$$\mu > \mu_0 := \max\left\{\frac{b}{r}, \frac{Lr+b}{Nr}, \frac{Lb}{N(\|\xi\|_0 - a)}, \frac{Lb}{N(b - \|\xi\|)}\right\},$$

where $\|\xi\|_0 = \min\{|\xi_i| : 1 \leq i \leq k\}$, and for any neighborhood U of $x^* = 0$, there exist a positive integer $n > 2$ and a Cantor set $\Lambda \subset U$ such that $F_\mu^n : \Lambda \to \Lambda$ is topologically conjugate to the symbolic dynamical system $\sigma : \Sigma_2^+ \to \Sigma_2^+$, where $F_\mu(x) = f(x) + g(\mu x)$. Consequently, there exists a compact and perfect invariant set $V \subset D$ containing a Cantor set such that the controlled system (3.4)-(3.5) is chaotic on V in the sense of both Devaney and Li-Yorke.

Theorem 3.9. Consider the controlled system (3.4) with controller (3.6) in Y_k with $k \leq \infty$. Assume that

(i) assumption (i) in Theorem 3.8 holds;
(ii) the map g satisfies the following conditions:

   (iia) g is continuous in $[-a,a]^k \cup [b,r]^k$ and continuously differentiable in $(-a,a)^k \cup (b,r)^k$ with $0 < a < b < r$;

   (iib) $x^* = 0$ is a fixed point of g and there exists a point $\xi \in (b,r)^k$ such that $g(\xi) = 0$;

(iic) $Dg(x)$ is an invertible linear operator for each $x \in (-a,a)^k \cup (b,r)^k$ and there exists a positive constant N such that
$$\|g(x) - g(y)\| \geq N\|x - y\|, \quad \forall\, x, y \in [-a, a]^k \text{ and } \forall\, x, y \in [b, r]^k.$$

Then, for each constant μ satisfying
$$\mu > \mu_0 := \max\left\{\frac{La+r}{Na}, \frac{Lr}{N(\|\xi\|_0 - b)}, \frac{Lr}{N(r - \|\xi\|)}\right\},$$
all the results in Theorem 3.8 hold for $F_\mu(x) = f(x) + \mu\, g(x)$ therein.

3.5. Chaotification in general Banach spaces

In this subsection, two chaotification algorithms are given for system (3.3) with a feedback controller u_n given by (3.5) and (3.6), respectively, where $(X, \|\cdot\|)$ is a general Banach space. The results can be proved by using Theorem 2.3 [Shi et al. (2006)].

Theorem 3.10 [Shi et al. (2006), Theorem 3.1]. Consider the controlled system (3.4) with controller (3.5). Assume that

(i) the origin $x^* = 0$ is a fixed point of f and there exist positive constants r and L such that f is continuous in $\bar{B}_r(0)$ and continuously differentiable in $B_r(0)$, satisfying
$$\|Df(x)\| \leq L, \quad \forall x \in B_r(0);$$

(ii) the map g satisfies the following conditions:

(iia) g is continuous in $\bar{B}_r(0) \cup \bar{\Omega}$ and continuously differentiable in $B_r(0) \cup \Omega$, where $\Omega = \{x \in X : a < \|x\| < b\}$ with $r < a < b$;
(iib) $x^* = 0$ is a fixed point of g and there exists a point $\xi \in \Omega$ such that $g(\xi) = 0$;
(iic) $Dg(x)$ is an invertible linear operator for each $x \in B_r(0) \cup \Omega$ and there exists a positive constant N such that
$$\|g(x) - g(y)\| \geq N\|x - y\|, \quad \forall x, y \in \bar{B}_r(0) \text{ and } \forall x, y \in \bar{\Omega}.$$

Then, for any constant μ satisfying
$$\mu > \mu_0 := \max\left\{\frac{b}{r}, \frac{Lr+b}{Nr}, \frac{Lb}{N(\|\xi\| - a)}, \frac{Lb}{N(b - \|\xi\|)}\right\}$$
and for any neighborhood U of $x^* = 0$, there exist a positive integer $n > 2$ and a Cantor set $\Lambda \subset U$ such that $F_\mu^n : \Lambda \to \Lambda$ is topologically conjugate to the symbolic dynamical system $\sigma : \Sigma_2^+ \to \Sigma_2^+$, where $F_\mu(x) = f(x) +$

$g(\mu x)$. Consequently, there exists a compact and perfect invariant set $V \subset D$ containing a Cantor set such that the controlled system (3.4)-(3.5) is chaotic on V in the sense of both Devaney and Li-Yorke.

The controller (3.5) in Theorem 3.10 can be easily designed such that the corresponding map g is simple. For example, g can be taken as one of the following four simple functions:

$$g_1(x) = \begin{cases} \pm x, & \text{if } \|x\| \leq r \\ \text{arbitrary}, & \text{if } r < \|x\| < a \\ \pm(x - \xi), & \text{if } a \leq \|x\| \leq b, \end{cases}$$

where $0 < r < a < b$ and $\xi \in X$ can be any given point satisfying $a < \|\xi\| < b$.

Theorem 3.11 [Shi et al. (2006), Theorem 3.2]. Consider the controlled system (3.4) with controller (3.6). Assume that

(i) assumption (i) in Theorem 3.10 holds;
(ii) the map g satisfies the following conditions:

   (iia) g is continuous in $\bar{B}_a(0) \cup \bar{\Omega}'$ and continuously differentiable in $B_a(0) \cup \Omega'$, where $\Omega' = \{x \in X : b < \|x\| < r\}$ with $0 < a < b < r$;
   (iib) $x^* = 0$ is a fixed point of g and there exists a point $\xi \in \Omega'$ such that $g(\xi) = 0$;
   (iic) $Dg(x)$ is an invertible linear operator for each $x \in B_a(0) \cup \Omega'$ and there exists a positive constant N such that

$$\|g(x) - g(y)\| \geq N\|x - y\|, \quad \forall x, y \in \bar{B}_a(0) \text{ and } \forall x, y \in \bar{\Omega}'.$$

Then, for each constant μ satisfying

$$\mu > \mu_0 := \max\left\{\frac{La + r}{Na}, \frac{Lr}{N(\|\xi\| - b)}, \frac{Lr}{N(r - \|\xi\|)}\right\},$$

all the results in Theorem 3.10 hold for $F_\mu(x) = f(x) + \mu g(x)$ therein.

The controller (3.6) in Theorem 3.11 can be easily designed such that the corresponding map g is simple. For example, g can be taken as one of the following four simple functions:

$$g_2(x) = \begin{cases} \pm x, & \text{if } \|x\| \leq a \\ \text{arbitrary}, & \text{if } a < \|x\| < b \\ \pm(x - \xi), & \text{if } b \leq \|x\| \leq r, \end{cases}$$

where $0 < a < b < r$ and $\xi \in X$ can be any point satisfying $b < \|\xi\| < r$.

3.6. *Examples with simulations*

To illustrate the chaotification algorithms established in this section, some examples are now given with computer simulations.

Consider the following two-dimensional linear system:
$$x_{n+1} = \begin{pmatrix} 0 & a \\ -a & 0 \end{pmatrix} x_n, \quad n \geq 0, \tag{3.18}$$

where a is a real constant.

System (3.18) is asymptotically stable in the case of $|a| < 1$; it is stable but not asymptotically stable in the case of $|a| = 1$; it is unstable in the case of $|a| > 1$.

Now system (3.18) is to be chaotified by means of feedback control with the mod-operation and the sawtooth function, respectively.

(1) First, consider the controlled system with mod-operation:
$$x_{n+1} = \begin{pmatrix} 0 & a \\ -a & 0 \end{pmatrix} x_n + \mu x_n \pmod{1}, \quad n \geq 0. \tag{3.19}$$

By Theorem 3.1, for each μ satisfying
$$\mu > \mu_0 = \max\left\{5(1+|a|), 10|a|\right\},$$

the controlled system (3.19) is chaotic in the sense of Devaney as well as Li-Yorke on a Cantor set contained in $[-\frac{5}{2}\mu^{-1}, \frac{5}{2}\mu^{-1}]^2$.

For computer simulation, take $a = 1/2$ and $\mu = 8$; and $a = 2$ and $\mu = 21$, respectively. The simulation results show that the controlled system has complicated dynamical behaviors even though the original system is either asymptotically stable or unstable (see Figs. 1 and 2).

(2) Next, consider the controlled system with the sawtooth function:
$$x_{n+1} = \begin{pmatrix} 0 & a \\ -a & 0 \end{pmatrix} x_n + \mathrm{Saw}_\varepsilon(\mu x_n), \quad n \geq 0. \tag{3.20}$$

By Theorem 3.4, for each μ satisfying
$$\mu > \mu_0 = \max\left\{\frac{5}{2}r^{-1}\varepsilon, 5(|a|+1)\right\},$$

the controlled system (3.20) is chaotic in the sense of Devaney as well as Li-Yorke on a Cantor set contained in $[-\frac{5}{2}\mu^{-1}\varepsilon, \frac{5}{2}\mu^{-1}\varepsilon]^2$.

For computer simulation, take $r = \varepsilon = 1$, $a = 1/2$, and $\mu = 8$. The simulation result shows that the controlled system has complicated dynamical behaviors even though the original system is asymptotically stable (see Fig. 3).

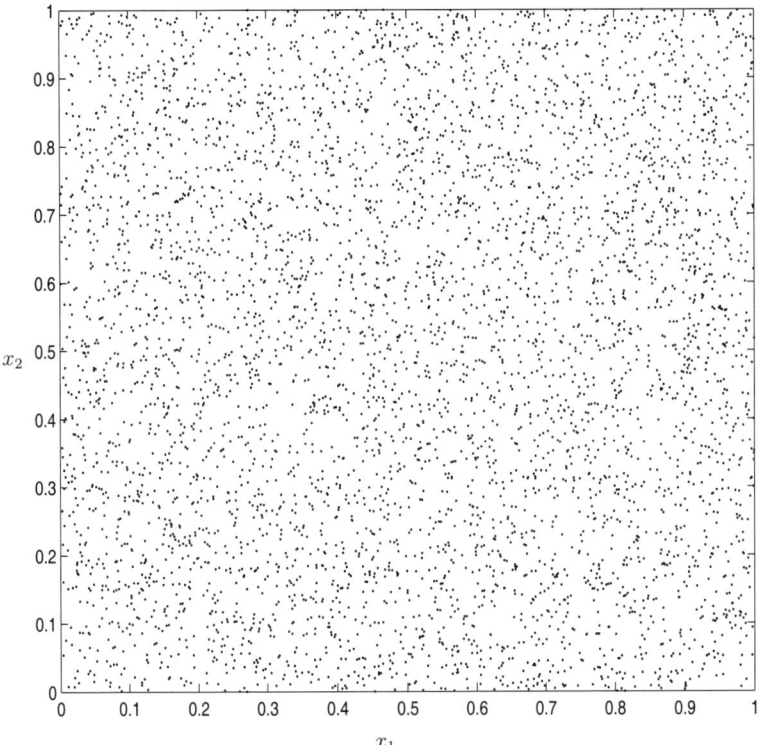

Fig. 1. 2-D computer simulation result with anti-controller via mod-operation, showing complex dynamical behaviors for the controlled system (3.19) with the system parameter $a = 1/2$, and the control parameter $\mu = 8$.

4. Chaotifying Partial Difference Equations

In this section, the chaotification algorithms established in the preceding section are applied to the following first-order partial difference equation:

$$x(n+1, m) = f(m, x(n, m), x(n, m+1)), \qquad (4.1)$$

where $n \geq 0$ is the discrete time step, m is the lattice point with $0 \leq m \preceq k \leq \infty$, $k+1$ is the system size, and $f : \{(m, x, y) : 0 \leq m \preceq k, (x, y) \in D\} \to \mathbf{R}$ is a map with $D \subset \mathbf{R}^2$.

A certain boundary condition is given to Eq. (4.1). In the case of $k < \infty$, for example, the following periodic boundary condition is imposed onto Eq. (4.1):

$$x(n, 0) = x(n, k+1), \quad n \geq 0. \qquad (4.2)$$

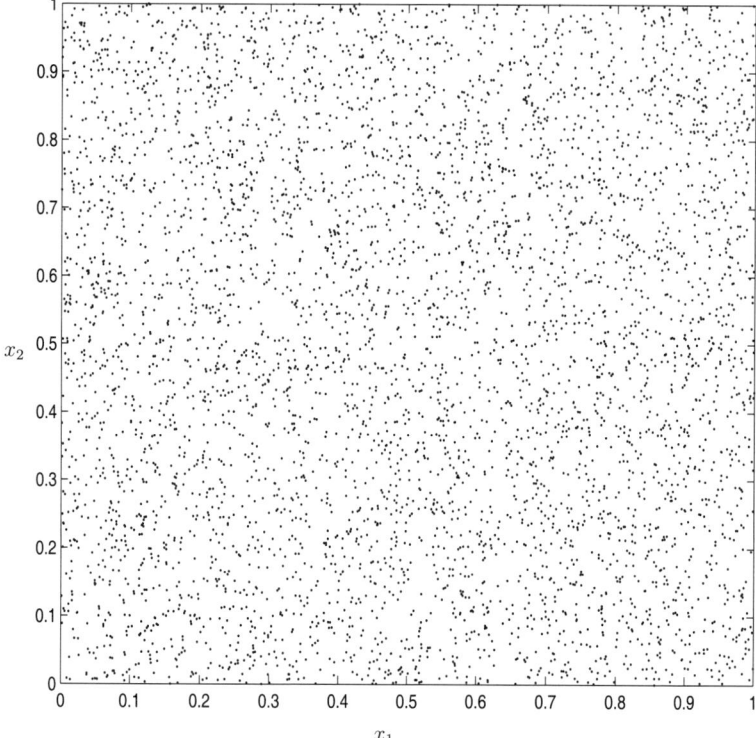

Fig. 2. 2-D computer simulation result with anti-controller via mod-operation, showing complex dynamical behaviors for the controlled system (3.19) with the system parameter $a = 2$, and the control parameter $\mu = 21$.

In the case of $k = \infty$, as another example, a suitable boundary condition is that the solution $\{x(n,m)\}_{n,m=0}^{\infty}$ of Eq. (4.1) is bounded for each $n \geq 0$; that is, $\{x(n,m)\}_{m=0}^{\infty} \in Y_\infty$ (as defined at the beginning of Sec. 3) for each $n \geq 0$.

For any given initial condition of Eq. (4.1), in the case of $k < \infty$,

$$x(0,m) = \phi(m), \quad 0 \leq m \leq k+1,$$

where ϕ satisfies the boundary condition (4.2), Eq. (4.1) has a unique solution $\{x(n,m) : n \geq 0, 0 \leq m \leq k\}$ satisfying the initial condition and the boundary condition (4.2), which can be verified by successive iterations. So, this type of initial-boundary problems is well-defined. In the case of $k = \infty$, on the other hand, for any given initial condition

$$x(0,m) = \phi(m), \quad m \geq 0,$$

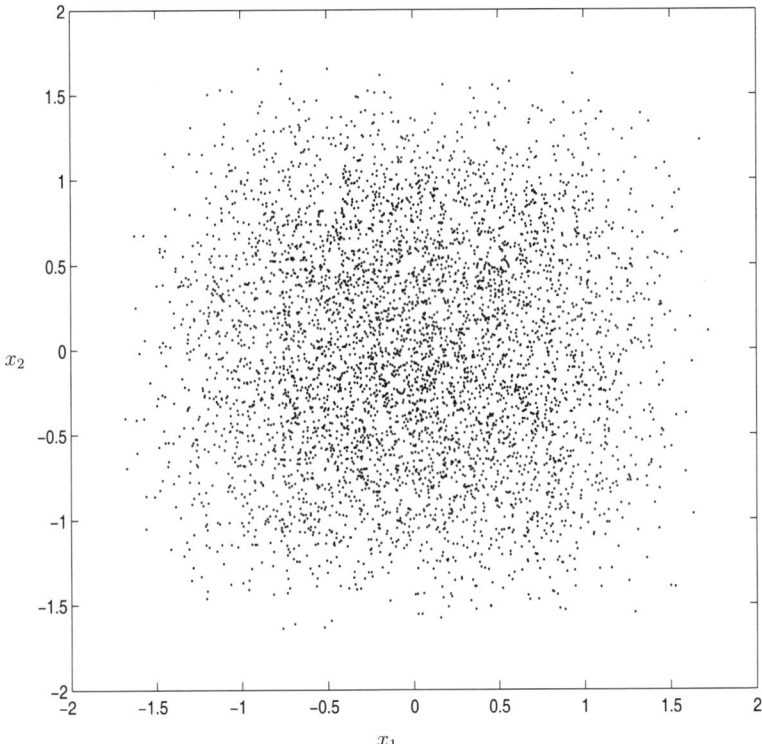

Fig. 3. 2-D computer simulation result with controller via the sawtooth function, showing complex dynamical behaviors for the controlled system (3.20) with the system parameter $a = 1/2$, and the control parameters: $r = \varepsilon = 1$, $\mu = 8$.

the initial value problem has a unique solution $\{x(n,m)\}_{n,m=0}^{\infty}$, again verified by successive iterations. It is noted that $\{x(n,m)\}_{m=0}^{\infty}$ may not be bounded for all $n \geq 0$ even if the initial-valued function $\phi = \{\phi(m)\}_{m=0}^{\infty}$ is bounded. The solution $\{x(n,m) : n \geq 0, 0 \leq m \preceq k\}$ is also called an orbit of system (4.1) (with (4.2) in the case of $k < \infty$) starting at ϕ.

The objective here is to design a control input sequence $\{u(n,m)\}$ or $\{u(n, m+1)\}$ such that the output of the controlled system

$$x(n+1,m) = f(m, x(n,m), x(n,m+1)) + u(n,m) \qquad (4.3)$$

or

$$x(n+1,m) = f(x(n,m), x(n,m+1)) + u(n,m+1) \qquad (4.4)$$

is chaotic in the sense of Devaney or Li-Yorke.

Similar to (3.5)-(3.6), the controller to be designed is in the forms of

$$u(n,m) = g(\mu\, x(n,m)), \qquad (4.5)$$

$$u(n,m) = \mu\, g(x(n,m)), \qquad (4.6)$$

and

$$u(n,m+1) = g(\mu\, x(n,m+1)), \qquad (4.7)$$

$$u(n,m+1) = \mu\, g(x(n,m+1)), \qquad (4.8)$$

where μ is a positive parameter and $g : J \to \mathbf{R}$ is a map, while J is an interval.

In the following, Sec. 4.1 reformulates the partial difference Eq. (4.1) into a special discrete system of the form (3.3) and then introduces some related concepts. Sections 4.2–4.4 give several chaotification algorithms via mod-operation, sawtooth function, and some general anti-controllers, respectively, by employing some chaotification algorithms established in Sec. 3, as discussed in [Shi et al. (2006)] and [Liang et al. (2008)].

4.1. *Problem reformulation and some basic concepts*

Instead of keeping the spatiotemporal setting of Eq. (4.1), the present attention is focused on the dynamical behaviors of the system with respect to the time evolution. To do so, Eq. (4.1) is first reformulated into a special discrete system in the form of (3.3).

By defining

$$x_n = \Big(x(n,0), x(n,1), \cdots\Big)^T \in \mathbf{R}^{k+1}, \quad n \geq 0,$$

Eq. (4.1) (with (4.2) in the case of $k < \infty$) can be written as

$$x_{n+1} = F(x_n), \quad n \geq 0, \qquad (4.9)$$

where

$$F(x_n) = \Big(f(x(n,0), x(n,1)), f(x(n,1), x(n,2)), \cdots, f(x(n,k), x(n,0))\Big)^T$$

in the case of $k < \infty$, or

$$F(x_n) = \Big(f(x(n,0), x(n,1)), f(x(n,1), x(n,2)), \cdots\Big)^T$$

in the case of $k = \infty$. This map F is called the induced map by f and the new system (4.9) is called the induced system by system (4.1) (with (4.2) in the case of $k < \infty$).

Similarly, the induced system by the controlled system (4.3) with (4.5) or (4.6) (with (4.2) in the case of $k < \infty$) can be written as

$$x_{n+1} = F(x_n) + G(x_n, \mu), \quad n \geq 0, \tag{4.10}$$

where G is the induced map by g; that is, $G(x_n, \mu) = \big(g(\mu\, x(n,0)), g(\mu\, x(n,1)), \cdots, g(\mu\, x(n,k))\big)^T$ or $G(x_n, \mu) = \mu\big(g(x(n,0)), g(x(n,1)), \cdots, g(x(n,k))\big)^T$ in the case of $k < \infty$; and $G(x_n, \mu) = \big(g(\mu\, x(n,0)), g(\mu\, x(n,1)), \cdots\big)^T$ or $G(x_n, \mu) = \mu\big(g(x(n,0)), g(x(n,1)), \cdots\big)^T$ in the case of $k = \infty$.

The controlled system (4.4) with (4.7) or (4.8) (with (4.2) in the case of $k < \infty$) can be written as

$$x_{n+1} = F(x_n) + G(\widehat{x}_n, \mu), \quad n \geq 0, \tag{4.11}$$

where

$$\widehat{x}_n := \big(x(n,1), x(n,2), \cdots, x(n,k), x(n,0)\big)^T \tag{4.12}$$

in the case of $k < \infty$; and

$$\widehat{x}_n := \big(x(n,1), x(n,2), \cdots\big)^T$$

in the case of $k = \infty$.

Next, some relative concepts are introduced for Eq. (4.1) (with (4.2) in the case of $k < \infty$).

Definition 4.1 [Shi et al. (2006), Definition 5.1]. A point $x \in Y_{k+1}$ is called a k_0-periodic point of Eq. (4.1) (with (4.2) in the case of $k < \infty$) if $x \in Y_{k+1}$ is a k_0-periodic point of its induced system (4.9); that is, $F^{k_0}(x) = x$ and $F^j(x) \neq x$ for $1 \leq j \leq k_0 - 1$. In the special case of $k_0 = 1$, x is called a fixed point or the steady state of Eq. (4.1) (with (4.2) in the case of $k < \infty$).

Definition 4.2 [Shi et al. (2006), Definition 5.2]. Equation (4.1) (with (4.2) in the case of $k < \infty$) is chaotic in the sense of Devaney (or Li-Yorke) on $V \subset Y_{k+1}$ if its induced system (4.9) is chaotic in the sense of Devaney (or Li-Yorke) on $V \subset Y_{k+1}$.

Remark 4.1. Several criteria of chaos for system (4.1) were established in [Shi (2008)].

4.2. Chaotification algorithms via mod-operation

It follows from Definition 4.1 that if $f(0,0) = 0$, then $\{x^*(m) = 0: 0 \leq m \preceq k\}$ is a fixed point of equation (4.1) (with (4.2) in the case of $k < \infty$) and consequently, $x^* := 0 \in Y_{k+1}$ is a fixed point of the induced system (4.9).

The following two results follows from Theorems 3.1 and 3.2, respectively.

Theorem 4.1 [Liang et al. (2008), Theorem 3.5]. Assume that $f(0,0) = 0$ and there exist positive constants r and L such that f is continuous in $[-r, r]^2$ and satisfies the Lipschitz condition

$$|f(x_1, y_1) - f(x_2, y_2)| \leq L \max\{|x_1 - x_2|, |y_1 - y_2|\}, \ \forall \ x_1, x_2, y_1, y_2 \in [-r, r]. \tag{4.13}$$

Then, for each constant μ satisfying

$$\mu > \mu_0 := \max\{5(1+L), 10L\}, \tag{4.14}$$

there exists a Cantor set $\Lambda \subset (-\frac{5}{2}\mu^{-1}r, \frac{5}{2}\mu^{-1}r)^{k+1}$ such that the controlled system

$$x(n+1, m) = f(x(n, m), x(n, m+1)) + \mu x(n, m) \pmod{r},$$
$$n \geq 0, \ 0 \leq m \preceq k \leq \infty,$$

(with (4.2) in the case of $k < \infty$) is chaotic on Λ in the sense of both Devaney and Li-Yorke.

Theorem 4.2 [Liang et al. (2008), Theorem 3.6]. Assume that $f(0,0) = 0$ and there exist positive constants r and L such that f is continuous in $[-r, r]^2$ and furthermore satisfies (4.13). Then, for each constant μ satisfying (4.14), there exists a Cantor set $\Lambda \subset (-\frac{5}{2}\mu^{-1}r, \frac{5}{2}\mu^{-1}r)^{k+1}$ such that the controlled system

$$x(n+1, m) = f(x(n, m), x(n, m+1)) + \mu x(n, m+1) \pmod{r},$$
$$n \geq 0, \ 0 \leq m \leq k < \infty,$$

with (4.2) is chaotic on Λ in the sense of both Devaney and Li-Yorke.

Suppose that f is continuously differentiable in $[-r, r]^2$ for some $r > 0$. Denote by $f_x(x, y)$ and $f_y(x, y)$ the first partial derivatives of f with respect to the first and the second variables at (x, y), respectively. Moreover, define

$$L := \max\{|f_x(x, y)| + |f_y(x, y)| : x, y \in [-r, r]\}. \tag{4.15}$$

Then, the induced map F by f is continuously differentiable in $[-r,r]^{k+1}$. Further, in the case of $k<\infty$, for any $x=\{x(j)\}_{j=0}^{k}\in[-r,r]^{k+1}$,

$$DF(x) = \begin{pmatrix} f_x(u(0)) & f_y(u(0)) & 0 & \cdots & 0 \\ 0 & f_x(u(1)) & f_y(u(1)) & \cdots & 0 \\ 0 & 0 & f_x(u(2)) & \cdots & 0 \\ \cdots & \cdots & \cdots & \cdots & \cdots \\ f_y(u(k)) & 0 & 0 & \cdots & f_x(u(k)) \end{pmatrix}_{(k+1)\times(k+1)},$$

that is,

$$DF(x)z = \Big(f_x(u(0))z(0) + f_y(u(0))z(1),\ f_x(u(1))z(1)$$
$$+ f_y(u(1))z(2), \cdots,\ f_x(u(k))z(k) + f_y(u(k))z(0)\Big)^T,$$
$$z = \{z(j)\}_{j=0}^{k} \in Y_{k+1},$$

where $u(j) = (x(j), x(j+1))$ for $0 \le j \le k$ with $x(k+1) = x(0)$, and in the case of $k = \infty$, for any $x = \{x(j)\}_{j=0}^{\infty} \in [-r,r]^{\infty}$,

$$DF(x) = \begin{pmatrix} f_x(u(0)) & f_y(u(0)) & 0 & \cdots & 0 & 0 & \cdots \\ 0 & f_x(u(1)) & f_y(u(1)) & \cdots & 0 & 0 & \cdots \\ \cdots & \cdots & \cdots & \cdots & \cdots & \cdots & \cdots \\ 0 & 0 & 0 & \cdots & f_x(u(k)) & f_y(u(k)) & \cdots \\ \cdots & \cdots & \cdots & \cdots & \cdots & \cdots & \cdots \end{pmatrix},$$

that is,

$$DF(x)z = (f_x(u(0))z(0) + f_y(u(0))z(1),\ f_x(u(1))z(1) + f_y(u(1))z(2), \cdots)^T,$$

where $u(j) = (x(j), x(j+1))$ for $j \ge 0$ and $z = \{z(j)\}_{j=0}^{\infty} \in Y_{\infty}$. It can be verified from the above relations that

$$\|DF(x)\| \le L,\quad x \in [-r,r]^{k+1}.$$

Based on the above discussion, the following results are direct consequences of Theorems 4.1 and 4.2.

Corollary 4.1 [Liang et al. (2008), Corollary 3.1]. Assume that $f(0,0) = 0$ and f is continuously differentiable in $[-r,r]^2$ for some $r > 0$. Then, all the results in Theorems 4.1 and 4.2 hold, respectively, where L is defined by (4.15).

4.3. *Chaotification algorithms via sawtooth functions*

The following two results follow from Theorems 3.4 and 3.5, respectively.

Theorem 4.3 [Shi et al. (2006), Theorem 5.3]. Assume that f is continuously differentiable in $[-r, r]^2$ for some $r > 0$ and $f(0,0) = 0$. Then, for each constant μ satisfying

$$\mu > \mu_0 = \max\left\{\frac{5}{2}r^{-1}\varepsilon, 5(L+1)\right\}, \tag{4.16}$$

there exists a Cantor set $\Lambda \subset [-\frac{5}{2}\mu^{-1}\varepsilon, \frac{5}{2}\mu^{-1}\varepsilon]^{k+1}$ such that the controlled system

$$x(n+1, m) = f(x(n,m), x(n, m+1)) + \text{Saw}_\varepsilon(\mu\, x(n,m)), \quad n \geq 0,\ 0 \leq m \preceq k$$

(with (4.2) in the case of $k < \infty$) is chaotic on Λ in the sense of both Devaney and Li-Yorke for any given $\varepsilon > 0$, where L is defined in (4.15).

Theorem 4.4 [Liang et al. (2008), Theorem 3.7]. Assume that $f(0,0) = 0$ and there exist positive constants r and L such that f is continuous in $[-r, r]^2$ and furthermore satisfies (4.13). Then, for each constant μ satisfying (4.16), there exists a Cantor set $\Lambda \subset [-\frac{5}{2}\mu^{-1}\varepsilon, \frac{5}{2}\mu^{-1}\varepsilon]^{k+1}$ such that the controlled system

$$x(n+1, m) = f(x(n,m), x(n, m+1)) + \text{Saw}_\varepsilon(\mu\, x(n, m+1)),$$
$$n \geq 0,\ 0 \leq m \leq k < \infty.$$

with (4.2) is chaotic on Λ in the sense of both Devaney and Li-Yorke.

The following two results do not require $f(0,0) = 0$, and they can be proved by Theorems 3.6 and 3.7, respectively.

Theorem 4.5 [Liang et al. (2008), Theorem 3.8]. Assume that there exist positive constants r and L such that f is continuous in $[-r, r]^2$ and furthermore satisfies (4.13). Then, for each constant μ satisfying

$$\mu > \mu_0 := \max\{1 + L, 5 + 6(L + |f(0,0)|r^{-1})\},$$

there exists a Cantor set $\Lambda \subset [-r, r]^{k+1}$ such that the controlled system

$$x(n+1, m) = f(x(n,m), x(n, m+1)) + \mu\, \text{Saw}_{\frac{r}{3}}(x(n,m)),$$
$$n \geq 0,\ 0 \leq m \preceq k \leq \infty,$$

(with (4.2) in the case of $k < \infty$) is chaotic on Λ in the sense of both Devaney and Li-Yorke.

Theorem 4.6 [Liang et al. (2008), Theorem 3.9]. Consider the controlled system

$$x(n+1,m) = f(x(n,m), x(n,m+1)) + \mu \, \text{Saw}_{\frac{r}{3}}(x(n,m+1)), \quad (4.17)$$
$$n \geq 0, \ 0 \leq m \leq k < \infty,$$

with (4.2). If all assumptions in Theorem 4.5 hold, then all the results in Theorem 4.5 hold for the controlled system (4.17).

Corollary 4.2 [Liang et al. (2008), Corollary 3.2]. If there exists a positive constant r such that f is continuously differentiable in $[-r, r]^2$, then all the results in Theorems 4.5 and 4.6 hold, where L is defined by (4.15).

4.4. *Chaotification algorithms via some general anti-controllers*

The following four results from [Liang et al. (2008), Theorems 3.1-3.4] follow from Theorems 3.8 and 3.9, respectively.

Theorem 4.7. Consider the controlled system (4.3) with controller (4.5), where $k \leq \infty$. Assume that

(i) f is continuously differentiable in $[-r, r]^2$ for some $r > 0$ and $f(0,0) = 0$;
(ii) g satisfies the following conditions:
   (iia) g is continuously differentiable in $[-r, r] \cup [a, b]$ with $r < a < b$ and $g'(x) \neq 0$ for all $x \in [-r, r] \cup [a, b]$;
   (iib) $g(0) = 0$ and there exists a point $\xi \in (a, b)$ such that $g(\xi) = 0$.

Then, for any constant μ satisfying

$$\mu > \mu_0 := \max\left\{\frac{b}{r}, \frac{Lr+b}{Nr}, \frac{Lb}{N(\xi-a)}, \frac{Lb}{N(b-\xi)}\right\}, \quad (4.18)$$

there exists a compact and perfect invariant set $E \subset Y_{k+1}$ such that the controlled system (4.3) with controller (4.5) (with 4.2) in the case of $k < \infty$) is chaotic on E in the sense of both Devaney and Li-Yorke, and has a dense orbit in E, where E contains the origin $x^* = 0$ and a Cantor set, L is defined by (4.15), and $N = \min\{|g'(x)| : x \in [-r, r] \cup [a, b]\}$.

Theorem 4.8. Consider the controlled system (4.3) with controller (4.6), where $k \leq \infty$. Assume that

(i) f is continuously differentiable in $[-r, r]^2$ for some $r > 0$ and $f(0,0) = 0$;

(ii) g satisfies the following conditions:

(iia) g is continuously differentiable in $[-a,a]\cup[b,r]$ with $0<a<b<r$ and $g'(x)\neq 0$ for all $x\in[-a,a]\cup[b,r]$;

(iib) $g(0)=0$ and there exists a point $\xi\in(b,r)$ such that $g(\xi)=0$.

Then, for any constant μ satisfying

$$\mu>\mu_0:=\max\left\{\frac{La+r}{Na},\frac{Lr}{N(\xi-b)},\frac{Lr}{N(r-\xi)}\right\}, \tag{4.19}$$

there exists a compact and perfect invariant set $E\subset Y_{k+1}$ such that the controlled system (4.3) with controller (4.6) (with (4.2) in the case of $k<\infty$) is chaotic on E in the sense of both Devaney and Li-Yorke, and has a dense orbit in E, where E contains the origin $x^*=0$ and a Cantor set, L is defined by (4.15), and $N=\min\{|g'(x)|:x\in[-a,a]\cup[b,r]\}$.

Theorem 4.9. Consider the controlled system (4.4) with controller (4.7) with $k<\infty$. Under assumptions (i) and (ii) in Theorem 4.7, for any constant μ satisfying (4.18), there exists a compact and perfect invariant set $E\subset Y_{k+1}$ such that the controlled system (4.4) with controller (4.7) and the periodic boundary condition (4.2) is chaotic on E in the sense of both Devaney and Li-Yorke, and has a dense orbit in E, where E contains the origin $x^*=0$ and a Cantor set, L is defined by (4.15), and $N=\min\{|g'(x)|:x\in[-r,r]\cup[a,b]\}$.

Theorem 4.10. Consider the controlled system (4.4) with controller (4.8), where $k<\infty$. Under assumptions (i) and (ii) in Theorem 4.8, for any constant μ satisfying (4.19), there exists a compact and perfect invariant set $E\subset Y_{k+1}$ such that the controlled system (4.4) with controller (4.8) and the periodic boundary condition (4.2) is chaotic on E in the sense of both Devaney and Li-Yorke, and has a dense orbit in E, where E contains the origin $x^*=0$ and a Cantor set, L is defined by (4.15), and $N=\min\{|g'(x)|:x\in[-a,a]\cup[b,r]\}$.

Remark 4.2. References [Shi et al. (2006)] and [Liang et al. (2008)] provide some chaotification examples with computer simulations about partial difference equations.

5. Conclusions

The emerging field of anti-control of chaos (chaotification) is very important and challenging, but also stimulating and promising. This new direction of research has gradually started to make an impact on modern engineering and technology, with enormous opportunities in academic, medical, industrial, and commercial applications. Achievements notwithstanding, there are still many fundamental and challenging issues on chaos theory and its applications ahead. In retrospect, new theories for dynamical analysis, effective methodologies for controls and anti-controls, and practical design for circuit implementation altogether have already emerged as a typical interdisciplinary field which, still yet, is calling for further efforts and endeavors from the scientific communities of applied mathematics especially nonlinear dynamics, automatic controls, circuits and systems, and biological sciences alike.

Acknowledgments

This research was supported by the NNSF of Shandong Province (Grant Y2006A15) and by the Hong Kong Research Grants Council under Grant CityU 1117/08E.

References

Banks, J., Brooks, J., Cairns, G. Davis, G. and Stacey, P. (1992). On Devaney's definition of chaos, *Amer. Math. Monthly* **99**, pp. 332-334.
Block, L. S. and Coppel, W. A. (1992). *Dynamics in One Dimension*, Lecture Notes in Mathematics **1513** (Berlin/Heidelberg, Springer-Verlag).
Chen, G. (2003). Chaotification via feedback: the discrete case, in *Chaos Control* (Chen, G. and Yu, X., Eds.) (Springer, Berlin), pp. 159-178.
Chen, G. and Dong, X. (1998). *From Chaos to Order: Methodologies, Perspectives and Applications* (World Scientific, Singapore).
Chen, G. and Lai, D. (1997) Anticontrol of chaos via feedback, in *Proc. of IEEE Conference on Decision and Control* (San Diego, CA), pp. 367-372.
Chen, G. and Lai, D. (1998). Feedback anticontrol of discrete chaos, *Int. J. Bifurcation and Chaos* **8**, pp. 1585-1590.
Chen, G. and Shi, Y. (2006). Introduction to anti-control of chaos: theory and applications, *Philos. Trans. Roy. Soc. A* **364**, pp. 2433-2447.
Chen, G. and Wang, X. F. (2006) *Chaotification of Dynamical Systems – Theory, Methods and Applications* (in Chinese) (Shanghai Jiao Tong University Press, China)
Devaney, R. L. (1989). *An Introduction to Chaotic Dynamical Systems* (Addison-Wesley, New York).
Huang, W. and Ye, X. (2002). Devaney's chaos or 2-scattering implies Li-Yorke's chaos, *Topology and its Applications* **117**, pp. 259-272.

Li, T. and Yorke, J. A. (1975). Period three implies chaos, *Amer. Math. Monthly* **82**, pp. 985–992.

Liang, W., Shi, Y. and Zhang, C. (2008). Chaotification for a class of first-order partial difference equations, *Int. J. Bifurcation and Chaos* **18**, pp. 717-733.

Marotto, F. R. (1978). Snap-back repellers imply chaos in $\mathbf{R}^n$, *J. Math. Anal. Appl.* **63**, pp. 199-223.

Shi, Y. (2008). Chaos in first-order partial difference equations, *J. Differ. Equ. Appl.* **14**, pp. 109-126.

Shi, Y. and Chen, G. (2004a). Chaos of discrete dynamical systems in complete metric spaces, *Chaos, Solitons and Fractals* **22**, pp. 555-571.

Shi, Y. and Chen, G. (2004b). Discrete chaos in Banach spaces, *Science in China Ser A: Mathematics*: Chinese version 2004, **34**, pp. 595–609; English version 2005, **48**, pp. 222-238.

Shi, Y. and Yu, P. (2006). Study on chaos induced by turbulent maps in noncompact sets, *Chaos, Solitons and Fractals* **28**, pp. 1165-1180.

Shi, Y. and Chen, G. (2006). Some new criteria of chaos induced by coupled-expanding maps, in *Proc. the 1st IFAC Conference on Analysis and Control of Chaotic Systems* (Reims, France), pp. 157-162.

Shi, Y. and Yu, P. (2008). Chaos induced by regular snap-back repellers, *J. Math. Anal. Appl.* **337**, pp. 1480-1494.

Shi, Y., Ju, H. and Chen, G. (2009). Coupled-expanding maps and one-sided symbolic dynamical systems, *Chaos, Solitons and Fractals* **39**, pp. 2138–2149.

Shi, Y., Yu, P. and Chen, G. (2006). Chaotification of discrete dynamical systems in Banach spaces, *Int. J. Bifurcation and Chaos* **16**, pp. 2615-2636.

Wang, X. F. (2003). Generating chaos in continuous-time systems via feedback control, in *Chaos Control* (Chen, G. and Yu, X., Eds.) (Springer, Berlin), pp. 179-204.

Wang, X. F. and Chen, G. (2000). Chaotification via arbitrary small feedback controls, *Int. J. Bifurcation and Chaos* **10**, pp. 549-570.

Wiggins, S. (1990). *Introduction to Applied Nonlinear Dynamical Systems and Chaos* (Springer-Verlag, New York).

Chapter 5

DELAYED FEEDBACK CONTROL TECHNIQUES

Kestutis Pyragas* and Viktoras Pyragas
Semiconductor Physics Institute, LT-01108 Vilnius, Lithuania
pyragas@pfi.lt

1. Introduction

An idea of controlling chaos has been first formulated by Ott, Grebogi, and Yorke [1] and attracted great interest among physicists over the past one and half decade. Why are chaotic systems interesting subjects for control theory and applications? The major key ingredient for the control of chaos is the observation that a chaotic set, on which the trajectory of the chaotic process lives, has embedded within it a large number of unstable periodic orbits (UPOs). In addition, because of ergodicity, the trajectory visits or accesses the neighborhood of each one of these periodic orbits. Some of these periodic orbits may correspond to a desired system's performance according to some criterion. The second ingredient is the realization that chaos, while signifying sensitive dependence on small changes to the current state and henceforth rendering unpredictable the system state in the long time, also implies that the system's behavior can be altered by using small perturbations. Then the accessibility of the chaotic system to many different periodic orbits combined with its sensitivity to small perturbations allows for the control and manipulation of the chaotic process. These ideas stimulated a development of rich variety of new chaos control techniques (see Refs. [2–7] for review), among which the delayed feedback control (DFC) method [8] has gained widespread acceptance.

Although the delayed feedback control (DFC) method [8] has been introduced more than one and half decade ago it is still one of the most active fields in applied nonlinear science [4, 9]. The method allows a noninvasive stabilization of unstable periodic orbits (UPO's) of dynamical systems in

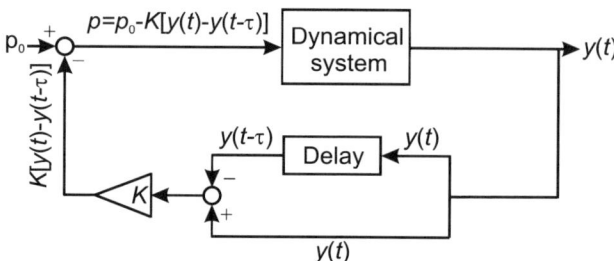

Fig. 1. Block diagram of the delayed feedback control method. $y(t)$ is an output variable, p is a control parameter, p_0 is its value at which the dynamical system possesses an unstable periodic orbit with a period τ, and K is the feedback gain.

the sense that the control force vanishes when the target state is reached. The DFC algorithm is reference-free and makes use of a control signal obtained from the difference between the current state of the system and the state of the system delayed by one-period of the target orbit. The block diagram of the method is presented in Fig. 1. The method allows us to treat the controlled system as a black box; no exact knowledge of either the form of the periodic orbit or the system of equations is needed. The DFC algorithm is especially superior for fast dynamical systems, since it does not require any real-time computer processing.

Successful implementation of the DFC algorithm has been attained in diverse experimental systems, including electronic chaotic oscillators [10–13], mechanical pendulums [14, 15], lasers [16–18], gas discharge systems [19–21], a current-driven ion acoustic instability [22], a chaotic Taylor-Couette flow [23], chemical systems [24, 25], high-power ferromagnetic resonance [26], helicopter rotor blades [27], and a cardiac system [28].

The DFC method has been verified for a large number of theoretical models from different fields. The problem of stabilizing high-speed semiconductor lasers was considered in Refs. [29–31]. Batlle et al. [32] implemented the DFC in a model of buck converter. The problem of controlling chaotic solitons by a time-delayed feedback mechanism was considered in Ref. [33]. Galvanetto [34] demonstrated the delayed feedback control of chaotic systems with dry friction. Bleich & Socolar [35] showed that the DFC can stabilize regular behavior in a paced, excitable oscillator described by Fitzhugh-Nagumo equations. Rappel et al. [36] used the DFC for stabilization of spiral waves in an excitable media as a model of cardiac tissue in order to prevent the spiral wave breakup. The DFC was also implemented in a model of a car-following traffic [37] and in economical models [38, 39]. Tsui and Jones [40] investigated the problem of chaotic satellite attitude control. Mensour and Longtin [41] proposed the DFC as a method to store information in delay-differential equations. Mitsubori and Aihara [42] suggested rather exotic application of the DFC, namely, the control of chaotic roll motion of a flooded ship in waves. Rosenblum and Pikovsky [43, 44] considered the influence of the DFC on the synchronization in an ensemble of globally coupled oscillators and discussed a possibility of using this approach to suppression of pathological brain rhythms. Yamasue and Hikihara [45] used the DFC in order to eliminate chaotic oscillations of microcantilever sensors in a dynamic force microscope. Li et al. [46] demonstrated that chaotic motions of robots with kinematical redundancy can be turned into regular motion by using the delayed feedback method. Liu et al. [46] showed that the method can eliminate oscillations in the Internet [47].

A variety of modifications of the DFC have been suggested in order to improve its performance. Adaptive versions of the DFC with automatic adjustment of delay time [48–50] and control gain [51, 52] have been considered. Basso et al. [53–55] showed that for a Lur'e system (system represented as feedback connection of a linear dynamical part and a static nonlinearity) the DFC can be optimized by introducing into a feedback loop a linear filter with an appropriate transfer function. For spatially extended systems, various modifications based on spatially filtered signals have been considered [56–58]. The wave character of dynamics in some systems allows a simplification of the DFC algorithm by replacing the delay line with the spatially distributed detectors. Mausbach et al. [21] reported such a simplification for an ionization wave experiment in a conventional cold cathode glow discharge tube. Due to dispersion relations the delay in time is equivalent to the spatial displacement and the control signal can be constructed without use of the delay line. Socolar, Sukow, and Gauthier [59] improved an original DFC scheme by using an information from many previous states of the system. This extended DFC (EDFC) scheme achieves stabilization of UPOs with a greater degree of instability [60, 61]. In [62–64] there were considered unstable DFC (UDFC) schemes using an unstable degree of freedom in a feedback loop to overcome the so called odd number limitation from which usual delayed feedback control suffers [65–68]. The EDFC and UDFC presumably are the most important modification of the DFC and they will be discussed at greater length in this chapter.

2. Delayed Feedback Control for Discrete Maps

To provide a platform for gaining fundamental analytical insights into time delay control methods we start with the simplest dynamical systems described by discrete-time maps rather than differential equations. The trends discovered through analysis of discrete maps are a good starting point for developing intuition about the behavior of continuous systems. Moreover, in systems with slow dynamics, the schemes for controlling discrete maps may be directly implemented.

2.1. *Controlling single-variable maps*

The simplest systems one might wish to control is an unstable fixed point of a single-variable map. Let the dynamics of the system variable y be governed by a map

$$y_{n+1} = f(y_n, p_n), \tag{1}$$

where the integer index n represents the discrete time and p_n is a system parameter that can be altered by an external signal. We assume that for $p_n = p_0$ the system has an unstable fixed point y^*. If the value y^* of the fixed point were known, we could try to stabilize it by using a standard proportional feedback control, i.e., adjusting the control parameter by the law $p_n = p_0 - K(y_n - y^*)$. However, we suppose that the reference value y^* is unknown. The main idea of the DFC control is to construct a feedback signal that adjusts p_n based on the difference between the current and past values of y. The simple approach is to let [8]

$$p_n = p_0 - K(y_n - y_{n-1}), \tag{2}$$

where K is a feedback gain. Then the success of the controller is determined by linear stability analysis of Eqs. (1) and (2). We make the following definitions:

$$\delta y_n \equiv y_n - y^*, \quad u_n \equiv y_n - y_{n-1}, \quad \mu_s \equiv \left.\frac{df}{dy}\right|_{y=y^*}, \quad k \equiv K \left.\frac{\partial f}{\partial p}\right|_{p=p_0} \tag{3}$$

Here μ_s is the Floquet multiplier (FM) of the free system and k is the modified value of the control gain. Since the fixed point of the free system is unstable we take $|\mu_s| > 1$. In the vicinity of y^*, Eqs. (1) and (2) then take the form

$$\delta y_{n+1} = \mu_s \delta y_n - k u_n, \quad u_n = \delta y_n - \delta y_{n-1}, \tag{4}$$

or

$$\begin{pmatrix} \delta y_{n+1} \\ u_{n+1} \end{pmatrix} = \begin{pmatrix} \mu_s & -k \\ \mu_s - 1 & -k \end{pmatrix} \begin{pmatrix} \delta y_n \\ u_n \end{pmatrix}. \tag{5}$$

The characteristic equation for the Floquet multipliers μ of this two-variable map reads

$$\mu^2 - (\mu_s - k)\mu - k = 0. \tag{6}$$

The map is stable if its both Floquet multipliers lie inside the unit circle of the complex plane, i.e. if $|\mu| < 1$. This leads to the inequalities $-1 < k < (\mu_s + 1)/2$, $\mu_s < 1$. It follows that this technique can stabilize unstable fixed points with only negative Floquet multipliers $\mu_s < -1$. This fact represents the simplest example of the so called odd number limitation of the DFC, which will be discussed in greater details in the following sections. Moreover, another limitation of this simple DFC approach is that it cannot stabilize strongly unstable fixed points with the Floquet multipliers $\mu_s < -3$.

In order to extend the utility of DFC to stronger instabilities in a practical manner (and without resorting to standard feedback schemes that rely on knowledge of the fixed point), values of the system variable from further in the past can be incorporated into the feedback signal. The simplest way to do this, from both an analytical and experimental perspective, is to replace Eq. (4) with [59]

$$\delta y_{n+1} = \mu_s \delta y_n - k u_n, \tag{7a}$$

$$u_n = \sum_{m=0}^{\infty} R^m (\delta y_{n-m} - \delta y_{n-m-1}), \tag{7b}$$

where R is a real parameter we are free to choose. An advantage of this scheme over other methods of incorporating past values of δy is that the sum in Eq. (7b) can be formed recursively by rewriting u_n as

$$u_n = \delta y_n - \delta y_{n-1} + R u_{n-1}. \tag{8}$$

We refer to this scheme as an extended delayed feedback control (EDFC) algorithm. The stability analysis of the EDFC is then quite similar to the above analysis of the DFC. The characteristic equation of the linearized two-variable map now reads

$$\mu^2 - (\mu_s - k + R)\mu - k + \mu_s R = 0. \tag{9}$$

Originally the EDFC has been considered for the parameter values $|R| < 1$. It was assumed that this restriction is necessary to provide the convergence

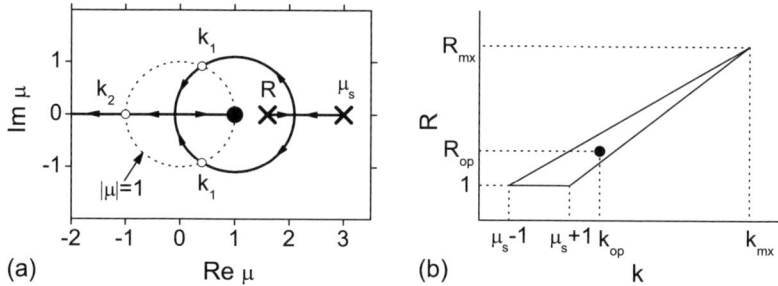

Fig. 2. (a) Root loci of Eq. (9) at $\mu_s = 3$, $R = 1.6$ as k varies from 0 to ∞. The crosses and black dot denote the location of roots at $k = 0$ and $k \to \infty$, respectively, (b) Stability region of Eqs. (7a) and (8) in the (k, R) plane; $k_{mx} = (\mu_s + 1)^2/(\mu_s - 1)$, $R_{mx} = (\mu_s + 3)/(\mu_s - 1)$.

of the sum in Eq. (7b). In other words, this condition guarantees the stability of the extended controller. The EDFC is superior to a simple DFC in the sense that it can stabilize fixed points with arbitrarily large negative Floquet multipliers $\mu_s < -1$. However, the stable EDFC suffers from the same odd number limitation as the simple DFC; it cannot stabilize the fixed points with positive Floquet multipliers $\mu_s > 1$.

In Ref. [62] it has been shown that the odd number limitation can be overcome by using an unstable EDFC (UEDFC) with the parameter $R > 1$. Although for $R > 1$ the controller is unstable, its connection with an unstable controlled system may make the whole closed loop system stable. The mechanism of such a stabilization for the positive μ_s is demonstrated in Fig. 2. The closed loop system is stable if both roots $\mu = \mu_{1,2}$ of Eq. (9) are inside the unit circle of the μ complex plain, $|\mu_{1,2}| < 1$. Figure 2(a) shows the characteristic root-locus diagram for $R > 1$, as the parameter k varies from 0 to ∞. For $k = 0$, there are two real eigenvalues greater than unity, $\mu_1 = \mu_s$ and $\mu_2 = R$, which correspond to two independent subsystems (7a) and (8), respectively; this means that both the controlled system and controller are unstable. With the increase of k, the eigenvalues approach each other on the real axes, then collide and pass to the complex plain. At $k = k_1 \equiv \mu_s R - 1$ they cross symmetrically the unite circle $|\mu| = 1$. Then both eigenvalues move inside this circle, collide again on the real axes and one of them leaves the circle at $k = k_2 \equiv (\mu_s + 1)(R+1)/2$. In the interval $k_1 < k < k_2$, the closed loop system (7a) and (8) is stable. By a proper choice of the parameters R and k one can stabilize the fixed point with an arbitrarily positive Floquet multiplier μ_s. Figure 2(b) shows the stability region in the plane of parameters (k, R). For a given value μ_s, there is an optimal choice of the parameters $R = R_{op} \equiv \mu_s/(\mu_s - 1)$, $k = k_{op} \equiv \mu_s R_{op}$ leading to zero eigenvalues, $\mu_1 = \mu_2 = 0$, such that the system approaches the fixed point in finite time.

2.2. Generalization for multi-variable maps

To stabilize systems consisting of many dynamical variables with multiple unstable directions, one may need to monitor several system variables and provide feedback signals to several parameters. Here we describe a generalization of EDFC called GEDFC. The GEDFC for a multi-variable map may be written as follows:

$$y_{n+1} = f(y_n, p_0 + B_0 G \cdot u_n), \tag{10a}$$

$$u_n = \sum_{m=0}^{\infty} R^m \cdot (y_{n-m} - y_{n-m-1}). \tag{10b}$$

Here $f(y_n, p_0)$ is the multi-variable uncontrolled map; u_n is the vector of control signal; p_0 is the vector of nominal values of the adjustable parameters. The EDFC control gain K is now substituted by the product of two matrices: a factor G that we are free to adjust and a factor B_0 that contains the information about which parameters are adjustable and how small changes in them affect the map. The EDFC scalar parameter R is also substituted by the matrix R that we are free to choose.

The linearized version of Eq. (10) in vicinity of fixed point y^* can be presented in the form

$$\delta y_{n+1} = A \cdot \delta y_n + B \cdot u_n, \tag{11a}$$

$$u_n = \sum_{m=0}^{\infty} R^m G \cdot (\delta y_{n-m} - \delta y_{n-m-1}) \tag{11b}$$

$$= G \cdot (\delta y_n - \delta y_{n-1}) + R \cdot u_{n-1}. \tag{11c}$$

Here $A = D_1 f(y^*, p_0)$ and $B = D_2 f(y^*, p_0) \cdot B_0$ are the Jacobi matrixes, where D_1 and D_2 are the vector-differential operators with respect to first and second arguments of the vector field f, respectively. For convenience, we included G in the definition of the control signal u, i.e. we substituted Gu by u. We assume A has full rank.

The design problem for such a controller is to choose G and R given A and B. One convenient approach is to make use of well-known design methods for standard proportional feedback control (PFC) systems in which the feedback signal is determined by the difference between the current variable values and their fixed point values. When the fixed point values are available for reference, the linearized controlled system takes the form

$$\delta y_{n+1} = A \cdot \delta y_n + B \cdot u_n, \tag{12a}$$
$$u_n = -K \cdot \delta y_n, \tag{12b}$$

where K is a matrix that can be chosen using techniques of discrete state optimal control theory if and only if A and B satisfy a controllability condition.

Equations (11) and (12) can both be cast in the form

$$\begin{pmatrix} \delta y_{n+1} \\ u_{n+1} \end{pmatrix} = Q \begin{pmatrix} \delta y_n \\ u_n \end{pmatrix}, \tag{13}$$

where we have

$$Q = \begin{pmatrix} A & B \\ G[A-I] & GB+R \end{pmatrix} \quad \text{or} \quad \begin{pmatrix} A & B \\ -KA & -KB \end{pmatrix}. \tag{14}$$

A comparison of these two forms immediately reveals that the GEDFC scheme has the same stability properties as the standard method if we choose

$$G = -K[A-I]^{-1}A \quad \text{and} \quad R = K[A-I]^{-1}B. \tag{15}$$

According to Eq. (15), the matrixes thus obtained satisfy the relation $R = -GA^{-1}B$. Any choice of of G and R that do not satisfy this relation corresponds to GEDFC schemes that are not equivalent to any standard proportional controller. Yamamoto [69] introduced a further generalization of DFC control in which a vector w of dynamical variables is added to the system as part of the control mechanism and the feedback signal is generated from it as follows:

$$u_n = G_1(\delta y_n - \delta y_{n-1}) + R_1 w_n, \tag{16a}$$
$$w_n = G_2(\delta y_{n-1} - \delta y_{n-2}) + R_2 w_{n-1}. \tag{16b}$$

The method is called dynamical delayed feedback control (DDFC). Nakajima [70] noted that GEDFC is a special case of DDFC in which $G_2 = I$ and $R_1 = G_1 R_2$, with the GEDFC parameters being $G = G_1$ and $R = R_2$.

3. Classification of Periodic Orbits in Continuous-Time Systems

In the rest of the chapter we discuss an application of the DFC methods to continuous-time systems. The DFC theory for continuous-time systems is much more difficult than for discrete maps. Continuous-time systems with time delay are hard to handle because the dynamics takes place in infinite-dimensional phase spaces. Even linear analysis of such systems is difficult due to the infinite number of Floquet exponents (FEs) characterizing the stability of controlled orbits. The linear and nonlinear analysis of such systems is usually performed numerically. In this context, a reasonable way for further development of the delayed feedback control theory is to look for problems allowing an analytical treatment. Our idea for the analytical approach is to consider dynamical systems close to bifurcation points of periodic orbits.

Most investigations in the theory of the DFC are devoted to the stabilization of unstable periodic orbits embedded in chaotic attractors of low-dimensional (usually three-dimensional) systems. The leading Floquet multipliers (FMs) of such orbits are real-valued and lie outside the unit circle in the complex plane (Figs. 3(a) and 3(b)). The orbits with the negative real multiplier arise from a period-doubling bifurcation and are typical, for example, for the Rössler system. The mechanism of stabilization of such orbits by delayed feedback is well understood [66, 71].

The orbits with the positive real multiplier come, for example, from a tangent or a subcritical Hopf bifurcation and are typical, for instance, for the Lorenz system. They satisfy the odd number limitation and cannot be stabilized by the usual delayed feedback method. The mechanism of stabilization of such orbits by the UDFC is described in Refs. [62–64]. In addition to the above mentioned orbits, there exists a large class of unstable periodic orbits with the complex conjugate pair of leading FMs [Fig. 3(c)].

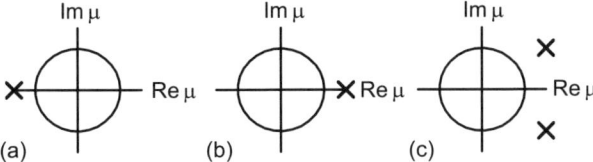

Fig. 3. Leading Floquet multipliers of unstable periodic orbits arising from different bifurcations: (a) period doubling, (b) tangent or subcritical Hopf, and (c) Nejmark-Sacker (discrete Hopf) bifurcations. The unit circle defines the region of stability.

Such orbits arise from a Nejmark-Sacker (discrete Hopf) bifurcation and cannot appear in low-dimensional chaotic attractors. However, such orbits may appear in low-dimensional non-chaotic systems.

Our aim is to describe an analytical treatment of the DFC algorithm for orbits arising in the three different bifurcations. Note that all three types of orbits have different topological properties. The orbits with the negative real multiplier flip their neighborhood during one turn. We consider them in Sec. 5. The orbits with the complex conjugate pair of the multipliers have a finite torsion and the orbits with the positive real multiplier have no torsion. We consider their stabilization in Secs. 6 and 7, respectively. In Sec. 4, we describe a relationship between the FEs of periodic orbits controlled by proportional and delayed feedback algorithms. This relationship is an essential tool of our analytical approach. We also utilize the methods of bifurcation theory and asymptotic methods of nonlinear dynamics, such as method of averaging, the center manifold theory, and a near identity transformation.

4. Proportional Versus Delayed Feedback

Consider a dynamical system described by ordinary differential equations

$$\dot{\boldsymbol{x}} = \boldsymbol{f}(\boldsymbol{x}, p, t), \tag{17}$$

where the vector $\boldsymbol{x} \in R^m$ defines the dynamical variables and p is a scalar parameter available for an external adjustment. We imagine that a scalar variable

$$y(t) = g(\boldsymbol{x}(t)) \tag{18}$$

that is a function of dynamic variables $\boldsymbol{x}(t)$ can be measured as the system output. Let us suppose that at $p = p_0 = 0$ the system has an UPO $\boldsymbol{x}_0(t)$ that satisfies $\dot{\boldsymbol{x}}_0 = \boldsymbol{f}(\boldsymbol{x}_0, 0, t)$ and $\boldsymbol{x}_0(t + T) = \boldsymbol{x}_0(t)$, where T is the period of the UPO. Here the value of the parameter p_0 is fixed to zero without a loss of generality. To stabilize the UPO we consider two continuous time feedback techniques, the proportional feedback control (PFC) and the DFC, both introduced by Pyragas [8].

The PFC uses the periodic reference signal

$$y_0(t) = g(\boldsymbol{x}_0(t)) \tag{19}$$

that corresponds to the system output if it would move along the target UPO. For chaotic systems, this periodic signal can be reconstructed from the chaotic output $y(t)$ by using the standard methods for extracting UPOs from chaotic time series data [72, 73]. The control is achieved via adjusting the system parameter by a proportional feedback

$$p(t) = G\left[y_0(t) - y(t)\right], \qquad (20)$$

where G is the control gain. If the stabilization is successful the feedback perturbation $p(t)$ vanishes. The experimental implementation of this method is difficult since it is not simply to reconstruct the UPO from experimental data.

More convenient for experimental implementation is the DFC method, which can be derived from the PFC by replacing the periodic reference signal $y_0(t)$ with the delayed output signal $y(t-T)$ [8]:

$$p(t) = K\left[y(t-T) - y(t)\right]. \qquad (21)$$

Here we exchanged the notation of the feedback gain for K to differ it from that of the proportional feedback. The delayed feedback perturbation (21) also vanishes provided the target UPO is stabilized. The DFC uses the delayed output $y(t-T)$ as the reference signal and the necessity of the UPO reconstruction is avoided. This feature determines the main advantage of the DFC over the PFC.

Hereafter, we consider a more general (extended) version of the delayed feedback control, the EDFC, in which a sum of states at integer multiples in the past is used [59]:

$$p(t) = K\left[(1-R)\sum_{n=1}^{\infty} R^{n-1} y(t-nT) - y(t)\right]. \qquad (22)$$

The sum represents a geometric series with the parameter $|R| < 1$ that determines the relative importance of past differences. For $R = 0$ the EDFC transforms to the original DFC. The extended method is superior to the original in that it can stabilize UPOs of higher periods and with larger FEs. For experimental implementation, it is important that the infinite sum in Eq. (22) can be generated using only single time-delay element in the feedback loop.

The success of the above methods can be predicted by a linear stability analysis of the target orbit. For the PFC method, the small deviations from the UPO $\delta \boldsymbol{x}(t) = \boldsymbol{x}(t) - \boldsymbol{x}_0(t)$ are described by variational equation

$$\delta \dot{\boldsymbol{x}} = [A(t) + GB(t)]\delta \boldsymbol{x}, \qquad (23)$$

where $A(t) = A(t+T)$ and $B(t) = B(t+T)$ are both T - periodic $m \times m$ matrices

$$A(t) = D_1 \boldsymbol{f}(\boldsymbol{x}_0(t), 0, t), \qquad B(t) = D_2 \boldsymbol{f}(\boldsymbol{x}_0(t), 0, t) \otimes Dg(\boldsymbol{x}_0(t)). \quad (24)$$

Here D_1 (D_2) denotes the vector (scalar) derivative with respect to the first (second) argument. The matrix $A(t)$ defines the stability properties of the UPO of the free system and $B(t)$ is the control matrix that contains all the details on the coupling of the control force.

Solutions of Eq. (23) can be decomposed into eigenfunctions according to the Floquet theory,

$$\delta\boldsymbol{x} = \exp(\Lambda t)\boldsymbol{u}(t), \quad \boldsymbol{u}(t) = \boldsymbol{u}(t+T), \quad (25)$$

where Λ is the FE. The spectrum of the FEs can be obtained with the help of the fundamental $m \times m$ matrix $\Phi(G, t)$ that is defined by equalities

$$\dot{\Phi}(G,t) = [A(t) + GB(t)]\Phi(G,t), \quad \Phi(G,0) = I. \quad (26)$$

For any initial condition $\boldsymbol{x}_{in}$, the solution of Eq. (23) can be expressed with this matrix, $\boldsymbol{x}(t) = \Phi(G,t)\boldsymbol{x}_{in}$. Combining this equality with Eq. (25) one obtains the system $[\Phi(G,T) - \exp(\Lambda T)I]\boldsymbol{x}_{in} = 0$ that yields the desired eigensolutions. The characteristic equation for the FEs reads

$$\det[\Phi(G,T) - \exp(\Lambda T)I] = 0. \quad (27)$$

It defines m FEs Λ_j (or Floquet multipliers $\mu_j = \exp(\Lambda_j T)$), $j = 1\ldots m$ that are the functions of the control gain G:

$$\Lambda_j = F_j(G), \quad j = 1, \ldots, m. \quad (28)$$

The values $F_j(0)$ are the FEs of the free system. By assumption, at least one FE of the free UPO has a positive real part. The PFC is successful if the real parts of all eigenvalues are negative, $\operatorname{Re} F_j(G) < 0$, $j = 1, \ldots, m$ in some interval of the parameter G.

Consider next the stability problem for the EDFC. The variational equation in this case reads

$$\delta\dot{\boldsymbol{x}} = A(t)\delta\boldsymbol{x}(t) + KB(t)\left[(1-R)\sum_{n=1}^{\infty} R^{n-1}\delta\boldsymbol{x}(t-nT) - \delta\boldsymbol{x}(t)\right]. \quad (29)$$

The delay terms can be eliminated due to Eq. (25), $\delta\boldsymbol{x}(t - nT) = e^{-n\Lambda T}\delta\boldsymbol{x}(t)$. As a result the problem reduces to the system of ordinary differential equations similar to Eq. (23)

$$\delta\dot{\boldsymbol{x}} = [A(t) + KH(\Lambda)B(t)]\delta\boldsymbol{x}, \quad (30)$$

where

$$H(\Lambda) = \frac{1 - \exp(-\Lambda T)}{1 - R\exp(-\Lambda T)} \qquad (31)$$

is the transfer function of the extended delayed feedback controller. Eqs. (23) and (30) have the same structure defined by the matrices $A(t)$ and $B(t)$ and differ only by the value of the control gain. The equations become identical if we substitute $G = KH(\Lambda)$. The price one has to pay for the elimination of the delay terms is that the characteristic equation defining the FEs of the EDFC depends on the FEs itself:

$$\det\left[\Phi(KH(\Lambda), T) - \exp(\Lambda T)I\right] = 0. \qquad (32)$$

Nevertheless, we can take advantage of the linear stability analysis for the PFC in order to predict the stability of the system controlled by time-delayed feedback. Suppose, that the functions $F_j(G)$ defining the FEs for the PFC are known. Then the FEs of the UPO controlled by time-delayed feedback can be obtained through solution of the transcendental equations

$$\Lambda = F_j(KH(\Lambda)), \quad j = 1\ldots m. \qquad (33)$$

We emphasize the physical meaning of the functions $F_j(G)$, namely, these functions describe the dependence of the Floquet exponents on the control gain in the case of the PFC.

5. Control of Periodic Orbits Arising from a Period-Doubling Bifurcation

In this section, we consider the theory of the DFC for periodic orbits arising from a period doubling bifurcation. Such orbits flip their neighborhood during one turn. More specifically, we consider UPOs whose leading Floquet multiplier is real and negative so that the corresponding FE obeys $\mathrm{Im}F_1(0) = \pi/T$. It means that the FE is placed on the boundary of the "Brillouin zone." Such FEs are likely to remain on the boundary under various perturbations and hence the condition $\mathrm{Im}F_1(G) = \pi/T$ holds in some finite interval of the control gain $G \in [G_{min}, G_{max}]$, $G_{min} < 0$, $G_{max} > 0$.

Let us introduce the dimensionless function

$$\phi(G) = F_1(G)T - i\pi \qquad (34)$$

that describes the dependence of the real part of the leading FE on the control gain G for the PFC and denote by

$$\lambda = \Lambda T - i\pi \qquad (35)$$

the dimensionless FE of the EDFC shifted by the amount π along the complex axes. Then from Eqs. (31) and (33) we derive

$$\lambda = \phi(G), \tag{36a}$$

$$K = G\frac{1 + R\exp(-\phi(G))}{1 + \exp(-\phi(G))}. \tag{36b}$$

These equations define the parametric dependence λ versus K for the EDFC.

To demonstrate the benefit of Eq. (36) let us derive the stability threshold of the UPO controlled by the extended delayed feedback. The stability of the periodic orbit is changed when λ reverses the sign. From Eq. (36a) it follows that the function $\phi(G)$ has to vanish for some value $G = G_1$, $\phi(G_1) = 0$. The value of the control gain G_1 is nothing but the stability threshold of the UPO controlled by the PFC. Then from Eq. (36b) one obtains the stability threshold for the EDFC:

$$K_1 = G_1(1 + R)/2. \tag{37}$$

It is interesting to note that the stability threshold for the DFC ($R = 0$) is equal to the half of the threshold in the case of the PFC, $K_1 = G_1/2$. In the following we demonstrate the use of the parametric Eqs. (36) for a specific example of the Rössler system.

5.1. *Example: Controlling the Rössler system*

Let us consider the problem of stabilizing the period-one UPO of the Rössler system [74]:

$$\begin{pmatrix} \dot{x}_1 \\ \dot{x}_2 \\ \dot{x}_3 \end{pmatrix} = \begin{pmatrix} -x_2 - x_3 \\ x_1 + ax_2 \\ b + (x_1 - c)x_3 \end{pmatrix} + p(t) \begin{pmatrix} 0 \\ 1 \\ 0 \end{pmatrix}. \tag{38}$$

Here we suppose that the feedback perturbation $p(t)$ is applied only to the second equation of the Rössler system and the dynamic variable x_2 is an observable available at the system output, i.e., $y(t) = g(\boldsymbol{x}(t)) = x_2(t)$.

For parameter values $a = 0.2$, $b = 0.2$, and $c = 5.7$, the free ($p(t) \equiv 0$) Rössler system exhibits chaotic behavior. Linearizing Eq. (38) around the UPO one obtains explicit expressions for the matrices $A(t)$ and $B(t)$ defined in Eq. (24).

First we consider the system (38) controlled by proportional feedback, when the perturbation $p(t)$ is defined by Eq. (20). By solving Eqs. (26)

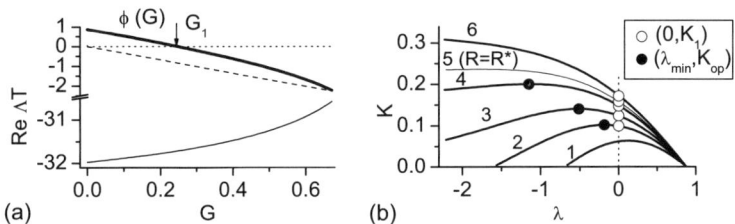

Fig. 4. (a) FEs of the Rösler system under PFC as functions of the control gain G. Thick solid, thin broken, and thin solid lines represent the functions $\Lambda_1 T - i\pi$, $\Lambda_2 T$ (zero exponent), and $\Lambda_3 T - i\pi$, respectively, (b) Parametric dependence K vs. λ defined by Eq. (36) for the EDFC. The numbers mark the curves with different values of the parameter R: (1) -0.5, (2) -0.2, (3) 0, (4) 0.2, (5) 0.28, (6) 0.4. Solid dots show the maxima of the curves and open circles indicate their intersections with the line $\lambda = 0$.

and (27) we obtain three FEs Λ_1, Λ_2 and Λ_3 as functions of the control gain G. The real parts of these functions are presented in Fig. 4(a). The values of the FEs of the free $(G = 0)$ UPO are $\Lambda_1 T = 0.876 + i\pi$, $\Lambda_2 T = 0$, $\Lambda_3 T = -31.974 + i\pi$. Thus the first and the third FEs are located on the boundary of the "Brillouin zone". In Fig. 4(a), we restricted ourselves with a small interval of the parameter $G \in [0, 0.67]$ in which all FEs do not change their imaginary parts. An information on the behavior of the leading FE Λ_1 or, more precisely, of the real-valued function $\phi(G) = \Lambda_1 T - i\pi$ in this interval will suffice to derive the main stability properties of the system controlled by delayed feedback.

The main information on the EDFC performance can be gained from parametric Eq. (36). They make possible a simple reconstruction of the relevant Floquet branch in the (K, λ) plane. This Floquet branch is shown in Fig. 4(b) for different values of the parameter R. Let us denote the dependence K versus λ corresponding to this branch by a function ψ, $K = \psi(\lambda)$. It reads

$$\psi(\lambda) = \phi^{-1}(\lambda)\frac{1 + R\exp(-\lambda)}{1 + \exp(-\lambda)}, \tag{39}$$

where ϕ^{-1} denotes the inverse function of $\phi(G)$. The maximum in the region $\lambda < 0$ defines the minimal value of the leading FE λ_{min} for the EDFC and $K_{op} = \psi(\lambda_{min})$ is the optimal value of the control gain at which the fastest convergence of the nearby trajectories to the desired orbit is attained. From Fig. 4(b) it is evident, that the delayed feedback controller should gain in performance through increase of the parameter R since the maximum of the $\psi(\lambda)$ function moves to the left. At $R = R^\star \approx 0.28$ the maximum

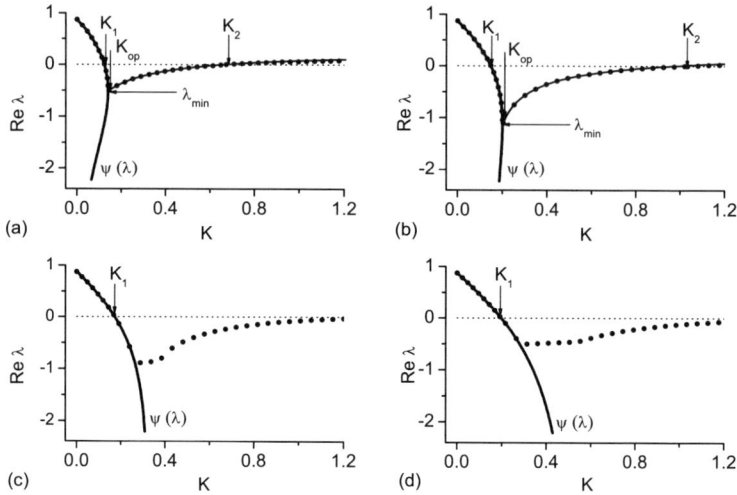

Fig. 5. Leading FEs of the Rösler system under EDFC as functions of the control gain K for different values of the parameter R: (a) 0.1, (b) 0.2, (c) 0.4, (d) 0.6. Thick solid lines symbolized by $\psi(\lambda)$ show the dependence $K = \psi(\lambda)$ for real λ. Solid lines in the region $K > K_{op}$ are obtained by analytical continuation of Eq. (40). The number of terms in series (40) is $N = 15$. Solid black dots denote the "exact" solutions obtained from complete system of Eqs. (26), (31) and (32).

disappears. For $R > R^\star$, it is difficult to predict the optimal characteristics of the EDFC. In this case the value λ_{min} is determined by the intersection of different Floquet branches.

An evaluation of the right boundary K_2 of the stability domain is a more intricate problem. Nevertheless, for the parameter $R < R^\star$ it can be successfully solved by means of an analytical continuation of the function $\psi(\lambda)$ on the complex region. For this purpose we expand the function $\psi(\lambda)$ at the point $\lambda = \lambda_{min}$ into power series

$$\psi(\lambda) = K_{op} + \sum_{n=2}^{N+1} \alpha_n (\lambda - \lambda_{min})^n. \tag{40}$$

We evaluate numerically the coefficients α_n by the least-squares fitting. Here we take the real values of λ. To extend the Floquet branch to the region $K > K_{op}$ we have to solve the equation $K = \psi(\lambda)$ for the complex argument λ.

Figure 5 shows the dependence of the leading FEs on the control gain K for the EDFC. The thick solid line represents the most important Floquet branch that conditions the main stability properties of the system. It is

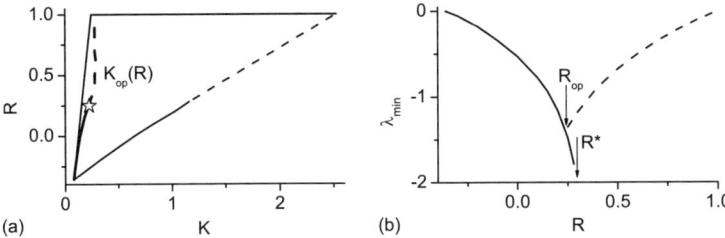

Fig. 6. (a) Stability region of the period-one UPO of the Rössler system under EDFC. The thick curve inside the region shows the dependence K_{op} versus R. The star marks the optimal point (K_{op}, R_{op}), (b) Minimal value λ_{min} of the leading FE as a function of the parameter R. In both figures solid and broken lines denote the solutions obtained from Eq. (36) and Eqs. (26), (31) and (32), respectively.

described by the function $K = \psi(\lambda)$ with the real argument λ. For $R < R^*$, this branch originates an additional sub-branch, which starts at the point (K_{op}, λ_{min}) and spreads to the region $K > K_{op}$. The sub-branch results from an analytical continuation of the function $\psi(\lambda)$ on the complex plane. As seen from the figures the Floquet sub-branches obtained by means of an analytical continuation are in good agreement with the "exact" solutions evaluated from the complete system of Eqs. (26), (31) and (32).

For $R > R^*$, the maximum in the function $\psi(\lambda)$ disappears and the Floquet branch originated from the eigenvalues $\lambda = \ln R \pm i\pi$ of the controller becomes dominant in the region $K > K_{op}$. This Floquet branch as well as the intersection point (K_{op}, λ_{min}) are unpredictable via a simple analysis. It can be determined by solving the complete system of Eqs. (26), (31) and (32). In Figs. 5(c) and 5(d) these solutions are shown by dots.

Figure 6 demonstrates how much of information one can gain via a simple analysis of parametric Eq. (36). These equations allow us to construct the stability domain in the (K, R) plane almost completely. Figure 6(b) shows of how the decay rate λ_{min} attained at the optimal value of the control gain K_{op} depends on the parameter R.

6. Control of Periodic Orbits Arising from a Nejmark-Sacker Bifurcation

In this section, we consider the problem of controlling UPOs with the complex conjugate pair of leading FMs (Fig. 3(c)). Such orbits may appear in low-dimensional non-chaotic systems, e.g., in a nonautonomous self-sustained oscillator exhibiting a quasiperiodic motion. We demonstrate an analytical treatment on the specific physical example of a weakly nonlinear van der Pol oscillator subjected to a periodic force and the DFC [75]. Without control, the oscillator can be synchronized by the periodic force only in a certain region of parameters. However, outside this region the system has UPOs that can be stabilized by the DFC. The feedback perturbation vanishes if the stabilization is successful and thus the region of synchronization can be extended noninvasively. By taking advantage of the fact that the system is close to a Hopf bifurcation, we derive a simplified averaged equation which can be treated analytically even in the presence of the delayed feedback. As a result we obtain simple analytical expressions defining the region of synchronization of the controlled system as well as an optimal value of the control gain.

6.1. *Example: Control of forced self-sustained oscillations*

Consider a weakly nonlinear van der Pol oscillator under action of external periodic force and delayed feedback perturbation

$$\ddot{x} + \omega_0^2 x + \varepsilon(x^2 - 1)\dot{x} = a\sin(\omega t) + k(x - x_T). \tag{41}$$

The parameter ω_0 is the characteristic frequency of self-sustained oscillations, and ε is responsible for the strength of nonlinearity of the oscillator. The first term in the right-hand side is an external periodic force and the second term describes the delayed coupling due to control. The parameter k is the feedback gain, $x_T \equiv x(t-T)$, and $T = 2\pi/\omega$ is the period of the external force. We suppose that ε is a small parameter, $\varepsilon \ll \omega_0$. Moreover we assume that the amplitude a, the frequency detuning $\omega - \omega_0$ as well as the control perturbation $k(x - x_T)$ are proportional to the small parameter ε.

We apply the method of averaging. First we rewrite Eq. (41) as a system

$$\dot{x} = y, \tag{42a}$$
$$\dot{y} = -\omega_0^2 x - \varepsilon(x^2 - 1)y + a\sin(\omega t) + k(x - x_T). \tag{42b}$$

As Eq. (41) or system (42) is close to that of linear oscillator, we can expect that the solution has a nearly harmonic form. Since there is a forced system we look for a solution with the characteristic frequency ω

$$x = (A(t)e^{i\omega t} + A^*(t)e^{-i\omega t})/2, \tag{43a}$$
$$y = i\omega(A(t)e^{i\omega t} - A^*(t)e^{-i\omega t})/2. \tag{43b}$$

Here $A(t)$ is a new variable, a slowly varying complex amplitude. Substituting Eq. (43) in system (42) we obtain the equation for the complex amplitude, which after averaging over the period T of fast oscillations takes the form

$$\dot{A} = \frac{\omega^2 - \omega_0^2}{2i\omega} A - \frac{\varepsilon}{2} A \left(\frac{|A|^2}{4} - 1 \right) - \frac{a}{2\omega} + \frac{k}{2i\omega}(A - A_T). \qquad (44)$$

By choosing an appropriate scale for the amplitude

$$A = 2z \qquad (45)$$

and introducing new parameters

$$\alpha = \frac{a}{2\varepsilon\omega}, \quad \nu = \frac{\omega^2 - \omega_0^2}{\varepsilon\omega} \approx 2\frac{\omega - \omega_0}{\varepsilon}, \quad \kappa = \frac{k}{\varepsilon\omega}, \qquad (46)$$

Eq. (44) can be simplified to

$$(2/\varepsilon)\dot{z} = -i\nu z - z(|z|^2 - 1) - \alpha - i\kappa(z - z_T). \qquad (47)$$

The parameters α, ν, and κ are proportional respectively to the amplitude of external force, the frequency detuning, and the delayed feedback gain.

The free system (47) (with $\kappa = 0$) has three periodic orbits provided

$$\alpha_1^2(\nu) < \alpha^2 < \alpha_2^2(\nu), \qquad (48a)$$

$$\alpha_{1,2}^2(\nu) = \frac{2}{27} \left[9\nu^2 + 1 \mp (1 - 3\nu^2)^{3/2} \right] \qquad (48b)$$

or one orbit otherwise. This result is obtained by solving the corresponding cubic equation.

The condition of the Hopf bifurcation defines the minimal amplitude of the stable orbit $A_{min} = \sqrt{2}$. The orbits with amplitude $|A_0| < A_{min}$ are unstable. In the (ν, α) plane, this condition defines the hyperbola

$$\alpha^2 = \nu^2/2 + 1/8, \qquad (49)$$

which is shown by a solid line in Fig. 7. Above this line the oscillator is synchronized with the external force. Below this line, in region of one periodic orbit, we usually have a quasiperiodic behavior. The orbits losing their stability through the Hopf bifurcation have a pair of complex conjugate exponents with the positive real part.

The delayed feedback changes the condition of the Hopf bifurcation, which now depends on the delayed feedback strength K. The relation between K, ν, and α is

$$\alpha^2 = \frac{1}{8}(1 - \nu K)\left[(1 + \nu K)^2 + 4\nu^2\right]. \qquad (50)$$

Here we use the notation

$$K = \kappa T \varepsilon / 2 = k\pi/\omega^2. \qquad (51)$$

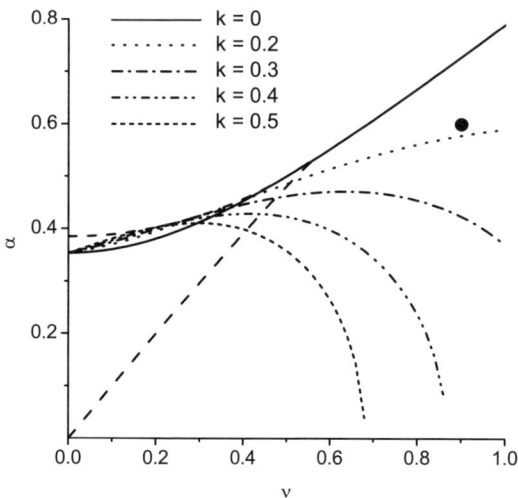

Fig. 7. The bifurcation diagram for van der Pol oscillator controlled by delayed feedback. In the region between two dashed lines there are three periodic orbits. Outside this region there is only one periodic orbit. The solid line defines the Hopf bifurcation for the uncontrolled system, and broken lines are defined by Eq. (50). Above these lines the oscillator is synchronized with the external force. The solid dot $(\nu, \alpha) = (0.9, 0.6)$ shows the set of parameters which we use to demonstrate the DFC performance.

In Fig. 7, the relations (50) are presented by curves in the (ν, α) plane for different fixed values of K. These curves define the boundaries of synchronization for the controlled oscillator. Above these curves the oscillator is synchronized with the periodic force. We see that the delayed feedback perturbation extends the synchronization region. The threshold of the Hopf bifurcation can be presented in the form

$$k_0 = \frac{\omega^2}{\pi} K_0 = \frac{\omega^2}{\pi \nu} \left(1 - \frac{|A_0|^2}{2} \right). \tag{52}$$

To demonstrate how the FEs depend on the control gain k we specify the parameters (ν, α) to be $(0.9, 0.6)$. This set of parameters is marked by a solid dot in Fig. 7. We have calculated the leading FEs of the initially unstable orbit using three different methods, namely, (i) solving transcendental equation (that is obtained by linerization of the averaged Eq. (47)), (ii) using quadratic equation (it is obtained by approximation $e^{-\lambda T} \approx 1 - \lambda T$ in the transcendental equation), and (iii) solving variational equations derived from the exact system (42). The results of the above analysis for $\varepsilon = 0.01$ are presented in Fig. 8. The exact values of the leading FE's are shown by

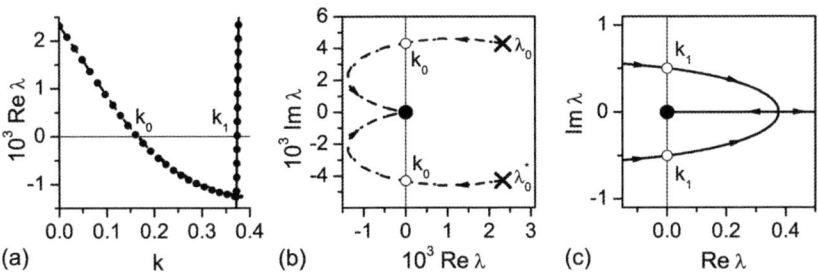

Fig. 8. Leading Floquet exponents as functions of the control gain for $(\nu, \alpha) = (0.9, 0.6)$. Here and in all numerical demonstrations below we take $\omega_0 = 1$. (a) Reλ vs k for $\varepsilon = 0.01$. For the given values of parameters ν, α, and ε, we have $\omega \approx 1.00451$ and $a \approx 0.01205$. The amplitude of the unstable orbit is $|A_0| \approx 1.034$ and its FE's for $k = 0$ are $\lambda_0 \approx (2.327 \pm 4.297i) \times 10^{-3}$. Solid dots are the values of the FEs obtained from variational equations derived from exact Eq. (42). The dashed and dotted lines calculated respectively from transcendental and quadratic equations [or Eq. (53)] approximate the left-hand branch. The solid line calculated from Eq. (56) approximates the right-hand branch, (b) Root loci of transcendental equation (dotted line) and quadratic equation (dashed line) as k varies from 0 to ∞ for the same values of parameters as in (a). Crosses and black dot denote the location of the roots for $k = 0$ and $k = \infty$, respectively, (c) Root loci of Eq. (54).

dots. There are two branches (the left-hand and the right-hand) defining the interval of stability $k_0 < k < k_1$ in which the real part of the leading FE is negative.

First we discuss the results for the left-hand branch. Figure 8(a) shows that all three above methods give quantitatively coinciding results. Thus for small ε the leading FE of the left-hand branch can be reliably obtained from the simple quadratic equation, which yields

$$\mathrm{Re}\lambda = \frac{\varepsilon}{2} \frac{1 - |A_0|^2/2 - \nu k\pi/\omega^2}{1 + (k\pi/\omega^2)^2}, \qquad (53)$$

and the threshold k_0 of the Hopf bifurcation is well described by Eq. (52).

For the right-hand branch, the nonlinear terms in Eq. (41) are small in comparison with the control term. Setting $\varepsilon = 0$ in the exact variational equations we obtain the characteristic equation

$$\lambda^2 + \omega_0^2 - k(1 - e^{-\lambda T}) = 0. \qquad (54)$$

Root loci diagram of the relevant branch for this equation when varying k is shown in Fig. 8(c). The pair of complex conjugate roots intersects the imaginary axes at the points $\lambda = \pm i\pi/T = \pm i\omega/2$. This intersection appears for $k = k_1$, where

$$k_1 = \frac{1}{2}\left(\omega_0^2 - \frac{\omega^2}{4}\right). \tag{55}$$

defines the upper threshold of stability. For $k = k_1$, the orbit loses stability by a period doubling bifurcation. Expanding the solution of Eq. (54) in Taylor series close to the threshold $k = k_1$, we obtain an approximate analytical expression

$$\mathrm{Re}\lambda = \frac{4\pi k_1/\omega}{\omega^2 + (2\pi k_1/\omega)^2}(k - k_1) \tag{56}$$

that describes well the Reλ vs k dependence for the right-hand branch (Fig. 8(a)).

To verify the validity of the linear theory we have numerically investigated the original nonlinear differential equations (42). Without control the system experiences a quasiperiodic motion (Fig. 9(a)). The DFC perturbation stabilizes an unstable UPO and we have a periodic motion synchronized with an external force (Fig. 9(b)). Whenever the synchronization is established the feedback perturbation vanishes (Fig. 9(c)). The envelopes

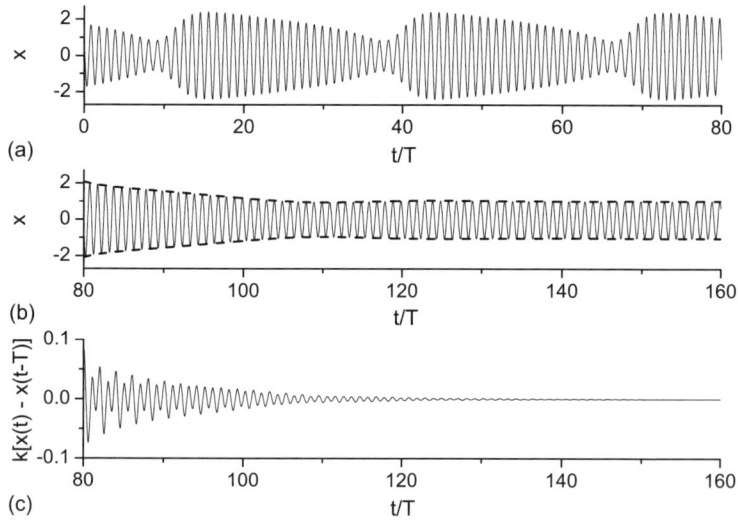

Fig. 9. Results of numerical integration of delay-differential Eq. (42) for $(\nu, \alpha) = (0.9, 0.6)$, $\varepsilon = 0.1$. (a) Dynamics of the x variable without control, (b) and (c) dynamics of the x variable and perturbation $k[x(t) - x(t-T)]$ when the control is switched on. The broken line (an envelop) in (b) is the dynamics of the complex amplitude $|A(t)| = 2|z(t)|$ obtained from averaged Eq. (47). The strength of the feedback gain is $k = 0.34$.

of the transient are well described by the averaged amplitude Eq. (47). This confirms the validity of the averaging procedure applied to the time-delay system (42).

Although here we have presented the analytical results only for the case of the DFC algorithm, the theoretical approach described in this section can be easily extended for the case of the EDFC [75].

7. Control of Torsion-Free Periodic Orbits in Autonomous Systems

It has been shown that the conventional DFC can stabilize only a certain class of periodic orbits characterized by a finite torsion. More precisely, the limitation is that any UPOs with an odd number of real Floquet multipliers greater than unity can never be stabilized by the DFC. This statement was first proved by Ushio [65] for discrete time systems. Just et al. [66] and Nakajima [67] proved the same limitation for the continuous time DFC, and then this proof was extended for a wider class of delayed feedback schemes, including the EDFC [68]. Hence it seems hard to overcome this inherent limitation, see, however, Sec. 10 for a recent correction of this view.

In Ref. [62], we have proposed to supplement the feedback loop with an additional unstable degree of freedom in order to overcome the odd number limitation. By including an additional unstable mode into the control loop one artificially enlarges the set of real multipliers greater than unity to an even number. The idea of an unstable controller we have demonstrated in Sec. 2.1 for a single-variable discrete map. We have shown that an unstable EDFC with the memory parameter $R > 1$ can stabilize an unstable fixed point satisfying conditions of the odd number limitation. Unfortunately, the idea of using an unstable EDFC with the parameter $R > 1$ fails for continuous time systems, since such a controller involves an infinite number of unstable modes. An appropriate controller for continuous time systems can be constructed on the basis of usual EDFC with the parameter $|R| < 1$. A required additional greater than unity real FM can be gained via supplementing the feedback loop by an additional unstable mode. The performance of such a controller has been numerically demonstrated for the Lorenz system [62] as a representative of systems with torsion free UPOs. Later on it has been realized that the unstable DFC allows an analytical treatment if a system is close to a subcritical Hopf bifurcation. First such a treatment has been performed for a simple nonlinear second order electronic circuit modeling a subcritical Hopf bifurcation [63]. Then this approach has been extended for wider class of dynamical systems with an arbitrary large phase space dimension [64]. Below we briefly describe this approach for the Lorenz system.

7.1. Example: Control of the Lorenz system at a subcritical Hopf bifurcation

We consider the Lorenz system [76]

$$\dot{x} = \sigma(y - x), \tag{57a}$$
$$\dot{y} = rx - y - xz, \tag{57b}$$
$$\dot{z} = xy - bz, \tag{57c}$$

for fixed values of the parameters $\sigma = 10$, $b = 8/3$, and variable parameter r. For $0 < r < 1$, the Lorenz system has a unique stable steady state (a stable node) at the origin $C^0 : (0,0,0)$. For $r = 1$, the origin becomes a saddle and two additional symmetrical stable fixed points $C^\pm$:

$$(x_f^\pm, y_f^\pm, z_f) = \left(\pm\sqrt{b(r-1)}, \pm\sqrt{b(r-1)}, r - 1\right) \tag{58}$$

appear. For $r = r_H$, the steady states $C^\pm$ become unstable. The value

$$r_H = \frac{\sigma(\sigma + b + 3)}{\sigma - b - 1} \tag{59}$$

represents the point at which the subcritical Hopf bifurcation occurs. Just below this bifurcation point, for

$$r = r_H - \Delta r, \quad 0 < \Delta r \ll r_H, \tag{60}$$

there are two small unstable limit cycles surrounding the stable steady states $C^\pm$. Moreover, at the same values of the parameter r there exists a strange attractor. Thus the system is multistable and depending on initial conditions the phase trajectory may either be attracted to the one of the steady states or exhibit a chaotic behavior on the strange attractor.

Our aim is to stabilize the unstable limit cycles arising at the Hopf bifurcation using the DFC technique. In particular, we are interested in analytical treatment of this problem. Note that the periodic orbits arising at this bifurcation are torsion-free and we need an unstable controller.

Specifically, we consider the following control algorithm:

$$\dot{x} = \sigma(y - x), \tag{61a}$$

$$\dot{y} = rx - y - xz + \varepsilon w(y - y_f), \tag{61b}$$

$$\dot{z} = xy - bz, \tag{61c}$$

$$\dot{w} = \varepsilon \lambda_c w + k\left[y - y(t - \tau)\right]. \tag{61d}$$

We suppose that y is an observable and apply the control perturbation $w(y - y_f)$ only to the second equation of the Lorenz system. We use a nonlinear perturbation. For definiteness, we consider the control of the periodic orbit surrounding the fixed point C^+ and take $y_f \equiv y_f^+ = \sqrt{b(r-1)}$.

Equation (61d) describes an unstable delayed feedback controller, which supplements the system with an additional unstable Floquet mode and eliminates the odd number limitation. The positive parameter $\lambda_c > 0$ defines the value of the additional FE introduced with the controller. The parameter k denotes the strength of the feedback gain. The delay time τ in Eq. (61d) is equal to the period of the unstable periodic orbit such that the controller does not change the periodic solution of the Lorenz system. Finally, the parameter

$$\varepsilon = \sqrt{b(r_H - 1)} - \sqrt{b(r - 1)} \approx \sqrt{b/(r_H - 1)}\Delta r/2 \tag{62}$$

defines the closeness of the system to the bifurcation point $r = r_H$. This is the main control parameter, whose smallness we exploit in the perturbation theory. Below we describe only the main steps of our analytical approach, for details we refer to our recent paper [64].

Generally Eq. (61) represent rather complicated system of nonlinear delay-differential equations. The dynamics of the system takes place in an infinite-dimensional phase space and reduction of the phase space dimension via the center manifold theory is a nontrivial task. To overcome the problem of an infinite-dimensional phase space we use the relationship between the FEs of the DFC and PFC algorithms described in Sec. 4. In Eq. (61d) we replace the delay term $y(t - \tau)$ with the periodic solution of the free Lorenz system $y_0(t)$ corresponding to the unstable limit cycle, which we intend to stabilize. Then instead of the DFC problem described by delay-differential Eq. (61) we get a nonautonomous system of four ordinary differential equations for the PFC problem. To transform the nonautonomous PFC problem to the autonomous we write an additional free Lorenz system with the initial conditions taken on the stable manifold of the desired UPO, such that it generates the signal $y_0(t)$ required for the PFC algorithm. As a result we get

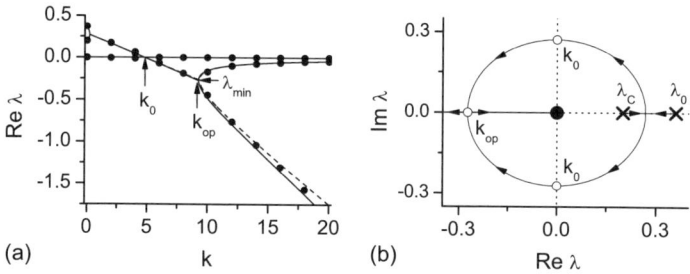

Fig. 10. (a) Real parts of leading Floquet exponents of the controlled limit cycle as functions of the control gain for $\varepsilon = 0.1$, $\lambda_c = 0.2$. Dashed and solid lines show the solutions of the characteristic Eqs. (63) and (64), respectively. Dots correspond to the values of Floquet exponents obtained from the exact variational equations derived from Eq. (61), (b) Root loci of Eq. (63) as k varies from 0 to ∞ for the same parameter values as in (a). Crosses and black dot denote the location of the roots for $k = 0$ and $k = \infty$, respectively.

an autonomous system of seven ordinary differential equations for the PFC problem. Using the closeness of the system to a subcritical Hopf bifurcation we apply the center manifold theory and reduce the system dimension. Then we simplify this system by using the near identity transformation and averaging. Linearization of the simplified system around the desired UPO leads to a characteristic equation of the PFC problem. Finally, using the relationship between the FEs of the PFC and DFC algorithms we derive an analytical characteristic equation for the DFC algorithm:

$$\lambda^2 - (\lambda_0 + \lambda_c)\lambda + \lambda_0\lambda_c + Qk[1 - \exp(-\varepsilon\lambda\tau)] = 0. \quad (63)$$

Here $\lambda_0 = \Lambda_0/\varepsilon \approx 0.360675991$ and $\lambda = \Lambda/\varepsilon$ are the rescaled values of the FEs Λ_0 and Λ of the free and DFC controlled UPO, respectively. An approximate value of the parameter Q is 1.743243862.

For $\varepsilon|\lambda|\tau \ll 1$, we can use an approximation $\exp(-\varepsilon\lambda\tau) \approx 1 - \varepsilon\lambda\tau$, which transforms Eq. (63) to the simple quadratic equation

$$\lambda^2 - (\lambda_0 + \lambda_c - kQ\varepsilon\tau)\lambda + \lambda_0\lambda_c = 0. \quad (64)$$

In Fig. 10(a), we compare the leading FEs of the controlled system determined by three different methods: (i) by solving the quasipolynomial Eq. (63), (ii) using the solutions of the quadratic Eq. (64), and (iii) by solving the exact system of variational equations derived from Eq. (61). All three above results are in good quantitative agreement, as viewed in Fig. 10(a). Thus the leading FEs are reliably predicted by the simple quadratic Eq. (64).

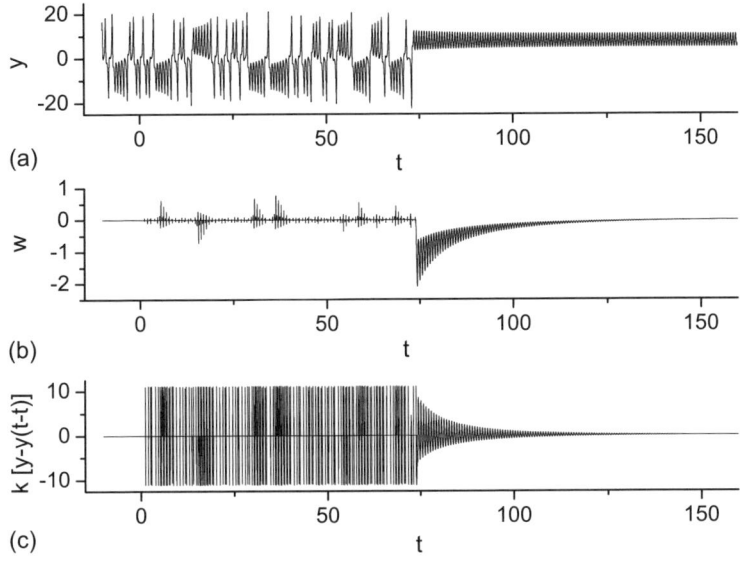

Fig. 11. Dynamics of (a) variable y, (b) controller variable w, and (c) delayed feedback perturbation $k[y-y(t-\tau)]$. The initial conditions are $x(-15\tau) = 8.109559459$, $y(-15\tau) = 13.03719946$, $z(-15\tau) = 14.27465065$, $w(-15\tau) = 0$. $y(t) = 0$ for $-15\tau < t \leq -14\tau$. The control is initiated at $t = \tau$. The values of the parameters are $\varepsilon = 0.1$, $\lambda_c = 0.2$, $\tau = 0.67398328$, $k = 0$ for $-15\tau \leq t < \tau$ and $k = 9.25$ for $t \geq \tau$. For $|y - y(t-\tau)| > Y_{max} = 1.2$, the controller is off.

The mechanism of stabilization is evident from Fig. 10(b). For $k = 0$, two real positive solutions of Eq. (64) $\lambda = \lambda_0$ and $\lambda = \lambda_c$ describe an unstable eigenvalue of the free system and the free controller, respectively. With increasing k, the eigenvalues approach each other on the real axis, then collide and pass to the complex plane. For $k = k_0$, where

$$k_0 = (\lambda_0 + \lambda_c)/Q\varepsilon\tau, \tag{65}$$

they cross the imaginary axis and move symmetrically into the left half plane, i.e., both the system and the controller become stable. An optimal value of the control gain is

$$k_{op} = k_0 + 2\sqrt{\lambda_0 \lambda_c}/Q\varepsilon\tau, \tag{66}$$

since it provides the fastest convergence to the stabilized limit cycle with the characteristic rate $\lambda_{min} = -\sqrt{\lambda_0 \lambda_c}$.

The validity of the linear theory is confirmed by numerical analysis of the original system of nonlinear delay-differential Eq. (61) presented in Fig. 11.

In numerical simulations, the controller is switched on only when the system is close to the desired periodic orbit and switched off when it is far away from the orbit. Without control ($t < \tau$), the Lorenz system demonstrates a chaotic behavior on the strange attractor. For $t > \tau$, the control algorithm starts to act and after a transient process the controlled system approaches a previously unstable limit cycle, and the feedback perturbation vanishes.

8. Control of Periodic Orbits without Torsion in Nonautonomous Systems

In this section we address the concept of the unstable delayed feedback controller to nonautonomous chaotic systems. We develop an analytical approach for the DFC algorithm that considerably differs from that considered in the previous section. Here we meet a favorable situation when a torsion-free periodic orbit is weakly nonlinear although it is embedded in a chaotic attractor for which the nonlinearity is an essential factor. A weak nonlinearity of the orbit allows us to develop a perturbation theory with respect to small nonlinear terms.

8.1. Example: Control of a forced double-well oscillator

As a paradigmatic model of nonautonomous chaotic system let us consider a forced double-well oscillator. If x and y denote the dynamical variables of the oscillator, A is the amplitude and ω is the frequency of the driving force then the equations of motion subjected to delayed feedback control read

$$\dot{x} = y, \tag{67a}$$
$$\dot{y} = \alpha x - \gamma x^3 - \beta y + A\cos(\omega t) - k(S + W), \tag{67b}$$
$$\dot{W} = \lambda_c W + kbS, \tag{67c}$$
$$S(t) = x(t) - (1 - R)B(t - \tau), \tag{67d}$$
$$B(t) = x(t) + RB(t - \tau), \tag{67e}$$

where $\alpha > 0$ and $\gamma > 0$ are the parameters of the double-well potential and $\beta > 0$ is the damping parameter. Here we use the extended version [12] of the DFC described by variables $S(t)$ and $B(t)$ with the delay time $\tau = 2\pi/\omega$ equal to the period of the driving force, and the memory parameter $0 < R < 1$. Note that quantity $B(t)$ can be excluded from Eqs. (67d) and (67e) by writing $S(t) = x(t) - x(t - \tau) + RS(t - \tau)$, but the latter

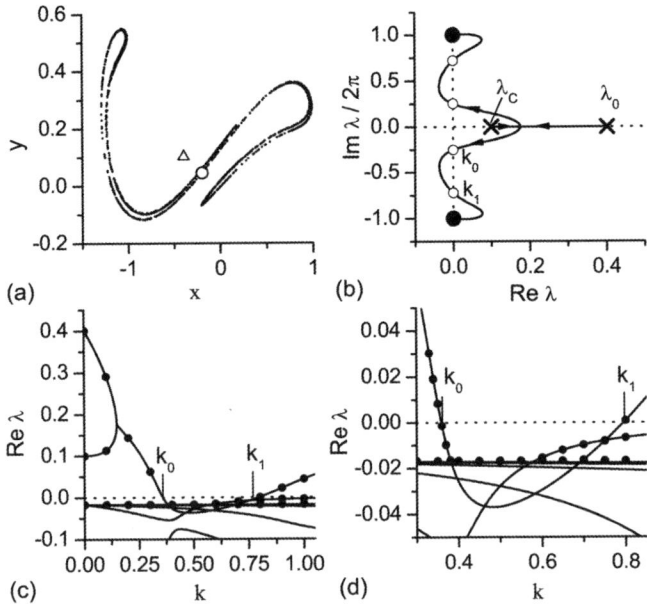

Fig. 12. (a) Stroboscopic map of the forced double-well oscillator for $\alpha = \beta = \gamma = 0.3$, $\omega = 1$, and $A = 0.27$. The open circle marks the target torsion-free UPO. The triangle shows the stable extraneous periodic orbit generated in the first step of control at $C = 0$ and $k = 0.5$, (b) Root loci of Eq. (73) for $b = 0.2$, $R = 0.9$, $\lambda_c = 0.1$ as k varies from 0 to ∞. Crosses and black dots show the location of roots for $k = 0$ and $k = \infty$, respectively, (c) Real parts of leading Floquet exponents vs k. Solid lines are the solutions of quasipolynomial Eq. (73) and dots show the solutions of the exact Eq. (70), (d) Enlarged essential part of (c).

equation contains two delayed variables $x(t - \tau)$ and $S(t - \tau)$. The form (67d) and (67e) with two variables $B(t)$ and $S(t)$ is more convenient from experimental point of view since it requires only one delay line for variable $B(t)$. To overcome the odd number limitation an unstable mode governed by the variable W with the parameter $\lambda_c > 0$ is incorporated. The strength of the feedback force is defined by parameters b and k. The τ-periodical solutions $x(t) = x_P(t) = x_P(t - \tau)$ of the uncontrolled oscillator (67a) and (67b) are also the solutions of the whole closed loop system (67), since they zero the control perturbation. Indeed, for such solutions Eqs. (67c)–(67e) are satisfied at $B(t) = x_P(t)/(1 - R)$, $S(t) = 0$, and $W(t) = 0$.

For a certain choice of the parameters, the free ($k = 0$) oscillator (67a) and (67b) exhibits chaotic behavior. An example of the stroboscopic map for $\alpha = 0.3$, $\beta = 0.3$, $\gamma = 0.3$, $\omega = 1.0$, and $A = 0.27$ is shown in Fig. 12(a).

The symmetric UPO marked by the circle is torsion-free; its largest Floquet exponent is $\lambda_0 \approx 0.401$ or Floquet multiplier $\mu_0 = e^{\lambda_0 \tau} \approx 12.423$. The latter value indicates that the orbit is highly unstable. This UPO is the subject of testing our control algorithm. It is interesting to note that the UPO is weakly nonlinear, although it is embedded in the chaotic attractor for which the nonlinearity is an essential factor. By a weakly nonlinear orbit we mean that the influence of the nonlinear term γx^3 to its solution is small and it can be found by a perturbation theory. In a zero approximation ($\gamma = 0$) the solution for this UPO is

$$x_P^{(0)}(t) = -A\mathrm{Re}[e^{i\omega t}/(\omega^2 + \alpha - i\omega\beta)]. \tag{68}$$

This approximation is good if $\gamma (x_P^{(0)})^2 << \alpha$, i.e. when

$$\gamma A^2 / [(\omega^2 + \alpha)^2 + \omega^2 \beta^2] << \alpha. \tag{69}$$

The set of parameters chosen above meets this inequality.

If the condition (69) is satisfied the Floquet exponents of the controlled UPO can be also obtained via a perturbation theory. To present the theory in a general form we introduce the vector notations. Let $z = (x\ y\ W)^T$ be the vector of dynamical variables of the system (67) and $z_P = (x_P\ y_P\ 0)^T$ be the corresponding UPO. Small deviations from the UPO $\delta z = z - z_P$ may be decomposed into eigenfunctions according to the Floquet theory, $\delta z = e^{\lambda t}u(t)$, $u(t) = u(t-\tau)$, where λ is the Floquet exponent. The equation for the periodic function $u(t)$ is

$$\lambda u + \dot{u} = Lu + \gamma N[z_P(t)]u - kK(\lambda)u. \tag{70}$$

The matrixes L and N,

$$L = \begin{pmatrix} 0 & 1 & 0 \\ \alpha & -\beta & 0 \\ 0 & 0 & \lambda_c \end{pmatrix}, \quad N[z_P(t)] = \begin{pmatrix} 0 & 0 & 0 \\ -3x_P^2(t) & 0 & 0 \\ 0 & 0 & 0 \end{pmatrix}$$

are respectively related to the linear and nonlinear terms of the free system (67), and $K(\lambda)$ is the control matrix

$$K(\lambda) = \begin{pmatrix} 0 & 0 & 0 \\ H(\lambda) & 0 & 1 \\ -bH(\lambda) & 0 & 0 \end{pmatrix}, \quad H(\lambda) = \frac{1 - e^{-\lambda\tau}}{1 - e^{-\lambda\tau}R}.$$

It depends on λ due to elimination of the delay terms.

We suppose that the target UPO is weakly nonlinear and seek solutions of Eq. (70) in the form of power series in γ: $\lambda = \lambda^{(0)} + \gamma\lambda^{(1)} + \ldots$, $u = u^{(0)} + \gamma u^{(1)} + \ldots$, and $z_P = z_P^{(0)} + \gamma z_P^{(1)} + \ldots$. In zero approximation, the right-hand side of Eq. (70) is independent of time and thus $u^{(0)}$ is independent of time as well. Therefore, the zero approximation gives rise to time-independent eigenvalue problem: $\lambda^{(0)} u^{(0)} = L u^{(0)} - kK(\lambda^{(0)}) u^{(0)}$. If we sought correction terms by the standard perturbation theory, we would come to intricate expressions. We show, however, that the Floquet exponents of the controlled UPO can be derived with accuracy $O(\gamma)$ from relatively simple time-independent eigenvalue problem. Let us average Eq. (70) over the period of the UPO

$$\lambda \bar{u} = L\bar{u} + \gamma \overline{N[z_P(t)] u} - kK(\lambda)\bar{u}, \qquad (71)$$

where $\bar{\varphi} \equiv (1/\tau) \int_0^\tau \varphi(t) dt$. One can easily verify that this equation can be transformed with accuracy $O(\gamma)$ to

$$\lambda \bar{u} = L\bar{u} + \gamma \overline{N[z_P^{(0)}(t)]} \bar{u} - kK(\lambda)\bar{u}, \qquad (72)$$

i.e. if we apply the perturbation theory to (71) and (72) we obtain equivalent results up to terms $O(\gamma)$. Equation (72) represents a time-independent eigenvalue problem and leads to a relatively simple quasipolynomial equation

$$\det(L + \gamma \overline{N[z_P^{(0)}(t)]} - kK(\lambda) - I\lambda) = 0, \qquad (73)$$

where I is the identity matrix.

The mechanism of stabilization is evident from the root loci diagram of Eq. (73) shown in Fig. 12(b). With the increase of k, the positive exponent λ_0 of the uncontrolled UPO and the eigenvalue λ_c of the unstable mode W approach each other on the real axes, then collide, pass to the complex plain and move to the stable region $\text{Re}\lambda < 0$. For $k \in [k_0, k_1] \approx [0.359, 0.796]$, the target orbit is stable. Figures 12(c) and 12(d) show that the solutions of the quasipolynomial Eq. (73) are in good quantitative agreement with the values of Floquet exponents obtained numerically from exact Eq. (70). Note that neglecting the nonlinear term $\gamma \overline{N[z_P^{(0)}(t)]}$ in Eq. (73) leads only to qualitative but not quantitative agreement with the exact results. Good quantitative results derived from quasipolynomial Eq. (73) confirm the validity of our analytical approach.

Successful stabilization of the torsion-free UPO is shown in Fig. 13. The results are obtained by numerical solution of the nonlinear delay-differential Eq. (67). When the system approaches the target orbit the variable W tends to zero and the control perturbation $S + W$ vanishes.

In Ref. [77] the above described unstable controller has been implemented experimentally ant the theoretical results of this section have

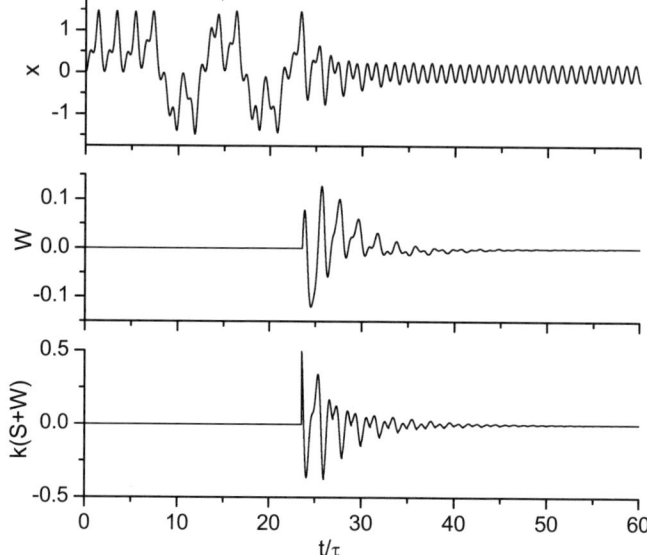

Fig. 13. Successful control of the torsion-free UPO. The results are obtained by numerical solution of Eq. (67). The control is activated at $t = 23.25\tau$, i.e. the control gain $k = 0$ for $t < 23.25\tau$ and $k = 0.5$ for $t \geq 23.25\tau$. Other parameters are the same as in Fig. 12.

been confirmed by electronic circuit experiments for a forced double-well oscillator.

9. On Global Properties of the Delayed Feedback Control

Due to complexity of the DFC theory, most investigations are restricted to a linear stability analysis. However, the linear stability of the target orbit does not guarantee successful control for any initial conditions. Unfortunately, the general results concerning global properties of DFC systems are still missing. It is completely unclear how large basins of attraction are, i.e. which initial condition is attracted towards a particular stabilized orbit. Basins of attraction may have an incredibly complicated topology. But for time delay systems phase spaces are infinite dimensional and even the visualization of trajectories becomes a challenge.

The importance of such global features of the DFC has been emphasized even in the original paper [8]. It has been shown that limiting the size of the control force by a simple cutoff increases the basin of attraction of the target state. This idea is based on the fact that chaotic dynamics is ergodic and a trajectory visits a neighborhood of each periodic orbit with finite probability. Turning on the control loop only when such a neighborhood is visited, i.e. when the control signal falls beyond a certain threshold one may obtain stabilization. The use of restricted perturbations is now popular in many theoretical and experimental investigations of the DFC schemes.

The analysis of basins of attraction in dynamical systems is usually a complicated task. Recently Just et al. [78] have presented a possible explanation of a generic mechanism, which would determine basins of attraction in DFC schemes. The key idea in [78] is to analyze the type of bifurcations on control boundaries, i.e. at values of the control gain K where the target orbit changes the stability. The authors state that a continuous (supercritical) transition at the control boundary should indicate a large basin of attraction and a discontinuous (subcritical) transition should lead to the small basin. The type of bifurcation on the control bondary can be easily checked by numerical simulations, whereas a proper theoretical approach calls for analyzing the instability in terms of weakly nonlinear analysis, a standard technique in bifurcation theory. An example of such a theoretical approach is presented in the recent paper [79], where various nonlinear DFC coupling schemes have been tested with regards to basins of attraction.

Along with usual restriction of the perturbation an alternative approach for enlarging basins of attraction of the DFC schemes has been proposed in Ref. [77]. Here the stabilization of the desired orbit is achieved in two steps. In the first step one seeks only a rough approach to the desired state. For this aim some of the system parameter is detuned artificially. In the second step the correct value of the parameter is returned and the target state is reached exactly.

The idea of the two-step algorithm is based on the observation that periodic orbits are robust [80]; their form and Floquet multipliers vary slowly with smooth parameter changes. An illustrative example is the period doubling bifurcation. The UPOs embedded in a chaotic attractor originated from this bifurcation do not differ considerably from stable periodic orbits, which were in the system below the critical value of the chaotic instability. Thus by switching a proper control parameter in the first step, one can expect a conversion of chaotic motion into a stable periodic motion close to the target UPO (cf. [81]). We refer to this generated periodic motion as an extraneous periodic orbit. In the second step we need only to move the system from the extraneous orbit to the target UPO. Such an algorithm may enlarge the basin of attraction of the target orbit if the extraneous orbit has a larger basin of attraction and if it lies in the basin of attraction of the target orbit.

The simplest dynamical toy models to which two-step delayed feedback control can be applied are time discrete maps. They are easier to handle since the dimension of phase space stays to be finite even if the control loop is included. Thus, visualization of global properties remains feasible in such cases. To illustrate the main idea of the two-step control algorithm we restrict ourselves to the minimal model. We consider the stabilization of a simple fixed point that does not require the use of the unstable controller and apply the simple DFC algorithm (not the extended version).

To be specific, we demonstrate the idea for the logistic map $x_{n+1} = f(x_n, a)$, where $f(x_n, a) = ax_n(1 - x_n)$. The map has a period-one orbit $x^* = 1 - 1/a$, which is unstable for $a > 3$. Our aim is to stabilize it via the two-step DFC algorithm. We suppose that a is an accessible control parameter. When increasing this parameter the system undergos the period doubling bifurcation and reaches a chaotic regime. The key idea of applying the two-step control to this system is as follows. In the first step, we shift the parameter to the interval $1 < a < 3$ where the period-one orbit is stable, and thus generate an extraneous orbit. In the second step, we return the original value of the parameter a and switch on the DFC algorithm in order to stabilize the target orbit. In other words, we adjust the control parameter a on each iteration by an amount $-(1 - C)\Delta a + Ck(x_n - x_{n-1})$ such that the controlled system is described by a two-dimensional map

$$x_{n+1} = [a - (1 - C)\Delta a + Ck(x_n - y_n)] x_n (1 - x_n), \qquad (74a)$$
$$y_{n+1} = x_n. \qquad (74b)$$

For $C = 1$ and $k = 0$, we have the free logistic map. In the first step, we take $C = 0$, such that parameter a is decreased by an amount Δa. As a result there appears an extraneous orbit $x_e = 1 - 1/(a - \Delta a)$ with the Floquet multiplier $\mu_e = 2 - a + \Delta a$. If $1 < a - \Delta a < 3$ the extraneous orbit is stable and the system approaches it during the characteristic time $\tau_e = 1/|\ln|\mu_e||$. Thus the duration t_C of the first step should satisfy $t_C > \tau_e$. In the second step, we return the original value of the parameter a by switching on $C = 1$ and activate the DFC control by switching on $k \neq 0$. The linear analysis of the system (74) at $C = 1$ shows that the target orbit $(x, y) = (x^*, x^*)$ is stable if the control gain is in the interval

$$\frac{a-3}{2}\frac{a^2}{a-1} < k < \frac{a^2}{a-1}. \qquad (75)$$

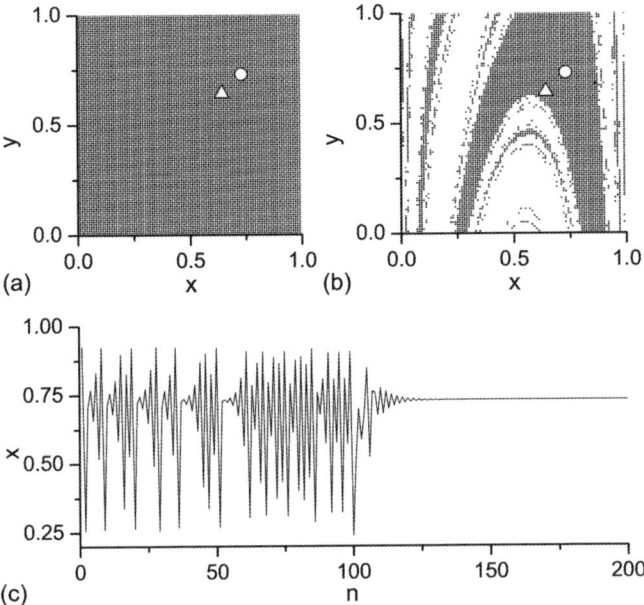

Fig. 14. Basins of attraction of (a) extraneus orbit ($C = 0$) and (b) target orbit ($C = 1$, $k = 2$) for $a = 3.7$ and $\Delta a = 0.9$. Extraneous and target orbits are shown respectively by triangle and circle. In (c) the dynamics of the two-step control is demonstrated. For $n < 100$ the control is off ($C = 1$, $k = 0$). The first step ($C = 0$) lasts 5 iterations in the time interval $100 \leq n \leq 104$. The second step ($C = 1$, $k = 2$) is activated for $n \geq 105$.

In Fig. 14 we demonstrate the numerical results for the parameters $a = 3.7$ and $\Delta a = 0.9$. We see that the basin of attraction of the extraneous orbit occupies the whole unity square $0 < x < 1$, $0 < y < 1$. The characteristic time of approaching this orbit is $\tau_e \approx 4.48$. The basin of attraction of the target orbit is shown in Fig. 14(b). It occupies only about 45% of the unity square. Thus the usual one-step DFC algorithm is successful only for 45% of initial conditions taken from the unity square. In contrast, the two-step control ensures 100% success for any initial conditions taken from the unity square. This is provided by two features of the extraneous orbit: (i) the extraneous orbit lies in the basin of attraction of the target orbit and (ii) the basin of attraction of the extraneous orbit occupies the whole unity square. In Fig. 14(c) we show the dynamics of the system controlled by the two-step algorithm when the duration of the first step $t_C = 5$ only slightly exceeds the characteristic time τ_e.

The two step algorithm works not only for discrete maps, but for more complex dynamical systems as well. It has been successfully implemented in the forced double well oscillator described in the previous section. Unfortunately, we do not have a general recommendation for selecting the control parameter which allows us to generate a proper extraneus orbit in the first step of control. For a given system the choice of such a parameter may depend on the convenience of experimental implementation and can be found by trial and error. For the double-well oscillator, we have found that the basin of attraction of the target orbit can be extended by a simple modification of Eq. (67e):

$$B(t) = x(t) + CRB(t - \tau), \qquad (76)$$

where C is an auxiliary parameter. When control is off ($k = 0$) we take $C = 0$. In the first step, which lasts some time interval t_C, we switch on $k \neq 0$ but hold C at zero. In the second step, we switch on $C = 1$.

First we have estimated the basin of attraction of the target orbit without using the two-step algorithm. For this aim we have calculated the statistics of the successful outcomes at different values of k taken from the stability interval $[k_0, k_1]$. The control has been activated from different 1000 points randomly chosen on the chaotic attractor of the free system. By way of illustration, one of numerical experiments for $k = 0.38, 0.54$, and 0.7 has shown respectively 846, 994, and 992 successful outcomes. The application of the two-step control algorithm for $t_C = 3\tau$ have shown 100% success rate at any values of $k \in [k_0, k_1]$.

For this system, rigorous consideration of a mechanism of the two-step algorithm is difficult, since one has to deal with the global dynamics in an infinite-dimensional phase space. However, a qualitative explanation of the mechanism is similar to that of the simple logistic map described above. For $C = 0$ and $k \neq 0$, the control force does not vanish; it generates an extraneous stable periodic orbit shown in Fig. 12(a) by the triangle. The bifurcation diagram presented in Fig. 15 indicates that the extraneous orbit is linked to the target state by a homotopy, i.e. the two orbits are "continuously connected". The generated orbit is close to the stabilized target UPO [cf. Fig. 12(a)] and lies in its basin of attraction. Numerical analysis shows that the extraneous orbit has better stability properties and larger basin of attraction than those of the stabilized UPO. If $t_C > 1/|\lambda_e|$, where λ_e is the leading exponent of the extraneous orbit, then in the first step the phase points located in the larger basin approach the extraneous orbit, and in the second step they approach the target orbit.

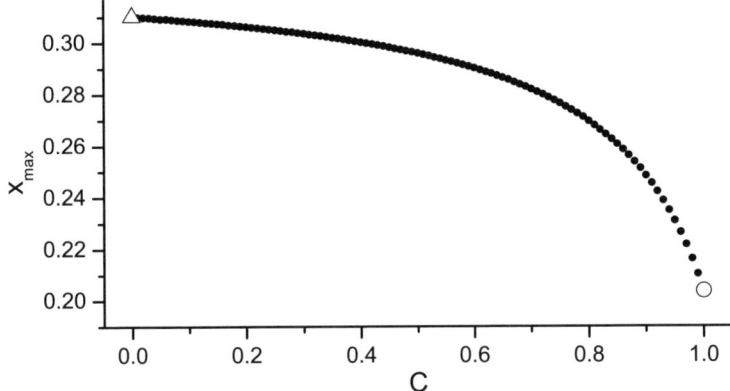

Fig. 15. The maxima of the x variable of delay differential Eq. (67) for $k = 0.5$ when one changes the parameter C continuously from 1 to 0. The initial value of x_{max} at $C = 1$ marked by open circle represents the target state and the final value at $C = 0$ marked by triangle corresponds to the extraneous orbit. Other parameters are the same as in Fig. 12.

10. The Odd-Number Limitation Revisited

The odd number limitation of the DFC techniques has been intensively discussed in the literature during the past decade. The main statement of the limitation is that any UPOs with an odd number of real Floquet multipliers greater than unity can never be stabilized by any DFC technique. The history of this limitation is dramatic. This statement was first proved by Ushio [65] for discrete time systems. Just et al. [66] and Nakajima [67] proved the same limitation for the continuous time DFC, and then this proof was extended for a wider class of delayed feedback schemes, including the EDFC [68]. However, Fiedler et al. [82] have recently shown by a simple example that this limitation does not hold in general.

The authors of Ref. [82] consider the normal form of a subcritical Hopf bifurcation, extended by a delayed feedback term

$$\dot{z}(t) = [\lambda + i + (1 + i\gamma)|z(t)|^2]z(t) + k[z(t - \tau) - z(t)], \qquad (77)$$

with complex z and real parameters λ and γ. Here the Hopf frequency is normalized to unity. The feedback amplitude k is represented by a complex number $k = k_R + ik_I = k_0 e^{i\phi}$ with real k_R, k_I, ϕ, and positive k_0. The presence of the phase $\phi \neq 0$ in the control amplitude provides a nontrivial bifurcation scenario.

Transforming Eq. (77) to amplitude and phase variables r, Θ using $z(t) = r(t)e^{i\Theta(t)}$, one obtains at $k = 0$

$$\dot{r}(t) = (\lambda + r^2)r, \tag{78a}$$
$$\dot{\Theta}(t) = 1 + \gamma r^2. \tag{78b}$$

At $\lambda = 0$ a subcritical Hopf bifurcation occurs. An UPO with $r^2 = -\lambda$ and period $T = 2\pi/(1 - \gamma\lambda)$ exists for $\lambda < 0$. This UPO has the only one real unstable Floquet multiplier and thus satisfies the condition of the odd number limitation. The DFC method chooses delay as $\tau = T$ such that the UPO remains unchanged in the presence of control $k \neq 0$. The analysis of Eq. (77) for fixed $\phi \neq 0$ and varying k_0 shows that for certain values of k_0 the UPO becomes stable. This means that the general statement of the odd number limitation does not hold for this system. The mechanism of stabilization is caused by a transcritical bifurcation in which the subcritical orbit [whose radius is given by $r = (-\lambda)^{1/2}$ independently of the control amplitude k_0] collides with a delay-induced periodic orbit [82]. This delay-induced orbit is generated at a finite value of the control amplitude k_0 by a saddle-node bifurcation (collision with another unstable delay-induced periodic orbit). At transcritical bifurcation, the subcritical orbit and the delay-induced orbit exchange stability. The delay-induced orbit has a period $T \neq \tau$, except at transcritical bifurcation. The possible existence of such delay-induced periodic orbits with $T \neq \tau$, which results in a Floquet multiplier $\mu = 1$ of multiplicity two at transcritical bifurcation, was overlooked in [68].

These ideas are further developed in the recent publication [83]. Using the Lorenz equations as an example the authors of Ref. [83] demonstrate that the stabilization mechanism identified by Fiedler *et al.* [82] for the Hopf normal form can also apply to unstable periodic orbits created by subcritical Hopf bifurcations in higher-dimensional dynamical systems. They suggest a possible strategy for choosing the feedback gain matrix in delayed feedback control of UPOs that arise from a subcritical Hopf bifurcation.

In conclusion of this section we would like to stress that despite very interesting findings presented in Ref. [82] the odd number limitation still exists as a real problem of the DFC techniques. Indeed, the proof of the odd number limitation originally presented by Nakajima [67] for nonautonomous systems seems to be true [note that Eq. (77) represents an autonomous system]. Thus all nonautonomous systems suffer from this limitation. The mistake was incorporated when attempting to extend the proof to autonomous systems. The mistake is related to the presence of the Floquet multiplier $\mu = 1$, which plays a crucial role in autonomous systems. Although the general proof for autonomous systems is false the experience shows that many autonomous systems suffer from this limitation as well. An example presented in Ref. [82] is in some sense exceptional. Moreover, it is unclear how to use the results of Ref. [82] in order to overcome the limitation in autonomous systems far away from subcritical Hopf bifurcation.

11. Conclusions

The aim of this paper was to review experimental implementations, applications for theoretical models and various modifications of the time-delayed feedback control method and to present some recent theoretical ideas in this field. We considered both discrete maps and continuous-time systems.

The main attention was concentrated to the problems allowing an analytical treatment. The key idea was to consider the control problems close to bifurcation points of periodic solutions. We managed to treat analytically the delayed feedback control algorithm at three different bifurcations, namely, the period doubling, the Nejmark-Sacker, and the subcritical Hopf bifurcations. All three types of bifurcations are characterized by different topological properties of periodic orbits. The orbits without torsion are particularly difficult to control due to the odd number limitation. The effective method to overcome the odd number limitation is to use an unstable controller. We also discussed the recent achievements in theory of the odd number limitation.

In our analytical approaches, we used the classical tools of nonlinear dynamics and bifurcation theory, such as the method of averaging, the center manifold theory, and the near identity transformation. We also utilized the relationship between Floquet exponents of the proportional and delayed feedback control algorithms, which allowed us to reduce the delayed feedback problems to finite dimensional phase spaces.

The analytical approaches were demonstrated for paradigmatic models of the Rössler system, the forced van der Pol oscillator, the Lorenz system, and nonautonomous double-well oscillator. We have obtained simple analytical expressions for the dependence of leading Floquet exponents on the control gain and determined the lower and upper thresholds of stability as well as optimal values of the control gain. The main results and the approaches are of general importance since they are relevant to any systems close to associated bifurcation points.

We believe that the developed analytical methods are important contributions to the theory of the delayed feedback control. They give not only a better insight into the mechanism of the delayed feedback control technique, but they are also important for optimizing the control algorithm. We hope that the ideas presented in this chapter will stimulate the search for further modifications of delayed feedback control technique aiming at the improvement of the control performance. We also believe in further experimental extension of the method to new fields of research.

References

[1] Ott, E., Grebogi, C. & Yorke, J. A. (1990). Controlling chaos. *Phys. Rev. Lett.*, **64**, p. 1196–1199.
[2] Shinbrot, T., Grebogy, C., Ott, E. & Yorke, J. A. (1993). Using smal perturbations to control chaos. *Nature*, **363**, p. 411–417.
[3] Kapitaniak, T. (1996) *Controlling chaos. Theoretical and practical methods in non-linear dynamics*, (London: Academic Press).
[4] Schoell, E., Shuster, H. G. (2008). *Handbook of Chaos Control*, (Wiley-VCH Verlag GmbH & Co. KGaA).
[5] Boccaletti, S., Grebogi, C., Lai, Y.-C. Mancini, H. & Maza, D. (2000). The control of chaos: theory and applications. *Physics Reports*, **329**, p. 103–197.
[6] Chen, G. & Raton, B. (eds.) (2000). *Controlling chaos and bifurcation in engeneering systems*, (FL: CRS Press).
[7] Chen, G. & Yu, X. (eds.) (2003). *Chaos control. Theory and applications*, (Springer).
[8] Pyragas, K. (1992). Continuous control of chaos by self-controlling feedback, *Phys. Lett. A*, **170**, p. 421–428.
[9] K. Pyragas, K. (2006). Philos. Trans. R. Soc. London, Ser. A 364, 2309–2334.
[10] Pyragas, K. and Tamasevicius, A. (1993). Experimental control of chaos by delayed self-controlling feedback, *Phys. Lett. A*, **180**, p. 99–102.
[11] Kittel, A., Parisi, J., Pyragas, K. and Richter, R. (1994). Delayed feedback control of chaos in an electronic double-scroll oscillator, *Z. Naturforsch.*, **49a**, p. 843–846.
[12] Gauthier, D. J., Sukow, D. W., Concannon, H. M. and Socolar, J. E. S. (1994). Stabilizing unstable periodic orbits in a fast diode resonator using continuous time-delay autosinchronization, *Phys. Rev. E*, **50**, p. 2343–2346.
[13] Celka, P. (1994). Experimental verification of Pyragas's chaos control method applied to Chua's circuit, *Int. J. Bifurcation Chaos Appl. Sci. Eng.*, **4**, p. 1703–1706.
[14] Hikihara, T. and Kawagoshi, T. (1996). Experimental study of on stabilization of unstable periodic motion in magneto-elastic chaos, *Phys. Lett. A* **211**, p. 29–36.
[15] Christini, D. J., In, V., Spano, M. L., Ditto, W. L. and Collins, J. J. (1997). Real-time experimental control of a system in its chaotic and nonchaotic regimes, *Phys. Rev. E*, **56**, p. R3749–3752.

[16] Bielawski, S., Derozier, D. and Glorieux, P. (1994). Controlling unstable periodic orbits by a delayed continuous feedback, *Phys. Rev. E*, **49**, p. R971–974.

[17] Basso, M., Genesio, R. and Tesi, A. (1997). Controller design for extending periodic dynamics of a chaotic CO2 laser, *Systems and Control Letters*, **31**, p. 287–297.

[18] Lu, W., Yu, D. and Harrison, R. G. (1998). Instabilities and tracking of travelling wave patterns in a three-level laser, *Int. J. Bifurcation Chaos Appl. Sci. Eng.*, **8**, p. 1769–1775.

[19] Pierre, T., Bonhomme, G. and Atipo, A. (1996). Controlling a chaotic regime of nonlinear ionization waves using the time-delay autosynchronization method, *Phys. Rev. Lett.*, **76**, p. 2290–2293.

[20] Gravier, E., Caron, X., Bonhomme, G., Pierre, T. and Briancon, J. L. (2000). Dynamical study and control of drift waves in a magnetized laboratory plasma, *Europ. J. Phys. D*, **8**, p. 451–456.

[21] Mausbach, Th., Klinger, Th., Piel, A., Atipo, A., Pierre, Th. and Bonhomme, G. (1997). Continuous control of ionization wave chaos by spatially derived feedback signals, *Phys. Lett. A*, **228**, p. 373–377.

[22] Fukuyama, T., Shirahama, H. and Kawai, Y. (2002). Dynamical control of the chaotic state of the current-driven ion acoustic instability in a laboratory plasma using delayed feedback, *Physics of Plasmas*, **9**, p. 4525–4529.

[23] Lüthje, O., Wolff, S. and Pfister, G. (2001). Control of chaotic Taylor-Coutte flow with time-delayed feedback, *Phys. Rev. Lett.*, **86**, p. 1745–1748.

[24] Parmananda, P., Madrigal, R., Rivera, M., Nyikos, L., Kiss, I. Z. and Gaspar, V. (1999). Stabilization of unstable steady states and periodic orbits in an electrochemical system using delayed feedback control, *Phys. Rev. E*, **59**, p. 5266–5271.

[25] Guderian, A., Munster, A. F., Kraus, M., Schneider, F. W. (1998). Electrochemical chaos control in a chemical reaction: Experiment and simulation, *J. of Phys. Chem. A*, **102**, p. 5059–5064.

[26] Benner, H. and Just, W. (2002). Control of chaos by time-delayed feedback in high-power ferromagnetic resonance experiments, *J. Korean Pysical Society*, **40**, p. 1046–1050.

[27] Krodkiewski, J. M. and Faragher, J. S. (2000). Stabilization of motion of helicopter rotor blades using delayed feedback - modelling, computer simulation and experimental verification, *J. Sound and Vibration*, **234**, p. 591–610.

[28] Hall, K., Christini, D. J., Tremblay, M., Collins, J. J., Glass, L. and Billette, J. (1997). Dynamic control of cardiac alternans, *Phys. Rev. Lett.*, **78**, p. 4518–4521.

[29] Simmendinger, C. and Hess, O. (1996). Controlling delay-induced chaotic behavior of a semiconductor laser with optical feedback, *Phys. Lett. A*, **21**, p. 97–105.

[30] Munkel, M., Kaiser, F. and Hess, O. (1997). Stabilization of spatiotemporally chaotic semiconductor laser arrays by means of delayed optical feedback, *Phys. Rev. E*, **56**, p. 3868–3875.

[31] Simmendinger, C., Munkel, M. and Hess, O. (1999). Controlling complex temporal and spatio-temporal dynamics in semiconductor lasers, *Chaos, Solitons and Fractals*, **10**, p. 851–864.
[32] Batlle, C., Fossas, E. and Olivar, G. (1999). Stabilization of periodic orbits of the buck converter by time-delayed feedback, *Int. J. Circuit Theory and Applications*, **27**, p. 617–631.
[33] Fronczak, P. and Holyst, J. A. (2002). Control of chaotic solitons by a time-delayed feedback mechanism, *Phys. Rev. E*, **65**, p. 026219.
[34] Galvanetto, U. (2002). Delayed feedback control of chaotic systems with dry friction, *Int. J. Bifurcation Chaos Appl. Sci. Eng.*, **12**, p. 1877–1883.
[35] Bleich, M. E. and Socolar, J. E. S. (2000). Delayed feedback control of a paced excitable oscillator, *Int. J. Bifurcation Chaos Appl. Sci. Eng.*, **10**, p. 603–609.
[36] Rappel, W. J., Fenton, F. and Karma, A. (1999). Spatiotemporal control of wave instabilities in cardiac tissue, *Phys. Rev. Lett.*, **83**, p. 456–459.
[37] Konishi, K., Kokame, H. and Hirata, H. K. (1999). Coupled map car-following model and its delayed-feedback control, *Phys. Rev. E*, **60**, p. 4000–4007.
[38] Holyst, J. A. and Urbanowicz, K. (2000). Chaos control in economical model by time-delayed feedback method, *Physica A*, **287**, p. 587–598.
[39] Holyst, J. A., Zebrowska, M. and Urbanowicz, K. (2001). Observations of deterministic chaos in financial time series by recurrence plots, can one control chaotic economy? *European Physical J. B*, **20**, p. 531–535.
[40] Tsui, A. P. M. and Jones, A. J. (2000). The control of higher dimensional chaos: comparative results for the chaotic satellite attitude control problem, *Physica D*, **135**, p. 41–62.
[41] Mensour, B. and Longtin, A. (1995). Controlling chaos to store information in delay-differential equations, *Phys. Lett. A*, **205**, p. 18–24.
[42] Mitsubori, K. and Aihara, K. U. (2002). Delayed-feedback control of chaotic roll motion of a flooded ship in waves, *Proceedings of the Royal Society of London Series A-Mathematical Phyzics and Engeneering Science*, **458**, p. 2801–2813.
[43] Rosenblum, M. and Pikovsky, A. (2004a). Controlling synchronization in an ensemble of globally coupled oscillators, *Phys. Rev. Lett.*, **92**, p. 114102.
[44] Rosenblum, M. and Pikovsky, A. (2004b). Delayed feedback control of collective synchrony: An approach to suppression of pathological brain rhythms, *Phys. Rev. E*, **70**, p. 041904.
[45] Yamasue, K. and Hikihara, T. (2006). Control of microcantilevers in dynamic force microscopy using time delayed feedback. *Review of scientific instruments*, **77**, p. 053703.
[46] Li, L., Liu, Z., Zhang, D., and Zhang, H. (2006). Controlling chaotic robots with kinematical redundancy. *Chaos*, **16**, p. 013132.
[47] Liu, C. L, Tian, Y. P. (2008). Eliminating oscillations in the Internet by time-delayed feedback control. *Chaos, Solitons & Fractals*, **35**, p. 878–887.
[48] Kittel, A., Parisi, J. & Pyragas, K. (1995). Delayed feedback control of chaos by self-adapted delay time. *Phys. Lett. A* **198**, p. 433–438.

[49] Nakajima, H., Ito, H. & Ueda, Y. (1997). Automatic adjustment of delay time acid feedback gain in delayed feedback control of chaos. *IEICE Transactions on Fundamentals of Electronics Communications and Computer Sciencies*, **E80A**, p. 1554–1559.

[50] Herrmann, G. (2001). A robust delay adaptation scheme for Pyragas' chaos control method. *Phys. Lett. A*, **287**, p. 245–256.

[51] Boccaletti, S., & Arecchi, F. T. (1995). Adaptive control of chaos. *Europhys. Lett.*, **31**, p. 127–132.

[52] Boccaletti, S., Farini, A. & Arecchi, F. T. (1997). Adaptive strategies for recognition, control and synchronization of chaos. *Chaos, Solitons and Fractals*, **8**, p. 1431–1448.

[53] Basso, M., Genesio, R. & Tesi, A. (1997). Stabilizing periodic orbits of forced systems via generalized Pyragas controllers. *IEEE Trans. Circuits Syst. I*, **44**, p. 1023–1027.

[54] Basso, M., Genesio R. & Tesi, A. (1997). Controller design for extending periodic dynamics of a chaotic CO2 laser. *Systems and Control Letters*, **31**, p. 287–297.

[55] Basso, M., Genesio, R., Giovanardi, L., Tesi, A. & Torrini, G. (1998). On optimal stabilization of periodic orbits via time delayed feedback control. *Int. J. Bifurcation Chaos Appl. Sci. Eng.*, **8**, p. 1699–1706.

[56] Bleich, M. E., Hochheiser, D., Moloney, J. V. & Socolar, J. E. S. 1997 Controlling extended systems with spatially filtered, time-delayed feedback. *Phys. Rev. E* **55**, 2119–2126.

[57] Hochheiser, D., Moloney, J. V. & Lega, J. 1997 Controlling optical turbulence. *Phys. Rev. A* **55**, R4011–R4014.

[58] Baba, N., Amann, A., Scholl, E. & Just, W. 2002 Giant improvement of time-delayed feedback control by spatio-temporal filtering. *Phys. Rev. Lett.*, **89**, 074101.

[59] Socolar, J. E. S., Sukow, D. W. and Gauthier, D. J. (1994). Stabilzing unstable periodic orbits in a fast diode resonator using continous time-delay autosinchronization, *Phys. Rev. E*, **50**, p. 2343–2346.

[60] Pyragas, K. (1995). Control of chaos via extended delay feedback, *Phys. Lett. A*, **206**, p. 323–330.

[61] Bleich, M. E. & Socolar, J. E. S. (1996). Stability of periodic orbits controlled by time-delay feedback. *Phys. Lett. A*, **210**, p. 87–94.

[62] Pyragas, K. (2001). Control of chaos via an unstable delayed feedback controller, *Phys. Rev. Lett.*, **86**, p. 2265–2268.

[63] Pyragas, K., Pyragas, V. and Benner, H. (2004). Delayed Feedback Control of Dynamical Systems at a Subcritical Hopf Bifurcation, *Phys. Rev. E*, **70**, p. 056222.

[64] Pyragas, V. and Pyragas, K. (2006). Delayed feedback control of the Lorenz system: An analytical treatment at a subcritical Hopf bifurcation,*Phys. Rev. E*, **73**, p. 036215.

[65] Ushio T. (1996). Limitation of delayed feedback control in nonlinear discrete-time systems, *IEEE Trans. Circuits Syst. I*, **43**, p. 815–816.

[66] Just, W., Bernard, T., Ostheimer, M., Reibold, E. and Benner, H. (1997). Mechanism of time-delayed feedback control, *Phys. Rev. Lett.*, **78**, p. 203–206.

[67] Nakajima, H. (1997). On analytical properties of delayed feedback control of chaos, *Phys. Lett. A*, **232**, p. 207–210.

[68] Nakajima, H. and Ueda, Y. (1998). Limitation of generalized delayed feedback control, *Physica D*, **111**, p. 143–150.

[69] Yamamoto, S., Hino, T., and Ushio, T. (2001). Dynamic Delayed Feedback Controlles for Chaotic Discrete-Time Systems, *IEEE Transactions on Circuits and Systems I*, vol. **48**, no. **6**, p. 785–789.

[70] Nakajima, H. (2000). A generalization of the extended delayed feedback control for chaotic systems, *Proc. Control Oscillations Chaos*, **2**, p. 209-212.

[71] Pyragas, K. (2002). Analytical properties and optimization of time-delayed feedback control, *Phys. Rev. E*, **66**, p. 026207.

[72] Lathrop, D. P. and Kostelich, E. J. (1989). Characterization of an experimental strange attractor by periodic orbits, *Phys. Rev. A*, **40**, p. 4028–4031.

[73] So, P., Ott, E., Schiff, S. J., Kaplan, D. T., Sauer, T. and Grebogi, C. (1996). Detecting unstable periodic orbits in chaotic experimental data, *Phys. Rev. Lett.*, **76**, p. 4705–4708.

[74] Rössler, O. E. (1976). An equation for continous chaos,*Phys. Lett. A*, **57**, p. 397–398.

[75] Pyragiene, T. and Pyragas, K. (2005). Delayed feedback control of forced self-sustained oscillations, *Phys. Rev. E*, **72**, p. 026203-9.

[76] Lorenz, E., N. (1963). Deterministic non-periodic flow. *J. Atmos. Sci.*, **20**, p. 130–141.

[77] Tamaševčius A., Mikolaitis G., Pyragas V., and Pyragas, K. (2007). Delayed feedback control of periodic orbits without torsion in nonautonomous chaotic systems: Theory and experiment, *Phys. Rev. E*, **76**, p. 026203-6.

[78] Just, W., Benner, H., and Loewenich, C. (2004). On global properties of time-delayed feedback control: weakly nonlinear analysis. *Physica D*, **199**, p. 33.

[79] Höhne, K., Shirahama, H., Choe, C.,-U., Benner, H., Pyragas, K., and Just, W. (2007). Global Properties in an Experimental Realization of Time-Delayed Feedback Control with an Unstable Control Loop, *Phys. Rev. Lett.*, **98**, p. 214102.

[80] Cvitanović, P. (1988). Invariant Measurement of Strange Sets in Terms of Cycles. *Phys. Rev. Lett.*, **61**, p. 2729.

[81] Kittel, A. Pyragas, K., and Richter, R. (1994). Prerecorded history of a system as an experimental tool to control chaos. *Phys. Rev. E*, **50**, p. 262.

[82] Fiedler, B., Flunkert, V., Georgi, M., Hoevel, P., and Schoell, E. (2007). Refuting the Odd-Number Limitation of Time-Delayed Feedback Control. *Phys. Rev. Lett.*, **98**, p. 114101.

[83] Postlethwaite Claire, M., and Silber, M. (2007). Stabilizing unstable periodic orbits in the Lorenz equations using time-delayed feedback control. *Phys. Rev. E*, **76**, p. 056214.

Chapter 6

PHASE CONTROL IN NONLINEAR SYSTEMS

Samuel Zambrano[1], Jesús M. Seoane[1], Inés P. Mariño[1],
Miguel A. F. Sanjuán[1] and Riccardo Meucci[2,*]

[1]*Department of Physics, Universidad Rey Juan Carlos,
Tulipán, s/n, 28933, Móstoles, Madrid, Spain*
[2]*CNR-Istituto Nazionale di Ottica Applicata,
Largo E. Fermi 6, 50125, Florence, Italy*
*riccardo.meucci@inoa.it

1. Introduction

Since the pioneering work on controlling chaos due to Ott, Grebogi and Yorke (OGY) [19], different control schemes have been proposed that allow to obtain a desired response from a dynamical system by applying some small but accurately chosen perturbations [23]. In this context, some techniques that allow avoiding escapes in open dynamical systems presenting transient chaos have been proposed, with applications to many different situations in physics and engineering (see Ref. [1] and references therein).

The methods stated to control chaos can be classified in *feedback* and *nonfeedback methods* [5], depending how they interact with the system. Feedback methods of chaos control, as the celebrated OGY [19], stabilize one of the unstable orbits that lie in the chaotic attractor by using small state-dependent perturbations into the system. However, in experimental implementations, the fast response that these methods require cannot usually be provided. For these situations, nonfeedback methods are more useful. Nonfeedback methods have been mainly used to suppress chaos in periodically driven dynamical systems.

$$\dot{\mathbf{x}} = \mathbf{f}(\mathbf{x}, \lambda) + \mathbf{F}\cos\omega t,\qquad(1)$$

where $\mathbf{x}, \mathbf{f}$ and $\mathbf{F}$ are vectors of the m-dimensional phase espace, and λ is a parameter of the system. The main idea of these nonfeedback methods is to apply a harmonic perturbation either to some of the parameters of the system

$$\dot{\mathbf{x}} = \mathbf{f}(\mathbf{x}, \lambda(1 + \epsilon \cos(r\omega t + \phi))) + \mathbf{F} \cos \omega t \qquad (2)$$

or as an additional forcing,

$$\dot{\mathbf{x}} = \mathbf{f}(\mathbf{x}, \lambda) + \mathbf{F} \cos \omega t + \epsilon \mathbf{u} \cos(r\omega t + \phi), \qquad (3)$$

where $\mathbf{u}$ is a conveniently chosen unitary vector.

The effectiveness of this type of methods has been tested experimentally in different works [12, 17]. In the first where these nonfeedback method was explored, the numerical and experimental explorations were essentially focused on the role played by the perturbation amplitude ϵ and the resonance condition r, but the role of the phase difference ϕ was hardly explored. However, in Ref. [17], it was observed that the phase difference ϕ between the periodic forcing and the perturbation had certain influence on the dynamical behavior of the system. Furthermore, in the Ref. [21], the authors have shown that ϕ plays a crucial role on the global dynamics of the system. Thus, it was clear that the role of the phase difference is important in the global dynamics of the system. The type of control based on varying the phase difference ϕ in search of a desired dynamical behavior is known as the *phase control* technique.

The aim of this chapter is to show that the phase control is very versatile and that it can be applied in many different contexts. As we said, this technique can be used to control chaos, but also to control other paradigmatic dynamical behaviors present in nonlinear dynamical systems. Thus, we are going to show in this Chapter that phase control can be used to suppress chaos [28], but also to control the phenomenon of crisis-induced intermittency [29] and to avoid escape in an open dynamical systems that presents transient chaotic behavior [22].

This Chapter is organized as follows. In Sec. 2 we present an application of the phase control method to control chaotic regimes by using as prototype model the Duffing oscillator. Section 3 explores how this technique can be used to control crisis-induced intermittency, taking as two paradigmatic models both, the Quadratic map and the CO_2 laser. Finally, in Sec. 4 we show, making use of the paradigmatic Helmholtz oscillator, that phase control can also be used to prevent divergences in open dynamical systems. Some conclusions and a discussion of the main results of this Chapter are presented in Sec. 5.

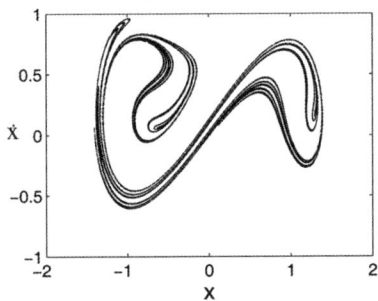

Fig. 1. Chaotic attractor observed for the orbits following the equation: $\ddot{x} + 0.15\dot{x} - x + x^3 = 0.258\cos(t)$. In this situation the value of the largest Lyapunov exponent is approx. 0.14

2. Phase Control of Chaos

2.1. *Description of the model*

In this Section we are going to explore how phase control can be used to suppress chaotic behavior in periodically driven chaotic system. A paradigmatic example of a system of this type is the two-well Duffing oscillator, whose equation of the motion reads

$$\ddot{x} + \delta\dot{x} - x + x^3 = F\cos(t). \tag{4}$$

A well known mechanical interpretation is the motion of a unit mass particle in a double well symmetric potential $V(x) = -\frac{x^2}{2} + \frac{x^4}{4}$ with dissipation and driven by an external periodic forcing. Depending on the values of F and δ, Eq. (4) yields a rich variety of dynamical solutions including stable equilibria, periodic oscillations and chaotic solutions. We tailor the parameters F, δ in such a way that the asymptotic state will be chaotic, and from then we study how to reach a periodic state under a suitable perturbation in our control method.

To analyze the dynamics of Eq. (4) we numerically integrate it by using a fourth order Runge Kutta algorithm with 500 integration steps per cycle, that is, with a time step $\Delta t = 2\pi/500$. Once the trajectories are integrated, Lyapunov exponents can be calculated using standard techniques. We thus find that for the set of parameters $F = 0.258$ and $\delta = 0.15$ the system is chaotic and the largest Lyapunov exponent is $\lambda \approx 0.14$. The Poincaré map of a trajectory is shown in Fig. 1. We keep these parameter values ($F = 0.258$, $\delta = 0.15$) all throughout the numerical part of this section.

Our aim here is to analyze the effectiveness of phase control for this system. To do this, we harmonically perturb one of the system's parameters, as in Eq. (2). In particular, we modulate periodically the cubic term of the restoring force. Hence the complete equation of our model is

$$\ddot{x} + \delta \dot{x} - x + (1 + \epsilon \cos(rt + \phi))x^3 = F \cos(t). \tag{5}$$

Notice that, as in Eqs. (2) and (3), and as in the remaining of this Chapter, the key parameters are the resonance between the frequency of the perturbation and that of the driving r, the perturbation amplitude ϵ and the phase difference ϕ. We will focus mainly on the role played by ϕ. Notice that, in this case, this modulation induces a slight harmonic variation of the width of the two wells and of the height of the potential barrier between the two minima of $V(x)$. Thus, ϕ can be interpreted also as the phase difference between this geometrical variation and the external forcing.

2.2. Numerical exploration of phase control of chaos

The equation we use for the numerical exploration of our technique once the parameters are fixed reads

$$\ddot{x} + 0.15\dot{x} - x + (1 + \epsilon \cos(rt + \phi))x^3 = 0.258 \cos(t), \tag{6}$$

where ϵ, ϕ and r will be used as free parameters. As shown in the previous subsection, when $\epsilon = 0$, this system is chaotic. In order to evaluate in a detailed way the role of ϕ, we calculate the largest Lyapunov exponent over every point in a 100×100 grid in the rectangle of the parameter plane $0 \leq \epsilon \leq 0.005$ $0 \leq \phi \leq 2\pi$, fixing r for each computation. As we are searching for areas in the parameter plane where a transition between chaotic and regular motion takes place, we take care of avoiding the transient states by waiting for a sufficiently long time to fix the corresponding stable regime.

In Fig. 2 we plot the results for several integer r values. The colour assigned to each point in the (ϵ, ϕ) plane indicates if for those values of the parameters Eq. (6) has a chaotic solution or a periodic solution. Instead of marking with two colors the points leading either to positive or negative Lyapunov exponents, we have chosen four different colors in order to better appreciate the structure of the chaos suppression regions. The black color denotes $\lambda < -0.025$, the grey color $-0.025 \leq \lambda < 0$, the silver color $0 < \lambda < 0.025$ and the white color $\lambda > 0.025$.

Figure 2 shows that there exist wide regions of the (ϵ, ϕ) plane where λ is smaller than zero, and therefore chaos is supressed. We note that the control

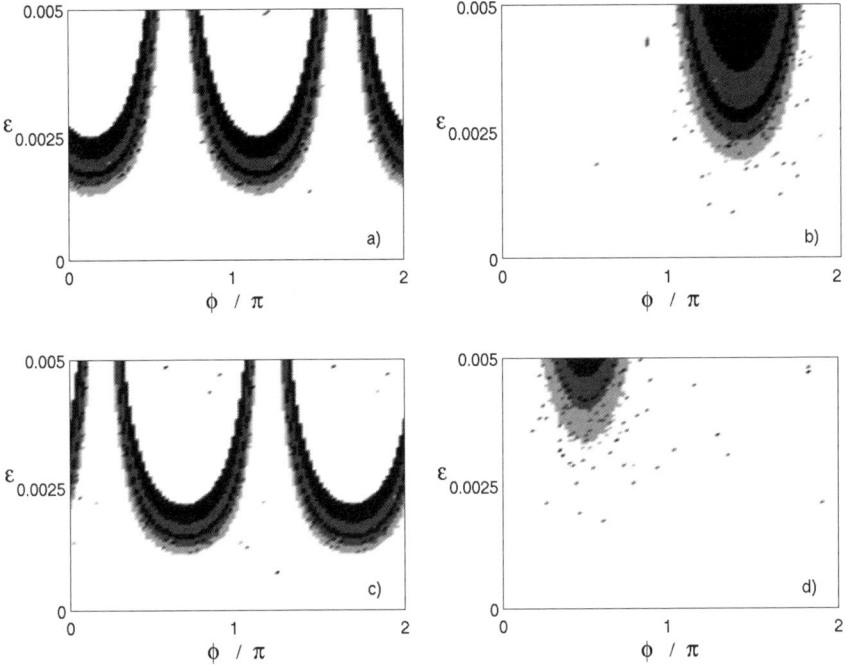

Fig. 2. Largest Lyapunov exponent λ computed at every point of a 100×100 grid of (ϵ, ϕ) values in the region $0 \leq \phi < 2\pi, 0 \leq \epsilon \leq 0.005$ for $\ddot{x} + 0.15\dot{x} - x + (1 + \epsilon \cos(rt + \phi))x^3 = 0.258 \cos(t)$, fixing (a) $r = 1$, (b) $r = 2$, (c) $r = 3$, (d) $r = 4$. The black color denotes $\lambda < -0.025$, the grey color $-0.025 \leq \lambda < 0$, the silver color $0 < \lambda < 0.025$ and the white color $\lambda > 0.025$. The control regions have a structure that follows the expected π symmetry for r odd and the trivial 2π symmetry for r even.

regions, far from having a trivial or irregular shape, present a symmetry that depends on the parity of the r parameter: π symmetry for odd r values (Figs. 2(a) and 2(c)) and the trivial 2π symmetry for even r values (Figs. 2(b) and 2(d)).

In order to explain this difference related to parity in the control areas, we note that there exists a wide area of the phase space which verifies that both the point $(x_0, \dot{x}_0, t_0)$ and $(-x_0, -\dot{x}_0, t_0 + \pi)$ belong to the same basin of attraction for a certain selection of the parameters (ϵ^*, ϕ^*). In this region, the invariance of Eq. (4) under the transformation $x \mapsto -x, t \mapsto t + \pi, \phi \mapsto \phi + r\pi$ allows us to infer that if the system is controlled for a certain pair of values (ϵ^*, ϕ^*), then chaos will also be controlled for $(\epsilon^*, \phi^* + r\pi \mod (2\pi))$, as observed in Fig. 2.

The most interesting feature is the role of the phase ϕ in selecting the final state of the system. Indeed, we shall see that the smallest ϵ values necessary to suppress chaos are obtained when we use the correct phase ϕ. Considering the $r = 2$ case (Fig. 2(b)), the maximal perturbation amplitude considered $\epsilon = 0.005$ would not lead to a regular motion at $\phi = 0$. Instead, at $\phi = 3\pi/2$, $\epsilon \approx 0.0025$ is enough to lead the system to a periodic state. For $r = 4$ (Fig. 2(d)) we can draw analogous considerations.

In the odd r cases ($r = 1$ and $r = 3$, see Figs. 2(a) and 2(c)) the selection of the ϕ value determines whether the final state of the system will be chaotic or periodic as in the even r values. However, in this odd case the control region is not unique as in the previous case, and there is another important difference to be pointed out: contrary to what intuition might say, for a fixed ϕ value a continuous increasing of the perturbation amplitude ϵ can lead from a chaotic state to a periodic state and then back to chaos. In addition, the elegible ϕ range is reduced as ϵ increases.

Up to this point we have not said a word to characterize the stable state that is reached by our system when chaos is controlled. The numerical calculations shown in Fig. 2 do not allow us to identify the orbits to which the system is led for each selection of parameters (ϵ, ϕ), but only to predict whether the system is chaotic or periodic. To distinguish the periodicity of the orbits, bifurcation diagrams will be computed. We evaluate bifurcation diagrams fixing two particular values of r and ϵ ($r = 2$, $\epsilon = 0.005$) and varying ϕ. The results are shown in Fig. 3, where for the sake of clarity we plot just the local negative maxima of x vs. ϕ. We note how the system can be directed to orbits of different periods by suitably adjusting ϕ. Thus, varying ϕ is like "tuning" the system towards a desired periodic orbit. Bifurcation diagrams also show the existence of small periodic windows within the chaotic sea. Thus, phase control enables the system to reach many different periodic orbits. However, considering that the distance between the successive bifurcations decreases very fast and that the periodic windows are quite narrow, severe limitations may occur in experimental devices affected by noise. In this case only a few periodic orbits will be accessible.

The selection of r is very important too, as we have seen that the ϕ values for which chaos is controlled drastically depends on this parameter. We can also consider what may happen if we take a non integer r value but close to an integer one. Let r be $r = k+\mu$ with k integer and $0 < \mu \ll 1$. In such a situation the parametric perturbation would be of the form: $\epsilon \cos(rt + \mu t + \phi) \equiv \epsilon \cos(rt + \phi(t))$ with $\phi(t) = \phi + \mu t$. Thus, this deviation from the resonance is equivalent to a slow variation of the phase ϕ, inducing an intermittent

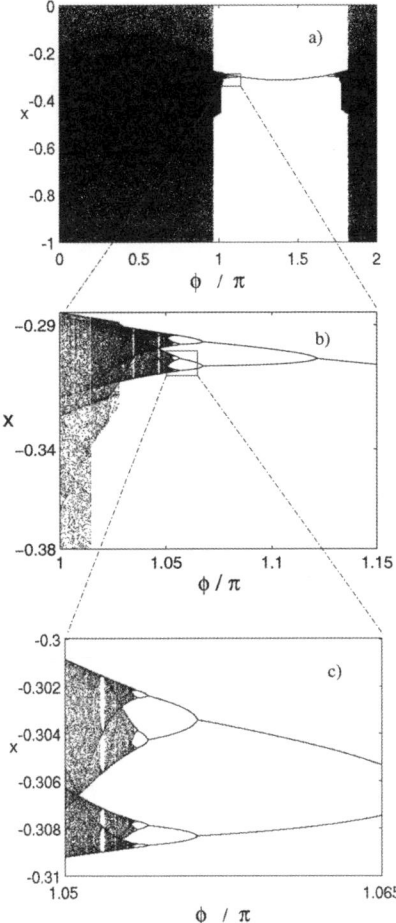

Fig. 3. Successive zooms of a bifurcation diagram the local maxima of the negative x values versus ϕ, in the equation $\ddot{x} + 0.15\dot{x} - x + (1 + 0.005\cos(2t + \phi))x^3 = 0.258\cos(t)$. The dependence of the periodic orbit on the ϕ value can be noted. The higher the period of the orbit, the narrower the ϕ interval where it can be reached. Periodic windows are also present inside the chaotic sea.

motion between chaotic and periodic solutions with a period $2\pi/\mu$. This phenomenon was previously referred to as a *breather* [21, 27]. In this situation the motion should be considered neither as a purely chaotic nor as a regular motion. Hence, if we are interested in a long lasting stable suppression of chaos it is important to be as close to the resonances as possible.

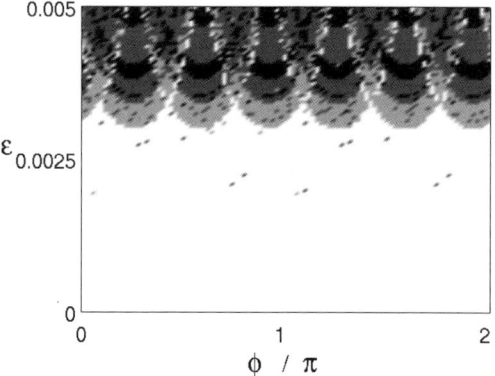

Fig. 4. Largest Lyapunov exponent λ computed at every point of a 100×100 grid of (ϵ, ϕ) values in the region $0 \leq \phi < 2\pi, 0 \leq \epsilon \leq 0.005$ for: $\ddot{x} + 0.15\dot{x} - x + (1 + \epsilon \cos(\frac{1}{3}t + \phi))x^3 = 0.258\cos(t)$. The black color denotes $\lambda < -0.025$, the grey color $-0.025 \leq \lambda < 0$, the silver color $0 < \lambda < 0.025$ and the white color $\lambda > 0.025$. Again, the control regions have very interesting structure that follow the expected $\frac{\pi}{3}$ symmetry.

2.3. *Experimental evidence of phase control of chaos*

Here we show a test of the phase control method in a laboratory system. This control technique was tested in the circuit that can be seen in Fig. 5. It consists of an electronic analog simulator implemented using commercial semiconductor devices. The variable V_x is the output of I_2, while V_y is the output of I_1. V_d is the driving voltage amplitude applied by means of the generator G_d, while V_c is the control voltage amplitude applied by G_c. The parameters ω_d and ω_c are the driving and control angular frequency respectively. The integrators I_1 and I_2 have been implemented using Linear Technology LT1114CN four quadrant operational amplifiers, while the multipliers are Analog Devices MLT04. The acquisition of the experimental data has been performed by means of a LeCroy digital oscilloscope and by means of a real time acquisition board connected to a personal computer provided with LabView software. G_c is a digital-to-analog arbitrary waveform generator SONY TEKTRONIX AWG420. It can provide two waveforms of whatever shape. The generator can also control the phase difference ϕ. Under a suitable normalization of the time scale the dynamics of this circuit is governed by Eq. (5), with a slight change in the values of the parameters $\delta = 0.1471$, $F = 0.262$, $\omega = 1.257$. In this way, the variables V_x, V_y can be associated to x, $\dot{x}$ respectively.

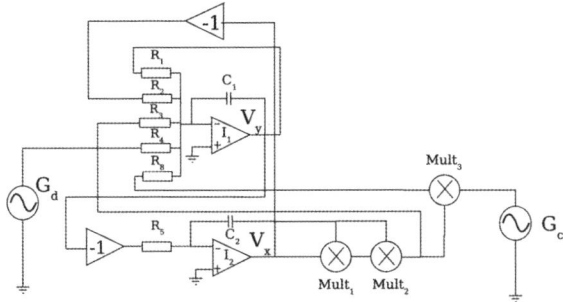

Fig. 5. Sketch of the electronic circuit. I: integrators; R: resistors; C: capacitors, Mult: multipliers; G_d: Sinusoidal wave generator; G_c: Arbitrary waveform generator. Under a suitable time normalization, the dynamics of the circuit is given by $\ddot{x} + 0.1471\dot{x} - x + (1+\epsilon\cos(r1.257t+\phi))x^3 = 0.263\cos(1.257t)$, where $x \propto V_x$ and $\dot{x} \propto V_y$. The ϵ, ϕ, r values can be fixed with the arbitrary waveform generator.

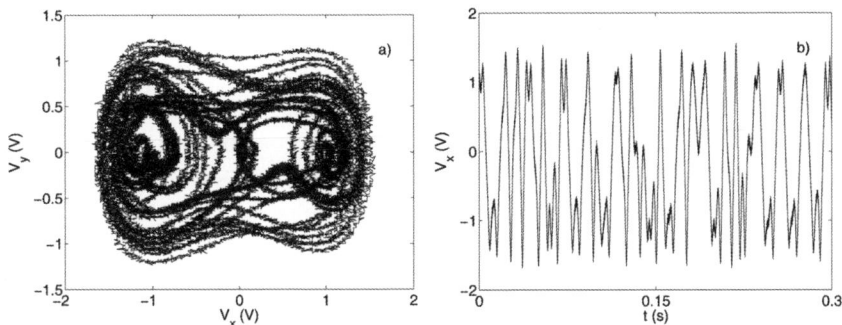

Fig. 6. (a) Experimental reconstruction of the chaotic attractor observed for the analog circuit, (b) Example of the experimental times series, where the asymmetry of the trajectory can be observed.

By observing the dynamics of the system with the aid of the oscilloscope we realized that the unperturbed ($\epsilon = 0$) system does not follow exactly the expected chaotic dynamics described by Eq. (5). We observe that a typical trajectory spends more time in the region of negative values of the x variable than in the region of the positive values, that is, it spends more time in the left well than in the right well, as is shown in Fig. 6. This is equivalent to an asymmetrical double-well potential.

An explanation of this phenomenon can be given considering how do multipliers work. Real multipliers differ from ideal ones because operations such as x^3 are done as $(x - \Delta_1)(x - \Delta_2)(x - \Delta_3) + C$, where $\Delta_i, C \ll 1$.

Thus, instead of doing x^3 our circuit made the following operation: $x^3 \mapsto x^3 + ax^2 + bx + c$. We have made experimental measurements of the x^3 term provided for the multipliers, and by polynomically fitting the results we obtain for the coefficients the values: $a = -0.014, b = 0.014, c = -0.12$.

Hence, the addition of these lower order terms besides x^3 in the restoring force makes the two-well potential $V(x)$ no longer symmetric, and it can be seen that now the left well becomes deeper than the right one. Thus, for the system it is more difficult to escape from the well at $x < 0$, where it spends more time. This is confirmed by Fig. 6(b), where an example of the experimental time series is shown.

Instead of considering this asymmetry as a problem we thought that this could be a good way to test the versatility and robustness of phase control of chaos. Following this strategy we repeat the numerical simulations considering the asymmetry in the potential. The new equation of the motion is:

$$\ddot{x} + \delta\dot{x} - x + (1 + \epsilon\cos(r\omega t + \phi))(x^3 + ax^2 + bx + C') = F\cos(\omega t). \quad (7)$$

In order to have a previous idea of the probable control areas, we test a wider region in the (ϵ, ϕ) plane than in the ideal case. Thus, we explore the range of values $0 \leq \epsilon \leq 0.03$ and $0 \leq \phi \leq 2\pi$. We compute Lyapunov exponents for Eq. (7) on each point (ϵ, ϕ) in a 100×100 grid in the considered region for the values $r = 1$, $r = 2$ and $r = 3$. The results are shown in Fig. 7.

Our numerical results confirm that some of the previously observed features also apply to the asymmetric case. The critical dependence on the phase ϕ to control chaos is preserved. However, in the considered region, control islands of chaos suppression coexist with narrow control areas, and their structure is much more intrincated and irregular than in the symmetric case. Indeed, due to the new terms added in the potential, the system does not present the invariance under the transformations described in the previous subsection for the symmetric case, so we should not be surprised to find that the $r\pi$ symmetry of the control regions is no longer present. Chaos is suppressed for a wide range of values of ϵ, even for values below $\epsilon \leq 0.005$.

Once we have a qualitative idea of the distortion induced by the potential asymmetry we can come back to the experiment. In our experiments we search for the control areas in the parameter space. We detect and reconstruct these control areas by doing bifurcation diagrams fixing r, ϵ (for different ϵ values) and varying the phase value ϕ. Such bifurcation diagrams are performed by searching for the maxima of the time series of the system

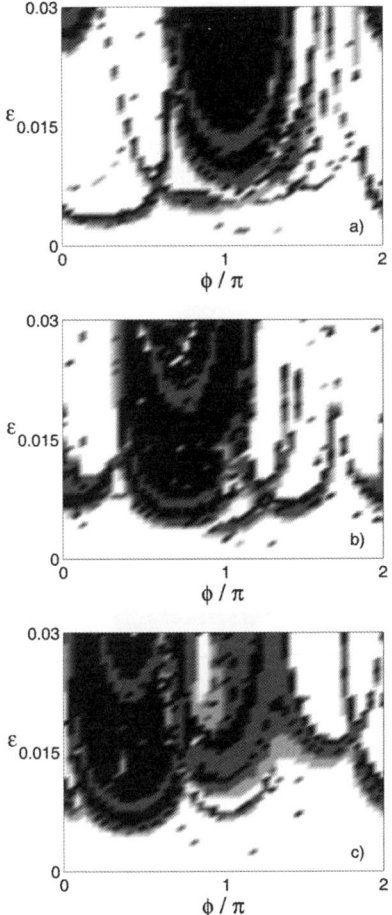

Fig. 7. Largest Lyapunov exponent λ computed at every point of a 100×100 grid of (ϵ, ϕ) values in the region $0 \leq \phi < 2\pi, 0 \leq \epsilon \leq 0.03$ considering the effective potential asymmetry: $\ddot{x} + 0.147\dot{x} - x + (1 + \epsilon \cos(1.257rt + \phi))(-0.12 + 0.014x - 0.014x^2 + x^3) = 0.262\cos(t)$. The calculations have been made for (a) $r = 1$, (b) $r = 2$, (c) $r = 3$. The black color denotes $\lambda < -0.025$, the grey $-0.025 \leq \lambda < 0$, the silver $0 < \lambda < 0.025$ and the white $\lambda > 0.025$. The control regions lack the symmetry of the former case, but wide zones of chaos suppression do still appear.

obtained when the phase ϕ is slowly varied to the characteristic time scale of our system, that is, we make $\phi = \mu t$ with $\mu \ll 1$.

The experimental control zones for $r = 1$, $r = 2$ and $r = 3$ can be observed in Figs. 8(a) to 8(c). They confirm the main features numerically

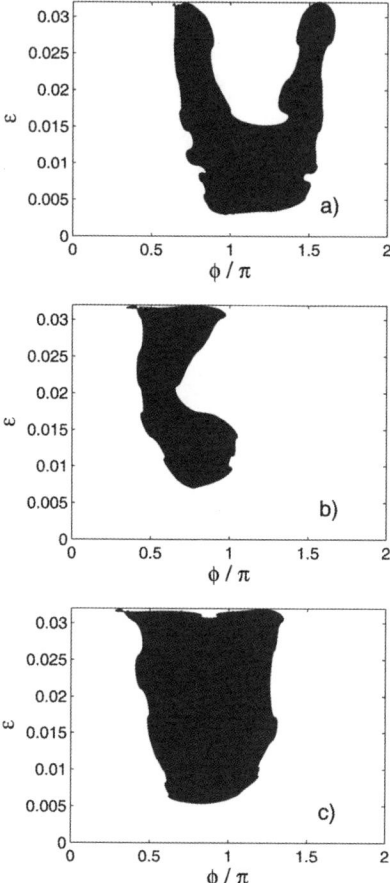

Fig. 8. Experimental control regions for the circuit of Fig. 5, reconstructed from experimental bifurcation diagrams, when (a) $r = 1$, (b) $r = 2$, (c) $r = 3$. In spite of the experimental errors there are still wide control zones that roughly coincide with those predicted by the numerical simulations.

predicted, that is, the crucial role of the phase ϕ on the final state and the smallness of ϵ needed to suppress chaos if ϕ is properly chosen. The latter is especially interesting as long as we did not consider any source of instability in our numerical simulations. We can also notice that the ϕ intervals where chaos is suppressed do roughly coincide with those predicted from numerical calculations, including the predicted symmetry breaking of the control areas.

The differences in shape between experimental and simulated control regions are due to the inherent noise of the circuit, which contributed to blurring the narrow control zones reported in the numerical simulations. Finally, we remark that the bifurcation in the control zone that appears in the $r = 1$ case (Fig. 8(a)), in spite of not being predicted numerically, was rather common in the symmetric case.

3. Phase Control of Intermittency in Dynamical Systems

3.1. *Crisis-induced intermittency and its control*

In the former section we studied a paradigmatic example of a chaotic system. For some of these systems it can be observed that, by modifying a control parameter, the chaotic attractor can touch an unstable periodic orbit inside its basin of attraction. This phenomenon, known as *interior crisis* [9], induces a sudden expansion of the attractor, after which most trajectories alternates periods of time in the region where the pre-crisis attractor lied with excursions out of it.

This type of intermittency, called *crisis-induced intermittency* is a widespread phenomenon [8], so it is common to find situations where a control of this type of behavior becomes desirable. Recently, [15] a feedback method to enhance or tame the intermittency has been devised. The strategy is to force the system with a feedback in which the "typical" frequency of the excursions, that is the frequency of the periodic orbit involved in the interior crisis, is either filtered or enhanced. This method has been shown to be effective in a periodically driven chaotic CO_2 laser, as the one described in Ref. [16].

However, as we said in previous section, feedback control methods might present some difficulties for their implementation. Thus, in some contexts non-feedback methods might be more useful. In last section we have shown that phase control of chaos [21, 27] is a powerful tool to control the dynamics of a periodically driven chaotic system. In this Section, we are going to show that the intermittent behavior of a dynamical system close to an interior crisis can be controlled by using the phase control scheme. We give experimental and numerical evidence of the validity of the method for the periodically driven CO_2 laser close to an interior crisis, and in order to have a deeper insight on the role of ϕ we also present an analysis of phase control of the quadratic map close to a crisis.

3.2. Experimental setup and implementation of the phase control scheme

We first address the experimental implementaion of the phase control scheme on a CO_2 laser, to control its intermittent behavior. The experimental setup consists of a single-mode CO_2 laser, as shown in Fig. 9. The laser cavity is defined by a totally reflecting grating and a partially reflecting mirror (G and M), and the gain medium is pumped by a constant electric discharge current. An electrooptic modulator (EOM) is inserted in the laser cavity in order to control the cavity losses by an external forcing, obtained from a sinusoidal generator (MD), that can be represented as

$$F(t) = \beta \sin(2\pi f_0 t) + b_0, \qquad (8)$$

where β is the amplitude of the external forcing, b_0 is a bias voltage and $f_0 = 100$ kHz is about twice the relaxation frequency of the laser.

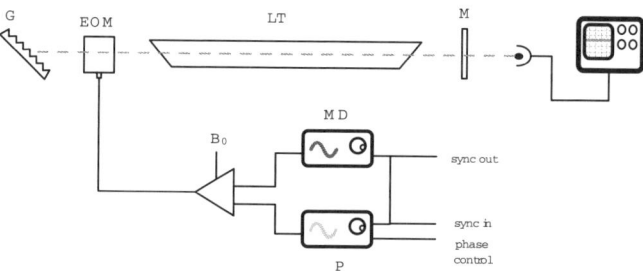

Fig. 9. Experimental setup for a single-mode CO_2 laser with modulated losses. EOM: intracavity electrooptic modulator, G: total reflecting grating, M: partial reflecting mirror, D: fast infrared detector, P: sinusoidal generator, MD: digital oscilloscope.

The CO_2 modulated laser is accurately described by the following model of five differential equations [14]:

$$\begin{aligned}
\dot{x}_1 &= kx_1(x_2 - 1 - \alpha \sin^2(F(t))), \\
\dot{x}_2 &= -\gamma_1 x_2 - 2kx_1 x_2 + gx_3 + x_4 + p, \\
\dot{x}_3 &= -\gamma_1 x_3 + gx_2 + x_5 + p, \\
\dot{x}_4 &= -\gamma_2 x_4 + zx_2 + gx_5 + zp, \\
\dot{x}_5 &= -\gamma_2 x_5 + zx_3 + gx_4 + zp.
\end{aligned} \qquad (9)$$

In the above equations, x_1 represents the laser output intensity, x_2 is the population inversion between the two resonant levels, and x_3, x_4 and

x_5 account for molecular exchanges between the two levels resonant with the radiation field and the other rotational levels of the same vibrational band. The parameters of the model are the following: k is the unperturbed cavity loss parameter, g is a coupling constant, γ_1 and γ_2 are population relaxation rates, z accounts for an effective number of rotational levels, α accounts for the efficiency of the electro-optic modulator and p is the pump parameter. The rest of the parameters are related to the external periodic forcing defined above.

By increasing the amplitude of the external forcing, the system undergoes a sequence of subharmonic bifurcations, and for $\beta < 0.1$ the dynamics is restricted to a certain region of the phase space, say $|x_1| < 0.013$, as shown in Fig. 10. Further increase of β induces an interior crisis, and the clear attractor expansion that can be observed in Fig. 10 for $\beta \approx 0.1$. This leads to the occurrence of a regime where there is an intermittency between orbits contained in the pre-crisis bounding region and excursions out of it, of period three and four. The set of parameters used in the numerical simulations are $k = 30$, $\alpha = 4$, $\gamma_1 = 10.0643$, $g = 0.05$, $p = 0.01987$, $\gamma_2 = 1.0643$, $z = 10$, $f_0 = 1/7$ and $b_0 = 0.1794$. The stability analysis provides a value of the relaxation oscillation frequency of 0.07, which is around half the frequency of the forcing signal.

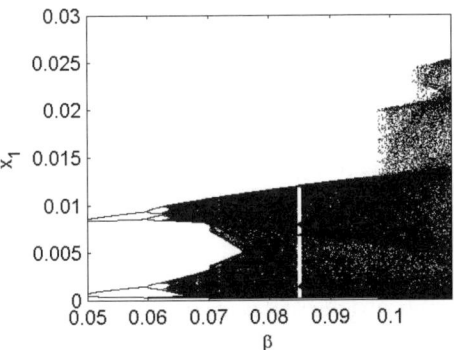

Fig. 10. Numerical bifurcation diagram for β. Two interior crisis are observed, but we are going to study the effect of harmonic perturbations on the laser around the first crisis.

The phase control *phase control scheme* is here implemeted as follows. We chose again to perturb harmonically one of the parameters of the system, b_0, because it is easily accessible in the experimental setup.

The perturbed parameter becomes a periodic function $b(t) = b_0(1 + \epsilon \sin(2\pi f_0 rt + \phi))$. The phase control scheme relies in an appropriate use of the phase ϕ, once ϵ and r are fixed.

Due to the fact that $F(t)$ depends linearly on the bias b_0, one can clearly see that adding a harmonic perturbation to the bias is equivalent to adding a second periodic forcing, which is one of the possible implementations of the phase control scheme [21, 27], corresponding to Eq. (3). Thus, for the perturbed system, the forcing term of Eq. (9) should read

$$F(t) = \beta \sin(2\pi f_0 t) + \epsilon' \beta \sin(2\pi f_0 rt + \phi) + b_0, \quad (10)$$

with $\epsilon' = b_0 \epsilon / \beta$. We consider ϵ' instead of ϵ, since it enables us to quantify the strength of the applied perturbation in terms of the main periodic forcing.

3.3. Phase control of the laser in the pre-crisis regime

We consider the role of the phase when the unperturbed laser is placed in the situation previous to the interior crisis, so that no intermittency takes place (not even induced by noise, since we choose to be quite far from the interior crisis). In this case, we characterize the effect of ϕ for fixed values of ϵ' and r by taking records of very long time series where ϕ is slowly varied $\phi \mapsto \phi(t) = 2\pi \mu t$ where $\mu \ll 1/f_0$, i.e., the phase varies very slowly compared to the typical time scale of the laser. Thus, for $t = 0$ the phase difference is $\phi = 0$ and it increases until $t = 1/\mu$, where it is $\phi = 2\pi$. The dynamical state of the system at a certain time t' corresponds essentially to the expected behavior for $\phi = 2\pi \mu t'$.

Let us first analyze the case in which the frequency of the perturbation is the same as the frequency of the main driving, that is, $r = 1$. The experimental long time series for this case is plotted in Fig. 11. We can observe how there is an increase of the amplitude of the peaks when ϕ is close to 0 and 2π, and a depression as ϕ goes to π. This phenomenon has a simple explanation, indeed

$$F(t) = \beta \sin(2\pi f_0 t) + \epsilon' \beta \sin(2\pi f_0 t + \phi) = \beta' \sin(2\pi f_0 t + \phi_0), \quad (11)$$

where

$$\beta' = \beta \sqrt{1 + \epsilon'^2 + 2\epsilon' \cos \phi}. \quad (12)$$

Notice that we basically have a single forcing, so the resulting ϕ_0 plays an irrelevant role. However, the effective amplitude of the perturbation, β', depends on ϕ. Thus, by choosing $\phi = 0$, the effective amplitude of

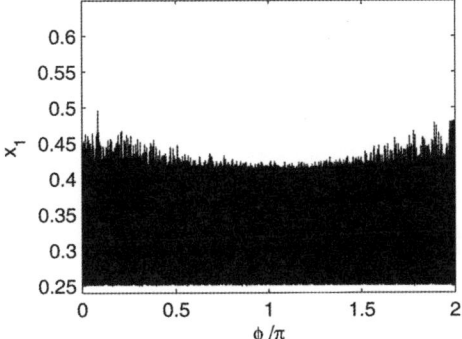

Fig. 11. Long time series varying the phase ϕ for $r = 1$ $\epsilon' = 0.1$. Note that for $\phi = 0$ and $\phi = 2\pi$ the maxima of the series are increased, as expected.

the periodic forcing is increased to a value closer to the critical value so the height of the peaks becomes bigger. Instead, by choosing $\phi = \pi$, β' becomes smaller, so the system is further away from the crisis, and the height of the peaks becomes smaller. We shall point out, in this figure, that with a perturbation of about 10% of the main forcing, the system is not led to the intermittent regime. This is not very relevant by itself, because Eq. (12) shows that the necessary amplitude of the perturbation to lead the system to the intermittent regime could be reduced just by placing the unperturbed system closer to the crisis. However, it is an important reference to evaluate the effectiveness of perturbations with different frequencies.

Now we consider the laser in the same unperturbed situation before the crisis and we apply a perturbation whose frequency is the same as the frequency of the unstable periodic orbit involved in the interior crisis [15], that is, $f_0/3$. The two main behaviors observed experimentally are summarized in the two diagrams shown in Fig. 12. We observe an evident $2\pi/3$ symmetry of the first diagram, which could be deduced from the invariance of Eq. (10) under the transformation $t \mapsto t + k/f_0$ and $\phi \mapsto \phi + 2\pi r k$, with k an integer. Figure 12(a) shows the crucial role played by the phase difference. For the same values of the perturbation amplitude, by adjusting the phase, the system can be placed either in an intermittent regime or in the pre-crisis regime. It is important to note that we observe experimentally this significant effect even if the amplitude of the perturbation applied is about 0.3% of the amplitude of the main forcing, which is much smaller than in the $r = 1$ case. However, as we observe in Fig. 12(b) for $\epsilon' = 0.006$,

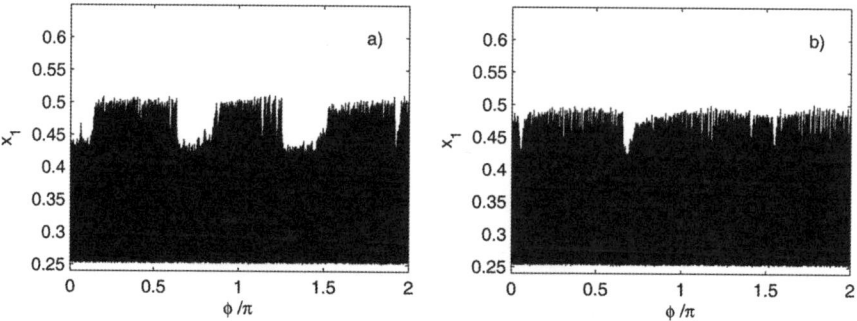

Fig. 12. (a) Long time series varying the phase ϕ for $r = 1/3$, $\epsilon' = 0.003$ and (b) $\epsilon' = 0.006$. The diagrams present the expected $2\pi/3$ symmetry, even if in (b) the intervals of ϕ leading to intermittency have merged and the behavior is nearly phase independent.

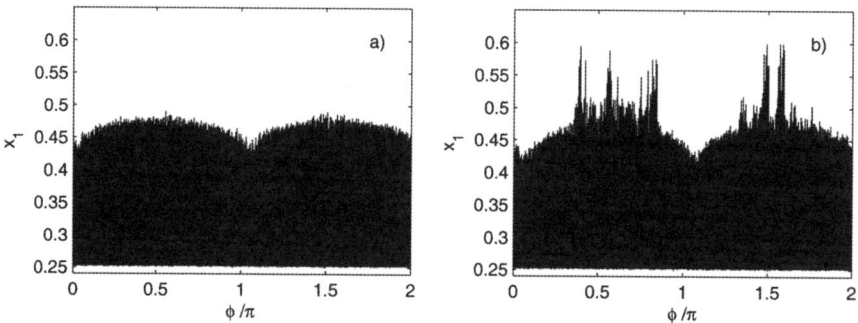

Fig. 13. (a) Long time series varying the phase ϕ for $r = 1/2$, $\epsilon' = 0.006$ and (b) $\epsilon' = 0.01$. We have a π symmetry, as expected. The dependence on the phase ϕ is clear, and we observe that in (b) a correct selection of the phase determines whether there is intermittency or not.

there is intermittency for nearly all values of ϕ. In summary, the $r = 1/3$ perturbation is much more effective than the $r = 1$ perturbation to control the intermittency.

New features arise when the system is perturbed with the frequency corresponding to the period doubling bifurcation of the system, $f_0/2$. Again, two experimental diagrams are presented to see the effect of the phase, shown in Fig. 13. We can observe the expected π symmetry in ϕ. On the other hand, Fig. 13(a) shows again that the phase difference modulates the maximum height of the peaks, but intermittency does not take place. However, when the perturbation is increased to $\epsilon' = 0.01$, Fig. 13(b), just 1% of the main forcing, the effect of the phase is even clearer: again, the phase enables us to place the system either in the intermittent regime or in the small chaos regime. In the intermittent regime observed in Fig. 13(b), the high amplitude orbits are related with the second interior crisis shown in Fig. 10, thus a variety of dynamical behaviors is accessible by varying ϕ.

Numerical calculations provide a confirmation of these results, together with a deeper insight on the role of the phase. A good indicator to discriminate between the different dynamical states of the laser for different values of the parameters is

$$< H > = < \max(x_1(t)) > |_{x_1(t) > x_0}, \qquad (13)$$

where $< \cdot >$ indicates the average over a long time series, and $\max()$ indicates the relative maximum of the series. The value of x_0 is chosen in such a way that $< H >$ enables us to distinguish between the small chaos and the intermittent regime. In the numerical simulations we have observed that taking $x_0 = 10^{-5}$, that is, neglecting only the extremely small peaks of the signal is sufficient for this discrimination. We have observed that $< H > \leq 0.006$ corresponds to the pre-crisis chaotic regime, $0.006 << H > \leq 0.0074$ matches with the intermittent regime observed after the first crisis shown in Fig. 10 and $< H >> 0.0074$ corresponds to the regime in which there are high amplitude orbits, like those observed in Fig. 10 after the second crisis. $< H >$ can be easily computed by numerical integration of the equations of the laser. We study the dependence of the global dynamics on the parameters of the system by calculating $< H >$ as a function of ϵ and ϕ, fixing r.

Numerical calculations are presented in Fig. 14. As for the experimental results, we include the calculations for the trivial case $r = 1$ for the sake of clarity. In this case, Fig. 14(a), the color of the diagram and thus $< H >$ changes smoothly as the parameters vary, from a minimum at $\phi = \pi$ to a maximum at $\phi = 0.2\pi$, as observed in the experiment (Fig. 11).

For the $r = 1/3$ case, Fig. 14(b), $< H >$ presents the expected $2\pi/3$ symmetry. On the other hand, it can be clearly observed how the value of $< H >$ increases gradually with ϵ. For a narrow interval of ϵ, approximately $\epsilon \in [0.002, 0.003]$, depending on ϕ we have values of $< H >$ bigger than 0.006 intercalated with values of $< H >$ smaller than 0.006. This agrees with the phase-induced transitions between the intermittency and the small chaos regime observed experimentally. However, as in the experiment, we can see that if the perturbation amplitude ϵ is further increased the intervals of ϕ giving rise to intermittency merge, so intermittency is observed almost independently of the phase.

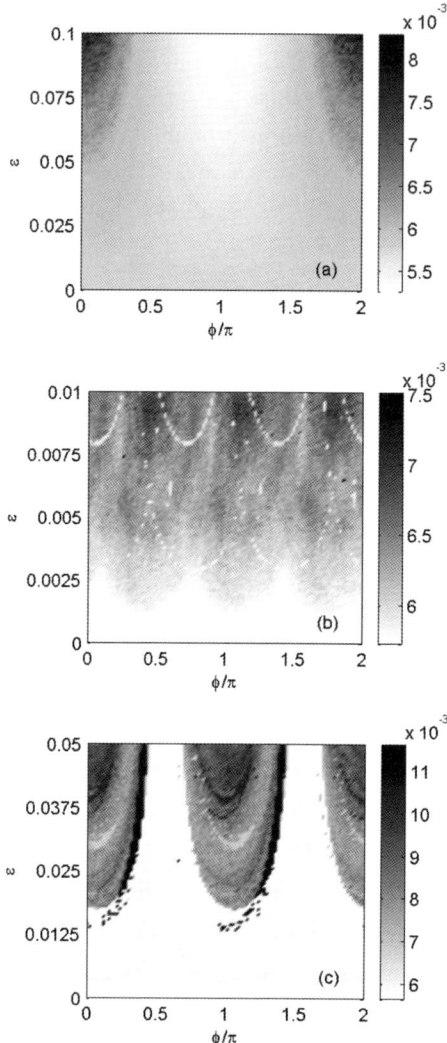

Fig. 14. The medium height of the maxima of x_1, $<H>$, as a function of ϵ and ϕ for $r = 1$ (a), $r = 1/3$ (b) and $r = 1/2$ (c).

Let us finally comment the results for the $r = 1/2$ case. For small values of ϵ, $\epsilon < 0.02$, the $<H>$ remains around $<H> \approx 0.005$ almost independently of the phase. However, when ϵ becomes bigger than a certain critical value $\epsilon_0 \approx 0.02$, there is a sudden change in the medium height of the peaks. This sudden transition to a high $<H>$ regime, which corresponds to

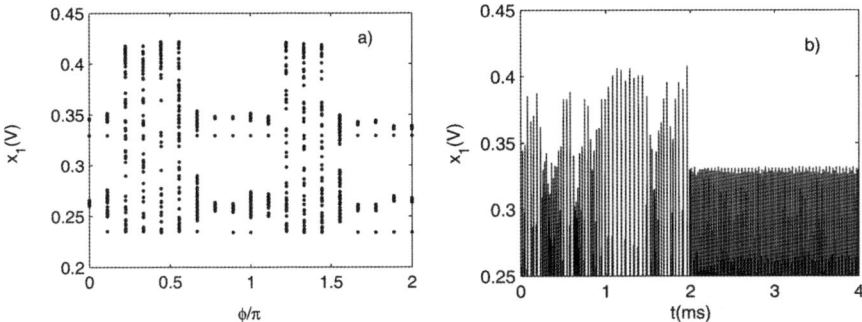

Fig. 15. (a) Experimental bifurcation diagram showing how an appropriate selection of the phase ϕ can take the laser from an intermittent regime to a small chaos regime. (b) A controlled time series of the laser, where control is applied at $t \approx 2$ ms and the intermittent behavior is nearly immediately suppressed.

the dynamical state observed after the second crisis of the laser, is evident from the drastic change of color that we can observe in the diagram of Fig. 14(c), which is fully consistent with the experiments on the laser. Thus, once again we see the important role played by ϕ in placing the system before or after the interior crisis.

3.4. *Phase control of the intermittency after the crisis*

Up to now we have shown that the intermittency of the CO_2 laser in the pre-crisis regime can be controlled, by just varying the phase ϕ. In this section, in analogy with [15], we show that phase control does also work when the unperturbed laser is placed in the post-crisis region.

In order to characterize the role of ϕ in this case we have opted to perform a bifurcation diagram by localizing the maxima of different time series of the laser with different values of ϕ, for $\epsilon = 0.01$ and $r = 1/2$, as shown in Fig. 15(a). We can clearly appreciate a π symmetry in the diagram as in the previous section. We can see how a variation of ϕ allows us to move from the intermittent regime to the small chaos regime. The action of the applied perturbation on the laser is illustrated by Fig. 15(b), where we can see how, once the perturbation is applied (for $t \approx 2$ ms), the system passes from an intermittent regime, with the characteristic large spikes, to a small chaos regime.

We have performed a numerical analysis to see this phenomenon in more detail. We characterize the role of the phase ϕ by calculating $<H>$,

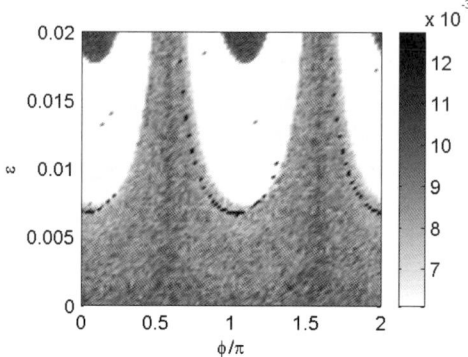

Fig. 16. The medium height of the maxima of x_1, $<H>$, as a function of ϵ and ϕ for $r = 1/2$.

defined as in Eq. (13), and the results are shown in Fig. 16. Again, the symmetry induced by our selection of r and the nontrivial role played by the phase ϕ are evident.

Thus, we have shown that we can use ϕ to control the intermittency after the interior crisis.

3.5. *Phase control of the intermittency in the quadratic map*

In order to gain a deeper insight of the role of ϕ in nonlinear systems, we study phase control in a paradigmatic nonlinear map close to an interior crisis. Our approach is quite different from that of other authors [13, 18], who have studied chaos control by harmonic perturbations in maps. Here we identify the key ingredients involved in the interior crisis for the unperturbed system. After this, we perform a perturbative analysis to estimate how these ingredients are affected by the presence of a periodic perturbation, thus emphasizing the nontrivial role played by ϕ.

We consider the unperturbed quadratic map given by

$$x_{n+1} = C - x_n^2 = F(C, x_n), \qquad (14)$$

a paradigmatic system that is conjugate to the well known logistic map [2], and whose interior crisis has been extensively studied [8–10] even when two of these maps are coupled, a situation that has been related with chaotic itinerancy [24]. For this system, at $C = C_s \approx 1.76$ a stable period three orbit is born, a phenomenon referred to as subduction. If we increase the parameter C, there is a period doubling bifurcation giving rise to three chaotic bands but, after a critical value, $C^* \approx 1.79$, the bands touch the

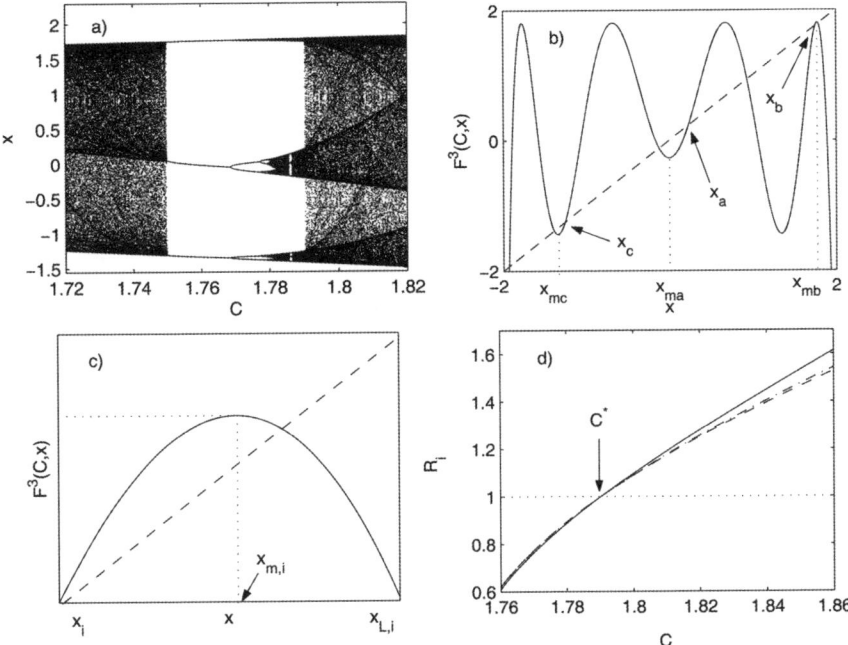

Fig. 17. (a) Bifurcation diagram showing the interior crisis for the quadratic map, (b) Plot of $F^3(C, x)$ as a function of x for $C \approx C^*$, where the unstable orbit x_a, x_b, x_c is marked, (c) Scheme of each of the three copies of a unimodal map present in $F^3(C, x)$, responsible for the three bands observed in the bifurcation diagram, (d) Plot of the numerical calculations of the three ratios R_i, $i \in \{a, b, c\}$, which are smaller than one for $C < C^*$ and bigger than one for $C > C^*$.

unstable period three orbit that lies in its basin of attraction and they disappear. As a consequence, a one-piece chaotic attractor is obtained. This process is summarized in the bifurcation diagram shown in Fig. 17(a).

We are interested in the form of $F^3(C, x) \equiv F(C, F(C, F(C, x)))$ close to the crisis. It is depicted in Fig. 17(b) for $C \approx C^*$, together with the three points of the unstable period-3 orbit involved in the crisis x_i which verify $F^3(C, x_i) = x_i$, where the subindex i will be a, b or c throughout this discussion. If we make a zoom of $F^3(C, x)$ close to each of the unstable periodic orbits we can see that we have three small unimodal maps (inverted or reflected, but equivalent), whose structure is sketched in Fig. 17(c). Each of them is associated with one of the three chaotic bands and they can be essentially characterized by the unstable orbit x_i, the limit point $x_{L,i}$ which verifies $F^3(C, x_{L,i}) = x_i$ (see Fig. 17(c)), and the maximum $x_{m,i}$, for which

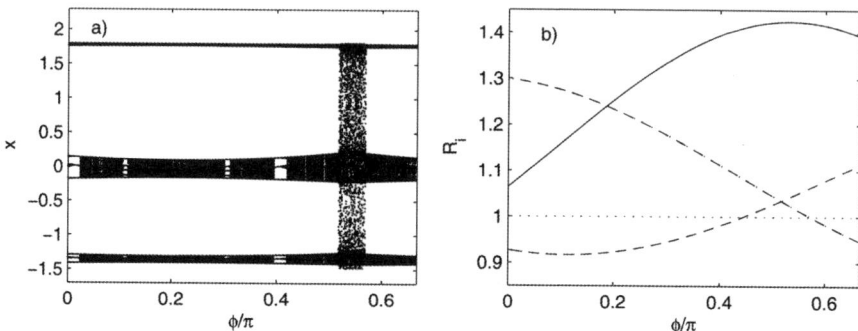

Fig. 18. (a) Bifurcation diagram of $G(x)$ for $C = 1.8$, $\epsilon = 0.005$ as a function of ϕ, showing that the value of the phase influences greatly in whether there is or not intermittency, (b) Graphs of our approximation of the perturbed ratios R'_i as a function of ϕ: the ϕ interval where the intermittency is observed corresponds roughly to the interval for which $R'_i > 1$, as claimed.

$\frac{dF^3(C,x)}{dx}|_{x_{m,i}} = 0$. The values of $x_{m,i}$, that can be calculated analytically, and the values of $x_i, x_{L,i}$, that can be calculated using the Newton-Raphson method [20], will depend on C. The key idea that we want to illustrate is that the existence of the three bands for $C \leq C^*$ and their disappearance at $C > C^*$ can be interpreted in terms of the evolution of the three unimodal maps that are present in $F^3(C,x)$. In fact, we claim that the bands will exist if these copies of a unimodal map present in $F^3(C,x)$ "trap" the orbits passing by $[x_i, x_{L,i}]$ (or $[x_{L,i}, x_i]$) under iterations of $F^3(C,x)$. Whether those three copies are "trapping" or not can be characterized by the three following ratios:

$$R_i(C) = \frac{|F^3(C, x_{m,i}) - x_i|}{|x_i - x_{L,i}|}, i \in \{a, b, c\}. \quad (15)$$

It is easy to see that the bands will exist if $R_i \leq 1$, and they will disappear for $R_i > 1$. Thus, we claim that the ratios $R_i(C)$ will go from values smaller than 1 for $C < C^*$, before the crisis, to values larger than 1 for $C > C^*$, after the crisis. Numerical calculations of these ratios R_i as functions of C have been carried out and are shown in Fig. 17(d). Notice that the value of C^* obtained in this way corresponds to the one obtained in [9, 10].

Bearing this in mind, we can now focus on the role of the phase ϕ when applying a harmonic perturbation to the system given by Eq. (14). Consider the perturbed map

$$x_{n+1} = C_n - x_n^2 = F(C_n, x_n), \quad (16)$$

where
$$C_n = C(1 + \epsilon \sin(2\pi r n + \phi)). \tag{17}$$

We will assume $\epsilon \ll 1$. In analogy with the laser, it is easy to see that the $r = 1$ case is quite trivial, because, by varying the phase, we are just moving in C in the interval $[C(1-\epsilon), C(1+\epsilon)]$.

Instead, in the $r = 1/3$ case the role of ϕ is far from being trivial. In this situation, the global dynamics will be governed by the autonomous map:

$$x_{n+3} = F(C_2, F(C_1, F(C_0, x_n))) \equiv G(x_n). \tag{18}$$

The smallness of ϵ will make $G(x_n)$ be quite similar to $F^3(C, x_n)$. Thus, if $C \approx C^*$, $G(x)$ will also contain three small unimodal maps, and our aim is to see how the perturbed ratios R'_i of each of them varies with ϕ, where i again can be a, b or c. First, it is straightforward to verify that the maxima of the perturbed system, $x'_{m,i}$, exist for ϵ sufficiently small, and can be written as,

$$x'_{m,a} = 0,$$
$$x'_{m,b} = \sqrt{C_1 + \sqrt{C_0}},$$
$$x'_{m,c} = -\sqrt{C_0}.$$

Thus,
$$G(x'_{m,a}) = C_2 - (C_1 - C_0^2)^2,$$
$$G(x'_{m,b}) = C_2,$$
$$G(x'_{m,c}) = C_2 - C_1^2.$$

The values of the perturbed periodic orbit x'_i, verifying $G(x'_i) = x'_i$, can be approximated starting from the unperturbed ones (x_i) by means of the Implicit Function Theorem (IFT) [11]. It is relatively simple to verify using this theorem that the order ϵ approximation is given by:

$$x'_i \approx x_i - \epsilon \frac{C \sin(\phi)}{2x_i}. \tag{19}$$

Analogously, we can calculate the values of the perturbed $x'_{L,i}$ from the unperturbed $x_{L,i}$ by making use of the IFT. It can be shown quite easily that:

$$x'_{L,i} \approx x_{L,i} - \epsilon \frac{C \sin(\phi)}{2x_{L,i}} \left(1 - \frac{1}{8F^2(C, x_{L,i}) F(C, x_{L,i})}\right), \tag{20}$$

for $i \in \{a, b, c\}$. With the exact expressions of obtained for $G(x'_{m,i})$ and the approximations of x'_i and $x_{L,i}$ shown above we can then approximate the new perturbed ratios:

$$R'_i(C, \epsilon, \phi) = \frac{|G(x'_{m,i}) - x'_i|}{|x'_i - x'_{L,i}|}. \tag{21}$$

Fixing ϵ and C, the three ratios will clearly depend on ϕ, so $R'_i = R'_i(\phi)$. We can see a bifurcation diagram and a plot of our approximations of $R'_i(\phi)$ in Fig. 18 for $C = 1.8$ (the unperturbed system displays intermittency) and $\epsilon = 0.005$. We can observe in the bifurcation diagram of Fig. 18(a) that the value of ϕ determines wether the system displays intermittency or not. On the other hand, we can see in Fig. 18(b) that the interval of ϕ for which there is intermittency corresponds roughly with the small interval around $\phi = \pi/2$ for which our approximations of the three ratios are bigger than one. In this region none of the copies of the unimodal map present in $G(x)$ are "trapping", so the orbits are not restricted to any band and there is intermittency. Thus, our calculations provide a way to understand the nontrivial role of ϕ in the global dynamics of the system that matches with the numerical simulations.

Summarizing, we have seen that varying ϕ modulates both the geometry of the system and the position of the periodic orbit involved in the interior crisis. Thus, when we are close to the crisis, its contribution becomes crucial.

4. Phase Control of Escapes in Open Dynamical Systems

4.1. *Control of open dynamical systems*

In this last section, we are going to show that phase control can be used to avoid escapes in an open dynamical system. Open dynamical systems are typical in nature. For systems of this kind, there is a region in phase space where nearly all the trajectories diverge asymptotically to infinity. They have attracted a great deal of attention in the context of transient chaos [25] and, particularly, in chaotic scattering problems [26], among others. In order to define what is an escape we can imagine the following scenario. We suppose that a particle is under the influence of some potential or massive object. Under this situation, we say that a dynamical system has an *escape* whenever this particle crosses a certain boundary and never comes back. As the reader might notice, this type of systems is related with the systems described in the former section. We shall see now that here the phase control technique can also be applied succesfully.

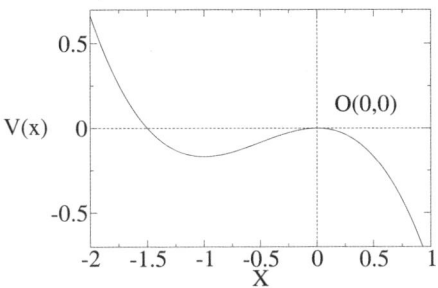

Fig. 19. Plot of the cubic potential $V(x) = -\frac{x^2}{2} - \frac{x^3}{3}$.

4.2. *Model description*

A paradigmatic example of a dynamical system with escapes is the Helmholtz oscillator, the model in which we will focus our attention. It provides probably the simplest way to model physical phenomena that present the ability to escape from a potential well. This nonlinear oscillator describes the motion of a unit mass particle in a cubic potential $V(x) = ax^2/2 + bx^3/3$, which eventually can be externally perturbed by a sinusoidal driving. By adding a linear dissipative force, the equation of motion is

$$\ddot{x} + \mu\dot{x} + ax + bx^2 = F\cos(\omega t), \tag{22}$$

where μ represents the damping coefficient, F the forcing amplitude, ω the forcing frequency and a and b determine the shape of the potential.

We fix the parameters all throughout this section to be $\mu = 0.1$, $\omega = 1$ and $a = b = -1$, for which the potential reads

$$V(x) = -\frac{x^2}{2} - \frac{x^3}{3}. \tag{23}$$

This potential has a maximum at $x = 0$ and a minimum at $x = -1$ as shown in Fig. 19. For this choice of parameter values the equation of motion is

$$\ddot{x} + 0.1\dot{x} - x - x^2 = F\cos t. \tag{24}$$

Note that the only free parameter is the forcing amplitude F. This simple system has been studied previously in several works. For instance, a thorough analysis about its dynamics can be found in Ref. [4] and work on

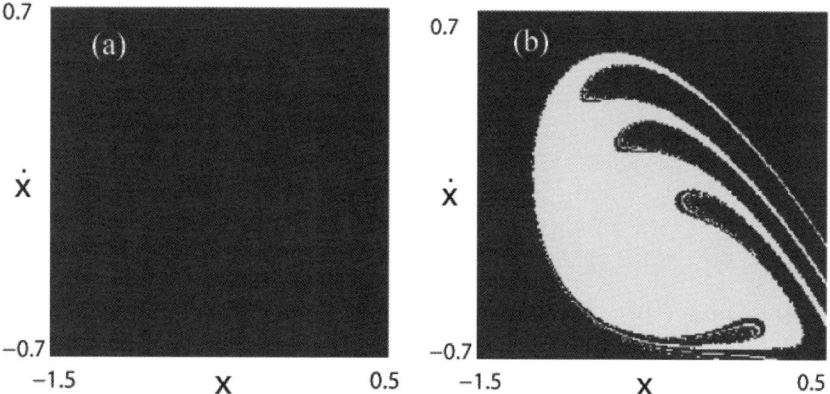

Fig. 20. (a) Basin of attraction of the Helmholtz oscillator, $\ddot{x}+0.1\dot{x}-x-x^2 = 0.21\cos t$. We denote black dots the set of points that escape from the potential well, and pale gray dots as the points that fall into the attractor(s). Note that in this picture all initial conditions escape after some period of time, (b) Basin of attraction of the Helmholtz oscillator, $\ddot{x} + 0.1\dot{x} - x - x^2 = 0.12\cos t$. Here the basins in phase space has a fractal structure where pale gray points denote the set of points falling into the attractors.

the integrability and symmetry-breaking of this oscillator is presented in Refs. [4, 6].

This system presents different behaviors depending on the value of the forcing amplitude F. For example, we can see a plot of the basins of attraction for this system for $F = 0.12$ and $F = 0.21$ in Fig. 20. In Fig. 20(b), which corresponds to the basin of attraction for $F = 0.12$, a bounded attractor (inside the cyan (pale gray) region) coexists with escaping orbits, (blue (cyan) dots). The attractor corresponds to a bounded orbit, that can be seen in Fig. 21. However, if we take an initial condition from any point of the basin of attraction of Fig. 20(a), the resulting trajectory diverges to infinity, or simply it *escapes*. In general, a trajectory escapes from the well when it crosses with positive velocity the maximum of the potential situated at $x = 0$ (see Fig. 19) and never comes back. The time that a certain particle spends from the well crossing the maximum of the potential is called *escape time T*.

The basins of attraction for $F = 0.21$, shown in Fig. 20(a), shows that for this value of the forcing all trajectories escape. This is the situation that we want to control, and this is the value of the forcing F that we consider in the remaining of this section. Our main goal here is to avoid escapes for the largest number of initial conditions using the phase control method.

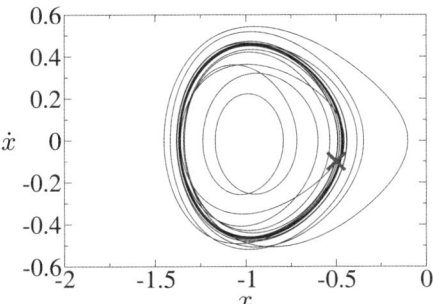

Fig. 21. This figure shows the phase space trajectory of the Helmholtz oscillator, $\ddot{x} + 0.1\dot{x} - x - x^2 = 0.12\cos t$, with an initial condition at the point $(x_0, \dot{x}_0) = (-0.5, -0.1)$, indicated by a cross. The dark color indicates the attractor.

We can now focus on how to implement the phase control scheme. Suppose that we have particles oscillating inside a potential well and an external forcing is acting. These particles can remain inside the well or escape from it depending on the amplitude forcing. A suitable way to modify the dynamical behavior of the particles inside the well is by modifying the shape of the potential well. Following this reasoning, we introduce, as in the former sections, a parametric perturbation. In this case, this perturbation is applied to the quadratic term of the equation of motion

$$\ddot{x} + 0.1\dot{x} - x - (1 + \epsilon\cos(t + \phi))x^2 = F\cos(\omega t), \tag{25}$$

where ϵ is the modulation amplitude and ϕ is the phase difference with the forcing that we simply call *the phase*. Note that we are using for the parametric perturbation a resonant frequency with the forcing amplitude, which is a common assumption for this type of nonfeedback control methods. The effects of a frequency mismatch, that might appear in some experimental situations, will be discussed in Sec. 4.4.

The modulation term $(1 + \epsilon\cos(t + \phi))$ can be interpreted as a modulation of the potential of the system, that can be rewritten as $V_{pert}(x,t) = -x^2/2 - (1 + \epsilon\cos(t+\phi))x^3/3$. In fact, this perturbed potential has a maximum on $x = 0$, for which $V_{pert}(0,t) = 0$. The potential is also zero for $x_{zero}(t) = -\frac{3}{2(1+\epsilon\cos(t+\phi))}$, so the width of the potential is $|\Delta x_{zero}(t)| = \frac{3}{2(1+\epsilon\cos(t+\phi))}$. This perturbed potential presents an oscillating minimum on $x_{min}(t) = -\frac{1}{1+\epsilon\cos(t+\phi)}$, for which the value of this perturbed potential is $\Delta V_{min}(t) = -\frac{1}{6(1+\epsilon\cos(t+\phi))^2}$, so it oscillates around the unperturbed value, $V(-1) = -1/6$.

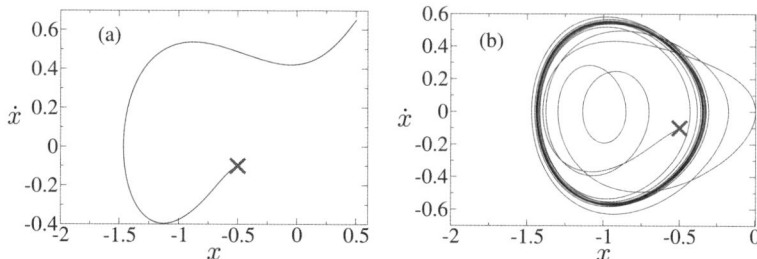

Fig. 22. (a) Single trajectory for the Helmholtz oscillator, $\ddot{x}+0.1\dot{x}-x-x^2=0.21\cos t$, with initial condition at the point $(x_0,\dot{x}_0)=(-0.5,-0.1)$ as indicated by the cross. The particle escapes after a lapse of time, (b) Single trajectory (as in Fig. 22(a)) for the Helmholtz oscillator with control, $\ddot{x}+0.1\dot{x}-x-(1+0.05\cos(t+\pi))x^2=0.21\cos t$. The perturbation keeps the particle in the well forever.

4.3. Numerical simulations and heuristic arguments

In this subsection we provide numerical evidence showing that by using an adequate value of ϕ and ϵ, we can avoid escapes for this system. A first numerical evidence about the effect of the control in the system can be observed in Fig. 22. Figure 22(a) shows a typical escaping trajectory for $F = 0.21$ with initial condition at $(x_0, \dot{x}_0) = (-0.5, -0.1)$, in absence of control. If we introduce the control, say $\epsilon = 0.05$, and a value of phase, $\phi = \pi$, the particle does not escape from the well, as is observed in Fig. 22(b).

We have explored this phenomenon in detail numerically. To do this, we have performed a numerical integration of trajectories whose initial conditions belong to a 60×60 grid in the phase space region, $x \in [-1.5, 0.5]$, $\dot{x} \in [-0.7, 0.7]$, for different combinations of ϵ and ϕ, and observed which of them escape. In the diagrams plotted in Figs. 23(a) to 23(c) the rate of particles that do not escape as a function of ϵ and ϕ are shown. Note that in some regions of these diagrams, for example for $\epsilon \approx 0.1$ and $\phi = \pi$, more than 50% of the particles are kept bounded. This is quite surprising since a value of ϵ of this order has a very small effect on the shape of the well. However, if we take another value of the phase ϕ, such as $\phi = 0$, nearly all trajectories escape. Thus the role of the phase ϕ is crucial if we want to keep the trajectories bounded.

This modulation can be quantified as follows. Assuming that if we set $\cos(t + \phi) = 1$, the depth of the well becomes $\Delta V = \frac{1}{6(1+\epsilon)^2}$ and the width $\Delta x = \frac{3}{2(1+\epsilon)}$. Note that the width of the potential (see Fig. 19 is the distance between the local maximum, x_{max}, and the point $x_{min} = -\frac{3}{2(1+\epsilon)}$. The effect of the modulation in these quantities is summarized in Table 1.

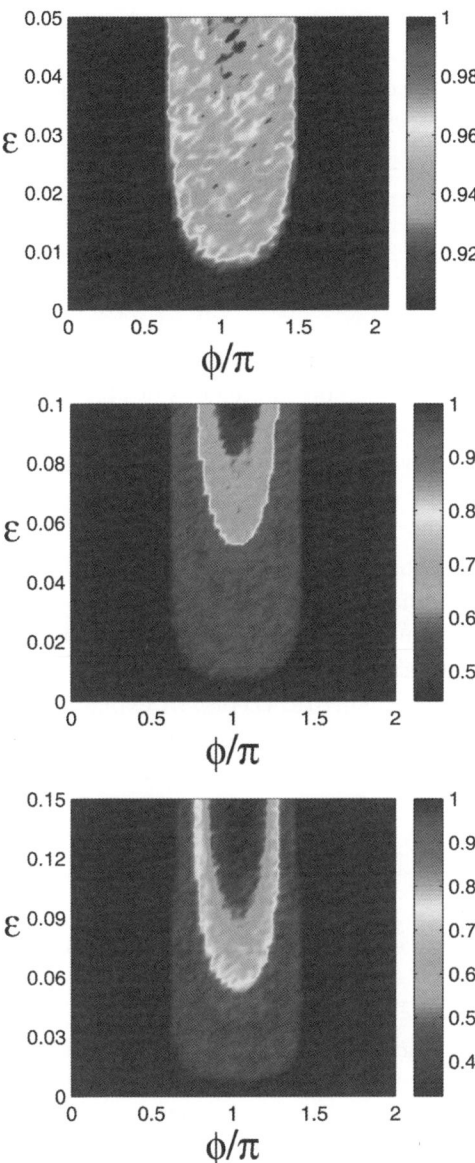

Fig. 23. A grayscale plot of the fraction of trajectories with initial conditions in the region $[-1.5, 0] \times [-0.7, 0.7]$ that do not escape to infinity for different values of ϵ and ϕ. Note that an adequate choice of the phase ϕ is critical if we are interested in avoiding escapes, and that apparently the optimal values of the phase for this purpose are located around $\phi = \pi$.

Table 1. Variation of the width, Δx, and depth, ΔV, of the well in function of ϵ. The percentages $\%\Delta x$ and $\%\Delta V$ indicate the percentage variation in the width and the depth of the potential well for the chosen values of ϵ.

ϵ	Δx	ΔV	$\%\Delta x$	$\%\Delta V$
0	1.50	0.16	0	0
0.05	1.43	0.15	4.5	9.5
0.10	1.36	0.14	8.8	15
0.15	1.31	0.13	11.6	21

We observe that by changing only a little bit the shape of the potential we can control escapes in some specific regions of phase space.

The typical basins of attraction of the Helmholtz oscillator after having applied the phase control for $F = 0.21$, $\phi = \pi$ and modulation amplitudes $\epsilon = 0.05$, $\epsilon = 0.1$, and $\epsilon = 0.15$, respectively, are plotted in Fig. 24. Observe the strong effects of this term, ϵ, and the phase, ϕ, on the basin of attraction of bounded orbits, whose area grows drastically with ϵ once we choose a suitable value of the phase $\phi = \pi$.

We have shown numerically that the harmonic modulation of the potential well with a suitable value of the phase ϕ are crucial to avoid escapes now. Now we provide a heuristic theory to explain the role of the phase ϕ in the control of escapes. We have seen that the effect of the perturbation in this system is analogous to applying a perturbation on the potential. In fact, the width of the perturbed potential well oscillates in time as $|\Delta x_{zero}(t)| = \frac{3}{2(1+\epsilon \cos(t+\phi))}$, and the depth of this perturbed potential well also oscillates around the unperturbed value, according to the expression $\Delta V_{min}(t) = -\frac{1}{6(1+\epsilon \cos(t+\phi))^2}$.

The maximum value of the forcing in the positive direction of the x-axis corresponds to $t = 0$, for which the value of the forcing is $F(t) = F_{max} \cos \omega t$, being $F = F_{max}$. For this value $F = F_{max}$, the deeper the potential well is, the more difficult it is for the particles to escape. It takes place when $\phi = \pi$ for which the minimum of the potential reaches its minimum value and the depth of the potential is maximum, $|\Delta V| = \frac{1}{6(1-\epsilon)^2}$. Furthermore, for this situation, the width of the potential reaches its maximum value $|\Delta x_{zero}| = \frac{3}{2(1-\epsilon)}$ and it makes as a consequence the possibility to escape more difficult. Summarizing, these simple arguments in which $\phi = \pi$ show that the potential well reaches the maximum values of both, depth and width, becoming more difficult for the particles to escape from the well and even keeping them inside it and they can never escape.

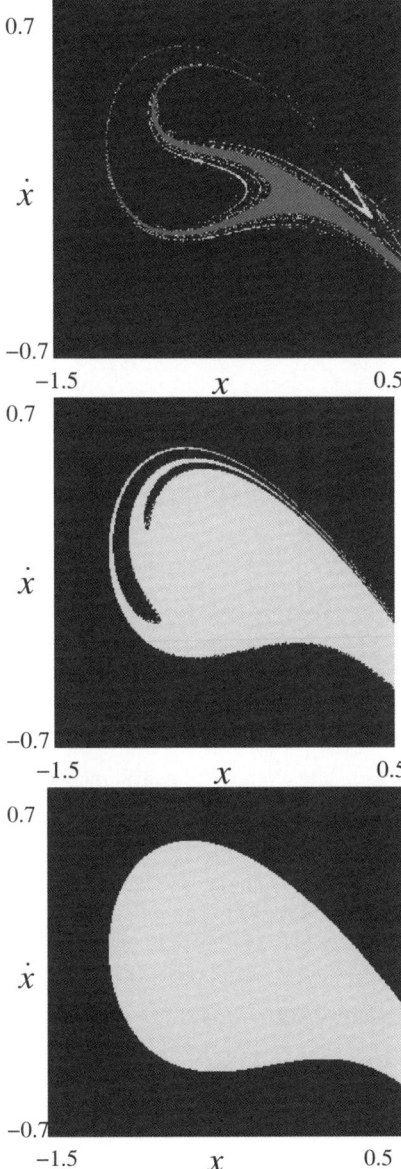

Fig. 24. Basins of attraction of the perturbed Helmholtz oscillator, $\ddot{x} + 0.1\dot{x} - x - (1 + \epsilon\cos(t + \pi))x^2 = 0.21\cos t$, with modulation amplitudes $\epsilon = 0.05$ (top left), $\epsilon = 0.1$ (top right), and $\epsilon = 0.15$ (bottom), respectively. Black dots denote the points that escape from the potential well and pale gray dots the points that fall into the attractor(s).

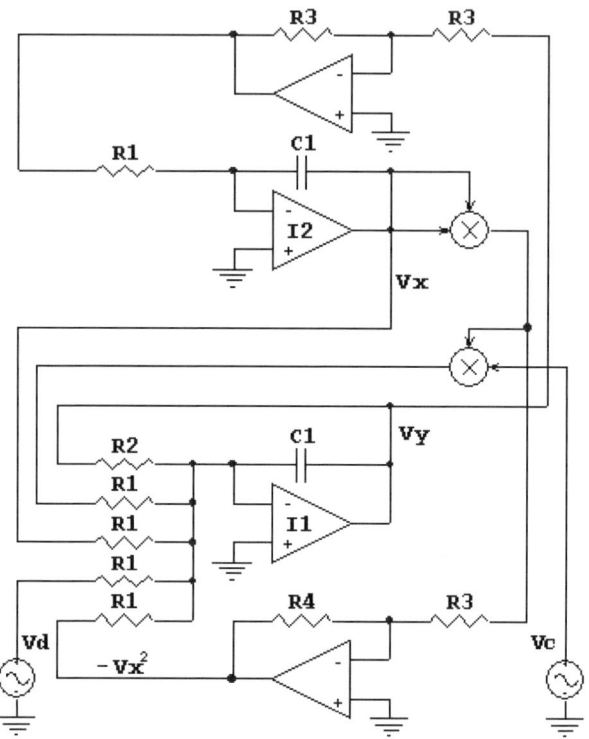

Fig. 25. Lay-out of the electronic circuit. I: integrators; R: resistors; C: capacitors; X: multipliers; V_d: sinusoidal driving signal; V_c: sinusoidal control signal. Under a suitable time normalization, the dynamics is ruled by the equation $\ddot{x} + 0.1\dot{x} - x - (1 + \epsilon \cos(t + \phi))x^2 = 0.21 \cos t$, where $x \propto V_x$ and $\dot{x} \propto V_y$. The parameters ϵ, ϕ are fixed by the arbitrary wave form generator V_c. The numerical values are: $R_1 = 100\ K\Omega$, $R_2 = 1\ M\Omega$, $R_3 = 2\ K\Omega$, $R_4 = 5\ K\Omega$ and $C_1 = 10\ nF$.

4.4. *Experimental implementation in an electronic circuit*

As we did in the two previous Sections, in this section we also provide the main features of a test of our control technique in a laboratory system. As we shall see, it confirms what we have observed numerically, i.e., that an adequate value of the phase difference ϕ between the main driving and the controlling perturbation can avoid escapes in the system.

With this purpose we have designed and built the electronic circuit sketched in Fig. 25. This circuit mimics the dynamics of the Helmholtz oscillator. A similar circuit was used in Ref. [7]. Our circuit consists of an electronic analog simulator implemented using commercial semiconductor devices. V_d is the driving voltage amplitude, applied by means of the generator G_d, while V_c is the control voltage amplitude applied by means of an arbitrary function generator TABOR 8024. The two sinusoidal signals, V_d and V_c, have the same frequency and the value of the phase ϕ is fixed for $t = 0$, so it is constant in the experiment. Thus, following the main idea of the phase control scheme, once the phase ϕ is chosen no further adjustment of its value needs to be done during the experiment. The integrators I_1 and I_2 have been implemented using Linear Technology LT1114CN four quadrant operational amplifiers, while the multipliers are Analog Devices MLT04. The acquisition of the experimental data has been performed by means of TEKTRONIX TDS 7104 digital oscilloscope connected to a personal computer. Under suitable time scale normalization the dynamics of this circuit is governed by Eq. (25), where the parameters are $\delta = 0.1$, $F = 0.2$, and $\omega = 1$, and the amplitude of the applied control is $\epsilon \approx 0.03$. The voltages of the circuit V_x and V_y can be associated to x and $\dot{x}$, respectively.

Different trajectories in phase space observed experimentally, corresponding to different values of the main driving amplitude V_d, are shown in Fig. 26. The chaotic attractor that can be observed for sufficiently low values of V_d is shown in Fig. 26(a). A diverging trajectory can be seen in Fig. 26(b). Finally, a periodic orbit observed for smaller values of the driving amplitude V_d smaller than the one leading to chaotic motion is shown in Fig. 26(c).

In order to test the validity of the control technique, we have performed different bifurcation diagrams of the system for different values of the amplitude of the perturbation ϵ and for a fixed value of the phase ϕ. In these bifurcation diagrams we capture the dynamics of the system for different values of the amplitude of the main forcing V_d. These diagrams have been performed by slowly varying the value of the forcing and by reporting the maxima of the resulting long time series. This way we can see the behavior of the system for different values of the forcing amplitude. In Fig. 27(a) we can see the experimental bifurcation diagram in absence of any external perturbations ($\epsilon = 0$). This diagram clearly shows a transition to chaos through a period-doubling cascade, after which there is a boundary crisis by which the chaotic attractor disappears and the particles escape to the infinity. The situation changes if we apply a perturbation with the suitable fixed value of the phase $\phi = \pi$. The resulting bifurcation diagram, that is shown in Fig. 27(b), shows that the boundary crisis and the divergence of the trajectories takes place for a value of the driving amplitude that is

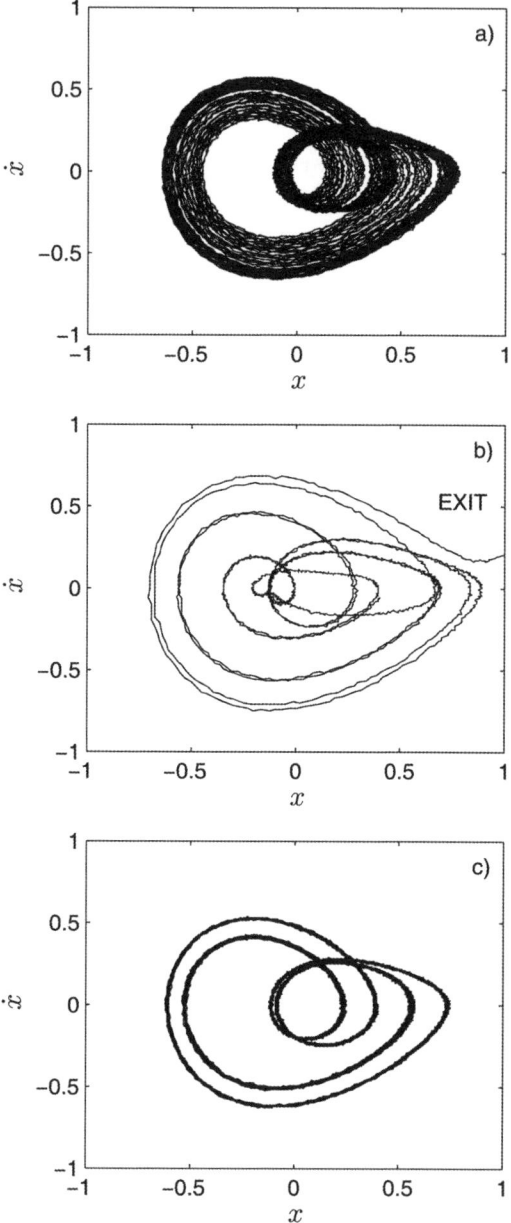

Fig. 26. Single trajectories in the experimental electronic circuit simulating the Helmholtz oscillator where we can find different dynamics. (a) Chaotic attractor, (b) Escaping trajectory, (c) Periodic orbit.

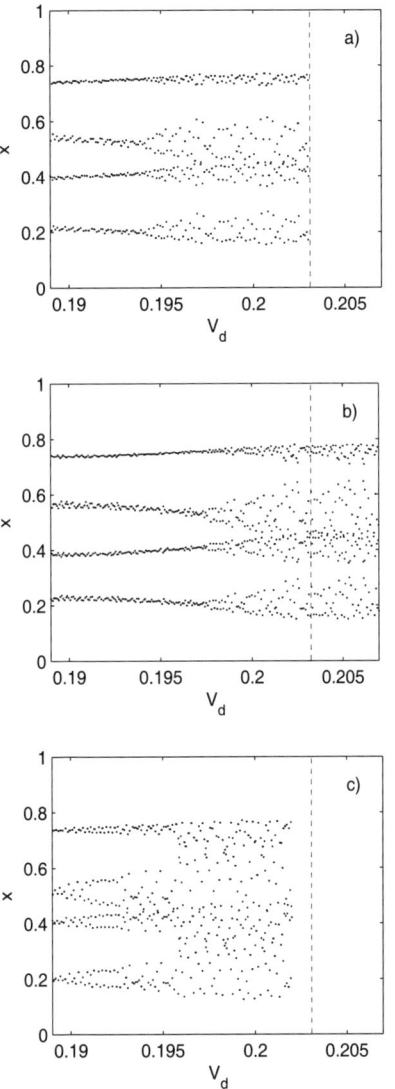

Fig. 27. Experimental bifurcation diagrams obtained from the circuit. These diagrams represent the variable x against the driving voltage V_d. We can observe different boundary crisis: (a) without control, (b) for $\epsilon = 0.05$ and $\phi = \pi$, (c) for $\epsilon = 0.05$ and $\phi = 0$. In every figure we can test that the modulation amplitude, ϵ, and an accurate value of the phase $\phi = \pi$ are fundamental for the control of escapes. Observe that in all figures dashed lines denote, in absence of control, the point in the bifurcation diagram in which the boundary crisis would take place.

higher than in the unperturbed case. This implies that trajectories that would typically diverge to infinity are kept bounded for this value of ϕ.

Thus, our experiment confirms that an adequate choice of the phase ϕ becomes crucial as well in this experiment. In Figure 27(c) we can see that with fixed value of the phase $\phi = 0$ the attractor is destroyed for a value of the main driving V_d sensibly smaller than the one reported in the unperturbed case. Thus, we can see once more that the value of ϕ plays a key role in the global dynamics of the system.

From an experimental point of view, it is interesting to consider which would be the effect of a mismatch between the frequency of the main driving and the frequency of the controlling perturbation, a situation that may arise in some experimental implementations. Considering that mismatch, the controlling perturbation applied to the system can be written as $\epsilon \cos((\omega + \delta\omega)t + \phi_{opt})$, where $\delta\omega << 1$ would be the value of the frequency mismatch. This perturbation can be rewritten as $\epsilon \cos(\omega t + \phi(t))$, where $\phi(t) \equiv \phi_{opt} + \delta\omega t$ is the new phase difference, that slowly varies in time. However, as long as the phase $\phi(t)$ remains in the interval for which trajectories remain bounded (detected numerically in Sec. 4.3) divergences are not expected, and the smaller $\delta\omega$ is the longer our phase control method will keep the trajectories bounded. Thus, in presence of a frequency mismatch our scheme allows to keep trajectories bounded for a period of time that scales as $1/\delta\omega$. This feature is shared by most control methods based on the application of a small harmonic perturbation to the system.

5. Conclusions and Discussions

In this Chapter we have made a thorough exploration of the applications of the phase control technique. Firstly, we have focused on its applications to control chaos in the paradigmatic Duffing oscillator. We have performed numerical simulations confirming the most important properties of this method: that only a correct choice of ϕ can lead the system to a periodic orbit and that, by adequately selecting the phase, the necessary amplitude to suppress chaos can be minimized. By using and extensive exploration of parameter space in search of zones of chaos suppression for different values of ϵ and ϕ, we have detected some interesting patterns. Most of the interesting patterns found numerically have been recovered in an experiment with an electronic circuit that mimics the dynamics of the Duffing oscillator with a slight potential asymmetry, even in the presence of noise. This fact suggests that phase control of chaos is robust even in the presence of distortions of the potential symmetry and that all the properties are of a quiet general nature, so they must be taken into account when applying this control method to the most diverse dynamical systems.

On the other hand, we have shown that the phase control scheme is able to control the intermittency in a chaotic system close to an interior crisis. First, we have shown both experimentally and numerically how, if we apply a harmonic perturbation to a chaotic CO_2 laser, it is possible to control the crisis-induced intermittency by accurately choosing ϕ. We have seen that this scheme is more effective when the frequency of the perturbation is equal to either the frequency of the unstable periodic orbit involved in the crisis or the frequency involved in the period doubling bifurcation, which can be obtained from experimental time series by using the Fourier transform. Furthermore, with the aid of the simple quadratic map, we have illustrated the nontrivial role that the phase difference ϕ plays when applying a harmonic perturbation to a dynamical system. Our analysis shows that the application of a periodic modulation to a system close to an interior crisis perturbs its geometry, and such perturbation depends strongly on ϕ, which becomes a key parameter for the global dynamics.

Concerning to the application to open dynamical systems, we have shown, by using as prototype model the Helmholtz oscillator, that an adequate parametric perturbation in the quadratic term of the equation of motion of this oscillator can avoid escapes in some regions in phase space. We provide numerical support and heuristic arguments for which we control the orbits in the well, avoiding the escapes from it by simply changing slightly the depth and width of the well and by using a suitable value of the phase ϕ. We have shown the robustness and the general nature of this method in the sense that the experimental implementation of a circuit confirms the same results. In the context of physical situations, the problems with escapes are typical in chaotic scattering problems, which have applications in many fields in physics. One important advantage of this scheme is its non-feedback nature. Furthermore, the key role of the phase in selecting the final dynamical state is very useful from a control point of view, since there is a large variety of situations in which the modulation of the accessible parameters might be limited, and ϕ is an additional degree of freedom that may be very useful.

In summary, we have shown that the phase control scheme is very versatile and useful in a wide variety of dynamical situations: to suppress chaos, to control the intermittency in chaotic systems close to a crisis, to avoid escapes in open dynamical systems.

Acknowledgments

The basins of attraction (Figs. 20 and 24) have been made using the software DYNAMICS [3]. This work was supported by the Spanish Ministry of Science and Technology under project number BFM2003-03081, by the Spanish Ministry of Education and Science under project number FIS2006-08525 and by Universidad Rey Juan Carlos and Comunidad de Madrid under projects number URJC-CM-2006-CET-0643 and URJC-CM-2007-CET-1601. R. Meucci also acknowledges financial support from Fondazione Ente Cassa di Risparmio di Firenze. Jesús M. Seoane and Samuel Zambrano acknowledge warm hospitality received at CNR-INOA where part of this work was carried out. The authors also acknowledge S. Euzzor (CNR-INOA) for technical assistance and competence in performing analog simulations.

References

[1] J. Aguirre, F. d'Ovidio, and M. A. F. Sanjuán. Controlling chaotic transients: Yorke's game of survival. *Phys. Rev. E*, 69:016203, 2004.
[2] K. T. Alligood, Tim D. Sauer, and J. A. Yorke. *Chaos, an introduction to dynamical systems*. Springer-Verlag, New York, 1996.
[3] Kathleen T. Alligood, Tim D. Sauer, and J. A. Yorke. *Dynamics: Numerical Exploration*. Springer, New York, 1994.
[4] J.A. Almendral, J. M. Seoane, and M. A. F. Sanjuán. The nonlinear dynamics of the Helmholtz oscillator. *Recent Res. Devel. Sound & Vibration*, 2:115, 2004.
[5] S. Boccaletti, C. Grebogi, Y. C. Lai, H. Mancini, and D. Maza. Control of chaos: theory and applications. *Phys. Rep.*, 329:103, 2000.
[6] H. Cao, J. M. Seoane, and M. A. F. Sanjuán. Symmetry-breaking analysis for the general Helmholtz-Duffing oscillator. *Chaos, Solitons and Fractals*, 34:197–212, 2007.
[7] E. del Río, A. R. Lozano, and M. G. Velarde. A prototype Helmholtz–Thompson nonlinear oscillator. *Rev.Sci. Instrum.*, 63:4208, 1992.
[8] C. Grebogi, E. Ott, F. Romeiras, and J. A. Yorke. Critical exponents for crisis induced intermittency. *Phys. Rev. A*, 36:5365–5380, 1987.
[9] C. Grebogi, E. Ott, and J. A. Yorke. Chaotic attractors in crisis. *Phys. Rev. Lett.*, 48:1507–1510, 1982.
[10] C. Grebogi, E. Ott, and J. A. Yorke. Crisis, sudden changes in chaotic attractors, and transient chaos. *Physica D*, 7:181–200, 1983.
[11] J. Hale and H. Koçak. *Dynamics and Bifurcations*. Springer-Verlag, Amsterdam, 1991.
[12] R. Lima and M. Pettini. Suppression of chaos by resonant parametric perturbations. *Phys. Rev. A*, 41:726, 1990.

[13] A. Y. Loskutov and A. I. Shishmarev. Control of dynamical systems behavior by parametric perturbations: An analytic approach. *Chaos*, 2:391–395, 1994.

[14] I. P. Marino, E. Allaria, M. A. F. Sanjuán, R. Meucci, and F. T. Arecchi. Coupling scheme for complete synchronization of periodically forced CO_2 lasers. *Phys. Rev. E*, 70:03628, 2004.

[15] R. Meucci, E. Allaria, F. Salvadori, and F. T. Arecchi. Attractor selection in chaotic dynamics. *Phys. Rev. Lett.*, 95:184101, 2005.

[16] R. Meucci, D. Cinotti, E. Allaria, L. Billings, I. Triandaf, D. Morgan, and I. B. Schwartz. Global manifold control in a driven laser: Sustaining chaos and regular dynamics. *Physica D*, 189:70–80, 2004.

[17] R. Meucci, W. Gadomski, M. Ciofini, and F. T. Arecchi. Experimental control of chaos by means of weak parametric perturbations. *Phys. Rev. E*, 49:R2528, 1994.

[18] K. A. Mirus and J. C. Sprott. Controlling chaos in low- and high-dimensional systems with periodic parametric perturbations. *Phys. Rev. E*, 59:5313, 1999.

[19] E. Ott, C. Grebogi, and J. A. Yorke. Controlling chaos. *Phys. Rev. Lett.*, 64:1196, 1990.

[20] W. H. Press, S. A. Teukolsky, W. T. Vetterling, and B. P. Flannery. *Numerical Recipes in C*. Cambridge University Press, Cambridge, 1988.

[21] Z. Qu, G. Hu, G. Yang, and G. Quin. Phase effect in taming nonautonomous chaos by weak harmonic perturbations. *Phys. Rev. Lett.*, 74:1736, 1995.

[22] J. M. Seoane, S. Zambrano, S. Euzzor, R. Meucci, F. T. Arecchi, and M. A. F. Sanjuán. Avoiding escapes in open dynamical systems using phase control. *Phys. Rev. E*, 78:016205, 2008.

[23] T. Shinbrot, C. Grebogi, E. Ott, and J. A. Yorke. Using small perturbations to control chaos. *Nature*, 363:411, 1993.

[24] G. Tanaka, M. A. F. Sanjuán, and K. Aihara. Crisis-induced intermittency in two coupled chaotic maps: Towards understanding chaotic itinerancy. *Phys. Rev. E*, 71:016219, 2005.

[25] T. Tél. *Directions in Chaos*. Edited by Bai-lin Hao, World Scientific, Singapore, 1990.

[26] T. Tél and E. Ott. Chaotic scattering: An introduction. *Chaos*, 3:417–425, 1993.

[27] J. Yang, Z. Qu, and G. Hu. Duffing equation with two periodic forcings: The phase effect. *Phys. Rev. E*, 53:4402, 1996.

[28] S. Zambrano, S. Brugioni, E. Allaria, I. Leyva, M. A. F. Sanjuán, R. Meucci, and F. T. Arecchi. Numerical and experimental exploration of phase control of chaos. *Chaos*, 16:013111, 2006.

[29] S. Zambrano, I. P. Marino, F. Salvadori, R. Meucci, M. A. F. Sanjuán, and F. T. Arecchi. Phase control of the intermittency in dynamical systems. *Phys. Rev. E*, 74:016202, 2006.

Chapter 7

RECENT ADVANCES IN CONTROL OF COMPLEX DYNAMICS IN MECHANICAL AND STRUCTURAL SYSTEMS

Giuseppe Rega

Dip. Ingegneria Strutturale e Geotecnica,
Sapienza Università di Roma,
via A. Gramsci 53, 00197, Roma, Italy
Giuseppe.Rega@uniroma1.it

Stefano Lenci

Dip. Architettura, Costruzioni e Strutture,
Università Politecnica delle Marche,
via Brecce Bianche, 60131, Ancona, Italy
Lenci@univpm.it

In the framework of an extended notion of control of chaos, recent research achievements on control of complex dynamics in the area of solid and structural mechanics are reviewed, by referring to some main classes of mechanical/structural systems and to a few paradigmatic problems. Then, a control method aimed at eliminating, or better shifting in parameter space, a global dynamic event (such as homoclinic or heteroclinic bifurcations) triggering unwanted complex dynamics, is illustrated. The elimination is obtained by modifying the shape of the excitation in an optimal way, through the solution of different optimization problems. This is why the proposed method can be named as "Optimal control of nonlinear dynamics and chaos". All of the main theoretical points of the overall research line are reviewed, by highlighting some general features of the method and by focusing on its application to several mechanical models and systems, starting with single degree-of-freedom, smooth or non-smooth, archetypal models up to the buckled beam, which is an infinite-dimensional system.

1. Background and Classification of Control of Chaos Methods

Control of nonlinear dynamics and chaos is a very active research area, as shown by the rapidly increasing number of publications on the subject. In the (possibly incomplete) references we quote 8 books[1-8], 9 journal special issues[9-17] and 12 review or survey papers[18-29] on the most renowned scientific journals. To give an idea of the number of publications in peer-reviewed journals, Fradkov *et al*[25] refer to 2700 papers published up to 2000, with more than half in 1997-2000, and to about 400 papers per year published in the period 1997-2002, with a year amount exceeding 500 in 2005.

Although some other works appeared at about the same time or earlier[88,113], it is believed that the starting point of "Controlling Chaos" is the celebrated work of Ott, Grebogi and Yorke[101], which is quoted in almost all papers on the subject, and which was successively reviewed[116], improved[63,75,76,126,127], and investigated with respect to its optimality[64]. The seminal idea is that of *exploiting* the chaotic behaviour of systems to control their dynamics, and this represents a major improvement with respect to independent (mathematicians', physicists' and engineers') points of view on the matter: one meaningful passage from analysis to synthesis of chaotic properties, based on the knowledge and exploitation of dynamical systems theory deeply studied in the past.

Later on several works merely aimed at *removing* chaos, by means of classical control techniques[117], empirical methods[118], or other keen approaches[106], have been also referred to as "Control of Chaos", although in this case some authors, and we agree with them, prefer the name "Suppression of Chaos" [26,65,74,131], which focuses on the effects of control rather than on the underlying skill of the control method. Thus, today, one refers to chaos control when:

(i) chaotic transients and/or attractors are eliminated "tout court," even if the tools employed to eliminate chaos have nothing to do with it;
(ii) typical properties of chaotic dynamics are involved in the control process, irrespective of the actual tools being employed to control the system and, indeed, of its actual response.

This twofold control aspect is clearly highlighted in Linder and Ditto[26], when they assert that "... some techniques merely suppress or remove chaotic behavior ... others actually exploit chaotic behavior...".

Within the latter, more general and modern, perspective, the capability to exploit typical properties of dynamical systems undergoing chaotic behaviours in order to *control* the system response refers to either *suppressing* or *enhancing* chaos (anti-control), or even *using* it, based on the specific goal of the considered method, system, and area of interest[28].

Another basic issue is concerned with whether the dynamical phenomenon to be controlled by whatever "chaos control" technique has to be intended as "chaos" in a strict sense or, rather, as any kind of *complex behaviour* of a dynamical system, which may have different aspects according to whether a theoretical or a practical viewpoint is adopted. In the former, dynamical systems oriented, perspective, one can think, e.g., of any global bifurcational event possibly entailing complex behaviour of the system, of the relevant escape from a safe subset (a potential well or a basin of attraction) in parameter control space, of the synchronization of different oscillators; in the latter, application-oriented, perspective, the kind and meaning of the "complex" phenomenon of interest is dictated by the practical goal to be attained.

Based on this extended notion of chaos control, the objective of this chapter is to provide the reader with an overview of the relevant advances in mechanical sciences. Yet, considering the strong cross-disciplinary character of the matter and the tight connections with on-going studies in companion sciences, it is suitable to shortly address the issue of the classification of many available control methods in order to establish an overall reference framework for successive discussions.

In view of the above mentioned extended notion of the matter, no relevant exhaustive classification seems to be yet available, although some earlier interesting attempts have been made. For example, Chen and Dong[1] proposed a classification based on various tools employed in the control process (parameter-dependent approaches, open-loop strategies, engineering feedback control, adaptive control, intelligent control, *etc.*), while Fradkov's "Chaos Control Bibliography (1997-2000)"[23] also contains a classification of the various applications in science and engineering. Recent classifications basically distinguish between feedback (or close-loop) and non-feedback (or open-loop) control techniques.

A more phenomenologically based classification, of major interest for the aim of this chapter, relies on the ascertainment of how one can statically/dynamically modify the system or the excitation to attain the control goal. For example, the OGY's method[101], some classical methods of control theory (CM)[82], the "control by system design" (CSD)[19], the "parametric variation methods" (PVM)[21] belong to the first class, while the second class includes other classical methods where a properly modified input (periodic or aperiodic, open-loop or feedback) is applied to the system (CM)[82,104,106,117], the "control through operating conditions" (COC)[19] based on modifying the frequency and/or the amplitude of the excitation, the methods based on either combining parametric and external excitations (PEE)[88], or applying weak periodic perturbations (WPP)[8], or modifying the shape of the excitation (SE)[113,132,139].

Obviously, different methods are expected to give different performances, at least theoretically. Thus, one has methods aimed at:

(i) stabilizing an unstable zone of parameter space (CSD, COC, PVM);
(ii) moving away from (previously known) chaotic zones (CSD, COC);
(iii) stabilizing a given, erratic solution (CM, OGY);
(iv) overall regularizing the system dynamics, irrespective of single solution behavior (PEE, WPP, SE).

It is the authors' opinion that this distinction is very important from a practical point of view because it suggests the use of the more appropriate control method fitting prescribed technical requirements.

Focusing the attention on the exploitation of chaotic behavior, there is even another possible classification, which seems not pursued in the literature, where various methods can be grouped according to the chaotic properties involved in the control. In this respect, one can distinguish methods based on the:

(i) properties of the saddles embedded in the chaotic attractor (OGY);
(ii) ergodicity of the chaotic attractor (belong to this class, for example, all methods where a preliminary targeting step[38,42,115] is followed by

the application of control tools in the neighborhood of the chosen control area);
(iii) sensitivity to initial conditions[114];
(iv) occurrence of homo/heteroclinic bifurcations (PEE, WPP, SE).

The chapter is organized as follows. In Sec. 2, a non-exhaustive overview of classes of systems and paradigmatic problems in solid and structural mechanics for which chaos control problems have been addressed in the recent literature, is provided. Section 3 is devoted to presenting a general approach to chaos control formulated by the authors with the underlying idea of controlling complex dynamics in mechanical and structural systems. The method is revisited through its chronological development and its progressive comprehension as it ensues from also an increasingly wider range of considered technical problems.

2. Control of Chaos in Solid and Structural Mechanics

Overall, considering the large amount of applications of chaos control to mathematical, physical, biological, and engineering systems, in different technical fields, it could appear that relatively few studies have dealt with the relevant applications in mechanics.

Up to the authors' knowledge, the earlier review paper[19] explicitly devoted to this topic dates back to 1993. The authors report on the newly developed (at that time) OGY method, and propose to control a mechanical system by either changing the forcing characteristics (e.g. the excitation frequency) – with the aim of moving away from a previously known chaotic region – or modifying any system property, such as inertia or stiffness. All of the proposed techniques fail in the area of "suppressing" chaos, and while being illustrated with reference to a Duffing oscillator (possibly equipped with also a tuned mass damper), they do not actually refer to specific issues associated with the mechanical nature of the systems.

A recent survey[25] of the application of various methods of chaos control to mechanical systems reports quite a long list of mechanical systems or processes for which control of nonlinear dynamics and chaos has been addressed in the literature, furnishing at least one reference for each of them. The list includes pendulums, beams and plates, friction, systems with impacts, spacecraft, vibroformers, microcantilevers, ship oscillations, tachometer, rate gyro, Duffing oscillator, robot-manipulator arm, earthquake civil engineering, milling process, whirling motion under mechanical resonance, and systems with clearance. The items in the list highlight the quite scattered nature of the matter due to the *mechanical complexity* and variety of the systems, of the involved dynamical processes, and of specific control goals.

Indeed, choosing a proper framework within which organizing, presenting and discussing the state of art in the field is not an easy task because, depending on various application fields, mechanics is concerned with both discrete (finite-dimensional) and continuous (infinite-dimensional) dynamical systems, with their possibly reliable low-dimensional models, and with a considerable richness of dynamical processes and phenomena to be possibly controlled. Though being traditionally a world of large scale systems whose dynamic analysis and design are mostly conducted via linear techniques, a large amount of research made in the last thirty years has highlighted a cornucopia of nonlinear dynamic phenomena, including of course chaotic response, along with their response in also mechanical and structural applications. Yet, regular nonlinear phenomena of such a variety of involved systems are already quite complicated in themselves, and such to exhibit a major importance in the actual real world of mechanics. As a matter of fact, the question of how important and how much pervasive the chaotic phenomena are in the behavior of mechanical/structural systems, is still to be answered. In the context of the present review, this entails that it is quite difficult to select studies devoted to "chaos control" in mechanics in a strict sense, whereas the topic has to be intended in the mentioned wider sense of *dynamical complexity*, and is often addressed in the literature within the more general matter of control of nonlinear, wanted or unwanted, phenomena.

On the other hand, the inherent complexity of systems from mechanical science and engineering poses the challenging problem of identifying proper reduced order models capable to reliably describe the ensuing rich nonlinear dynamics, an issue which becomes even more demanding if one is interested in dealing with possibly complex nonregular response of the actual system and on its control. This basically entails one major consequence and one question.

(i) A large majority of research works on control of chaos in mechanics is concerned with (relatively simple) archetypal nonlinear oscillators often representing idealizations of more involved discrete systems encountered in real mechanical engineering or minimal discretized representations – according to some reduction technique – of the infinite-dimensional continuous systems typical of structural engineering. This also entails that in most research papers the interest is focused on some characterizing dynamical or control aspect – transversal and unifying with respect to other scientific/technological fields, such as chaos control methods – rather than on the actual mechanical peculiarities of the considered systems.

(ii) Even being convinced of the significance of nonregular dynamics – and of their control – for real mechanical/structural systems, how much representative of their actual behavior are the chaotic regimes highlighted for such reference archetypal oscillators? Of course, this is a general point arising in whatever modeling and reduction problem, yet it is felt to have a special meaning in mechanics just owing to the above mentioned inherent complexity of the involved large-scale systems. The only possibility to clarify this issue consists in highlighting the actual occurrence, and the features, of chaotic response in experimental (mechanical and structural) systems, in developing refined theoretical and numerical models of the actual system, and in cross-validating the two approaches. In this respect, considerable research is going on to highlight the occurrence and the features of chaos in experimental and/or refined theoretical models, but quite few has been done as regards the relevant control.

In view of these issues and of the considerable amount of published works on chaos control in mechanical sciences – in the "extended" notion – no attempt is made herein to provide the reader with a comprehensive state of art of the research activity in the field. In contrast, a few, tentatively representative, samples, aimed at providing the reader with an overall flavor of the addressed topics, are shortly summarized, by grouping them in few main classes of systems, which is useful to help the reader get a not too scattered overview of the matter. It is hoped that the richness and the variety of the addressed systems are evidenced, along with some limitations of the present research in the area and with the ensuing advancement needs.

2.1. *An archetypal oscillator in mechanics: The mathematical pendulum*

The mathematical pendulum is the archetypal nonlinear oscillator *par excellence* in mechanics. Its chaos control has been studied through different techniques. Starrett and Tagg[121] and Starrett[120] used Bang-Bang methods (occasional, time proportioned perturbations), which essentially consist in applying a constant control for a certain time, whereas Corron *et al.*[53] aimed at experimentally showing how it is possible to build a simple chaos controller, in contrast to the very complicated controllers – even more complicated than the system to be controlled – usually adopted. Yagasaki and co-workers[123,124] applied feedback, Pyragas-like[106], techniques in a number of works containing important experimental results and interesting theoretical-experimental cross-validation, in addition to theoretical predictions and numerical results. Wang and Jing[129] applied the Lyapunov function method to design a controller capable to convert the chaotic motion of a pendulum to any periodic orbit in a shorter time than that required by OGY-based methods. Using a semi-continuous control method[79] built-up on the OGY technique, Pereira-Pinto *et al.*[103] considered stabilization of unstable periodic orbits in a simulated nonlinear pendulum based on parameters obtained from an experiment, with the attention being also paid to controlling chaotic behaviour using state space reconstruction. Several earlier[36,44,125] and recent[33,46,129] papers are devoted to studying control of chaos for a pendulum under external excitation in various forms whereas less attention has been paid to the presence of parametric excitation[39], whose effect on the inhibition of homoclinic/heteroclinic chaos in the externally excited pendulum has been addressed[128] through the Melnikov, non-feedback, control technique. In the more engineering-oriented context of control of dynamic phenomena, control of chaos in an autoparametric pendulum system has been recently considered[84] using a magneto-rheological damper.

Besides its own theoretical interest, the pendulum can be regarded as also a basic component of important engineering applications, like mechanical manipulators or robot arms. These are the underlying complex systems of a work on controllability conditions for (suppressing or inducing) chaos[47], which still makes use of the Melnikov technique and considers application of a weak resonant excitation along a research line followed by several authors in different contexts, see the general book by Chacon[8].

2.2. Non-smooth systems

Non-smooth or piecewise-smooth mechanical systems, which include frictional, clearance and vibro-impact oscillators, are a topic of considerable interest. Control of chaotic dynamics of a frictional oscillator has been investigated both theoretically[110], by Pyragas-like[106], or delayed-feedback, techniques, and experimentally[95], with the matter of anti-control (i.e., enhancing instead of eliminating chaos) being also addressed[95].

Vibroimpact systems have many applications in mechanics, ranging from drilling machinery and gearboxes to smart dampers for use in structural and automobile engineering. Control of chaos in oscillators with inelastic or elastic impacts has been investigated by Barreto et al.[35], Hu[78] and Bishop and co-workers[40,41]. Casas and Grebogi[51] applied the OGY method to standard impact oscillators, by using a fundamental discontinuity map – with a square root singularity in the Jacobian – previously introduced to study the grazing bifurcation[100]. A transcendental (impact) map was also used[54] in the implementation of the OGY method. Control of an unstable impact oscillator to a desired stable one is implemented by Lee and Yan[86] via a synchronization scheme, whereas a recent feedback technique[34,122] aimed at suppressing chaos by altering the system energy via a small-amplitude control signal has been implemented[56] to control an impact oscillator by varying the damping.

An inverted pendulum impacting on two lateral barriers has been the system initially considered[131,132] for developing a method of control of chaos based on modifying the shape of the excitation, with the relevant systematic application being reported successively[133,134]. The same technique has been applied by the authors for controlling the rocking dynamics of a rigid block[140], with special emphasis being paid to the mechanical phenomenon of overturning – which represents an escape phenomenon from a dynamical system viewpoint – whose anti-control has been also investigated[141]. Swing up control of a double-inverted pendulum has been addressed by di Bernardo and Stoten[59].

Overall, recent works on the dynamics of vibroimpact systems are reviewed in a paper by de Souza et al.[58], focusing on chaotic motion and its control. The systems considered are a gear-rattling model, active dampers, impact and liquid column dampers, as well as systems with...non-ideal energy source, represented by a limited power supply. Stabilization of the latter is obtained by using either impact dampers[55] or the above mentioned small-amplitude signal technique[57].

2.3. A few multidegree-of-freedom discrete systems

A few, more complicated, discrete systems have been studied, too. Ge and co-workers considered chaos control, anti-control and synchronization in several mechanical applications, by also applying different techniques from the literature, with special comments or comparisons. The systems include a rotational tachometer with vibrating support[72], an electro-mechanical gyrostat system[71], a rotational machine with centrifugal masses[68-69], a suspended track with moving loads[73], a two-degree-of-freedom louder speaker system[70], and single time scale[66] or three time scales[67] brushless DC motors. Anti-control of chaos in the 3D motion of a rigid body described by the Euler equations – possibly related to the Lorenz equations and to Chen system – has been studied[52] by considering suitable feedback gains.

A dynamical system constituted by coupled dynamos, and modelled by a three-degree-of-freedom system, is studied by Agiza[30], who applied various control techniques, including the Pyragas' one[106]. The chaotic fluctuations of temperature in a heater are experimentally controlled (i.e., eliminated) by an ad-hoc, simple but clever, algorithm which consists in incrementing the heating rate according to the difference of temperature between two points of the heater[118].

The relevance of the OGY[101] strategy of chaos control to spacecraft steering in different Hamiltonian situations is dealt with[90] by considering stabilization of chaotic trajectories and targeting, the latter being applied to some paradigmatic models in astrodynamics, namely the restrict circular three-body problem and the two-body Hill encounter problem.

2.4. *Smooth archetypal oscillators and refined models for continuous systems*

A large amount of works is being devoted to control of chaos for the large class of smooth archetypal nonlinear oscillators, which includes the Helmholtz oscillator, different Duffing-like oscillators, the Helmholtz-Duffing and higher-order polynomial oscillators. It is well-known how all of these systems may undergo chaotic, or anyway complex, dynamics in meaningful ranges of values of the relevant control parameters. In mechanics, these oscillators govern, to a different extent, the nonlinear dynamic behavior of various mechanisms or discrete systems, and, mostly, the non-internally resonant single mode dynamics of such continuous systems as, e.g., ships (Helmholtz), cables and arches (Helmholtz-Duffing), strings, beams, and plates (Duffing)[138,139]. Yet, in view of the general remarks made at the beginning of this section, two important requirements have to be fulfilled in this respect.

(i) Values of the coefficients of the various nonlinear oscillators have to be considered which are actually representative of system/problem parameters in technical situations of mechanical, naval, aerospace or structural engineering, and for which complex dynamics occur and there is need for their control.

(ii) The reliability of the considered (often minimal) reduced order models in describing the actual mechanical complexity of the reference system has to be estimated, along with the practical significance of the ensuing chaos control problem.

It is hard to state that these two points are satisfactorily accounted for, or even just mentioned, in all of the published research material in the field.

Control of chaos of Duffing-like oscillators has been studied by many authors. To name a few, a Duffing-Ueda (i.e., without linear term) system has been controlled by a classical Closed-Loop controller[32]. A simple, bang-bang like, control method utilizing a sequence of impulses has been applied to a Duffing oscillator with both linear and nonlinear positive terms[102]. The OGY method has been applied to a Duffing oscillator subjected to a partially stochastic excitation[60], while the application of Pyragas' method to the Duffing oscillator has been considered, e.g., by Mahmoud et al.[91]. Other works[43,99,108] developed control of chaos techniques and applied them to the Duffing oscillator.

A comparison of various control methods as applied to the Duffing equation has been proposed[31,117]. Sifakis and Elliott[117] considered four different control techniques. (i) *Open-loop periodic perturbation method*, which consists in adding a periodic perturbation to the system excitation. (ii) *continuous delayed feedback method*, i.e., the Pyragas method[106], (iii) the Hunt method[80,81], and (iv) the OGY method[101]. As to the first technique, the perturbation is chosen empirically by a trial-and-error procedure, whereas in other works it is optimally determined on the basis of a theoretical analysis relying on system dynamical properties. More classical control techniques, like the *optimal polynomial control* and the *robust sliding mode control*, have been also applied[31].

In turn, the theoretical features of a control method based on optimally modifying the shape of the excitation, as applied to different smooth archetypal oscillators, have been comparatively discussed by the authors[138,139], with the specific control performances being addressed elsewhere[135-137]. The underlying structural systems, with the associated values of the governing parameters, refer to technical applications from either macro-mechanics (cables, arches, beams) or micro-mechanics (beams in MEMS), with the latter involving nonlinearities from also non-mechanical sources[142]. Relevant advancements will be possibly concerned with somehow accounting for the coupled (thermo-fluid-electro-mechanical) effects which are likely to govern the features of actual occurrence of the so-called pull-in phenomenon (which is an escape event, in dynamical systems terminology) in these systems.

The effect of the excitation shape in modifying the homoclinic bifurcation, without any reference to control and in the restricted case of Jacobi elliptic functions, attracted the attention of researchers[111]. With reference to control, on the other hand, similar ideas, i.e. eliminating the homoclinic bifurcation by introducing an appropriate extra (controlling) excitation, called "function of stabilization," have been developed by Dzhanoev et al.[62]. Modified versions of the Lenci and Rega optimal control method have been also proposed and applied to smooth archetypal oscillators[48-50].

Control of chaos with some kind of nominal mechanical application in the background is considered by Ding[61] and Litak et al.[89] for escape systems representing, e.g., ship capsizing, and by Belhaq and Houssni[37] for a quasi-periodically driven system with quadratic and cubic nonlinearities. Within a more general framework of vibration control, inhibition of Melnikov chaos and/or escape via active control has been considered for a single-well Duffing oscillator[96], and for a two-well Duffing oscillator[97,98] representative of the single-mode dynamics of a buckled beam, by also paying attention to the problem of time-delay in the control action.

Experimental control of chaos in a continuous system has been addressed[94] by considering the same apparatus (a beam deformed by permanent magnets) already considered by the leading author in a number of pioneering papers[93] on experimental chaos in mechanical systems, with the control method being the Hunt's Occasional Feedback Control[80,81], which is quite similar to the Occasional Bang-Bang control.

Some results obtained in the control of regular nonlinear vibrations of a buckled beam via classical controllers[104,105] are felt to be easily extendable to control of chaos. In the same framework is a paper[45] dealing with the control of a shear-type framed structure such as those characterizing building skeletons in civil engineering. Here, the practical interest comes from the circumstance that real earthquake excitations (the El Centro accelerogram) are considered.

Zhang[130] has considered numerical control of chaotic motion in a 2:1 internally resonant two-degree-of-freedom model governing, according to a Galerkin reduction, the nonlinear nonplanar oscillations of a cantilever beam subjected to a harmonic axial excitation and transverse excitations at the free end. Numerical results show the usefulness of the transverse excitation for driving chaotic motion to a period *n* motion.

Within the more general framework of the reliability of reduced-order models in describing non-regular dynamic phenomena of infinite-dimensional systems[87], control of the homoclinic bifurcations triggering the complex dynamics of a buckled beam is dealt with[143] by considering different boundary conditions and the possibility to refer to the chaos control results obtained for the corresponding reduced-order models.

3. Control of Homo/Heteroclinic Bifurcations: The Authors' Contributions

The authors' control method reviewed in this chapter is based on the simple idea of eliminating global bifurcational events embedded in system dynamics, namely homoclinic or heteroclinic intersections, by varying the shape of the external excitation, i.e., for example, by adding appropriate superharmonic corrections to the basic harmonic excitation.

This approach was suggested by the well-established fact that homo/heteroclinic solutions are responsible for many chaotic behaviours, and "…it is not an exaggeration to claim that in virtually every manifestation of chaotic behaviour known thus far, some type of homoclinic behaviour is lurking in the background…"[85], a conjecture which has been repeatedly confirmed in various circumstances[83,92,119], both numerically and experimentally. Indeed, homo/heteroclinic intersections are certainly responsible for fractalization of boundaries of basins of attractions, for sensitivity to initial conditions, for chaotic transients and for appearing/disappearing of invariant chaotic sets constituted by infinitely many unstable periodic and aperiodic orbits, which largely influence the overall dynamics and often constitute the skeleton of chaotic attractors. Furthermore, they may be responsible for creation, widening or splitting of a chaotic attractor, for erosion of safe basin of attraction, and they are involved in the escape from potential wells. Thus, controlling homo/heteroclinic bifurcations permits to eliminate the previous phenomena, or at least to reduce or shift them in parameter spaces, and, accordingly, it has the objective of regularizing the *overall dynamics* rather then a given single orbit.

While the elimination of global bifurcations through modifications of the excitation represents a keystone of the method, modifying only the shape of the external excitation can be considered as a first step towards further improvements. In fact, one can consider both external and parametric harmonic excitations, as done by Lima and Pettini[88] or by Lenci and Rega[136] in the case of the Helmholtz equation, or general periodic external and parametric excitations, or quasi-periodic excitations, or aperiodic excitations, and so on. This would largely extend the generality of the method by increasing the range of performances obtained by modifying only the shape of a regular external excitation.

As far as possible applications of the method are concerned, we feel they are very large and simple, at least in principle: in fact, whatever is the dynamical system to be controlled – mechanical, physical, chemical, biological – it is sufficient to detect the homo/heteroclinic event responsible for the unwanted system response, if any. Then, one determines the dependence of this behavior on the excitation, and finally modifies it in such a way to eliminate as far as possible the homo/heteroclinic intersection and the related chaotic phenomena. The difficulties may possibly be of practical nature, as the detection of invariant manifolds may be difficult and challenging in dimensions greater than two, but this does not affect the theoretical framework of the method.

Contrary to other authors' papers[27,144] where the proposed method for eliminating homoclinic and heteroclinic bifurcations has been reviewed from a methodological point of view, herein we follow the successive chronological developments that the formulation, the implementation and the application of the method have actually undergone. Apart from better illustrating the main lines along which the research has been developed, this approach also permits to highlight the sequence of problems which have been addressed and solved, as well as a number of key issues to be dealt with for better clarifying the meaning and potentialities of the method as regards control of multi-degree-of-freedom and infinite-dimensional systems.

3.1. The initial works on the inverted pendulum

The whole research line started with the work published in 1998[132]. The original idea was to improve the results of Shaw[113], by considering an infinite number of controlling superharmonics (see the following) instead of the single one used by Shaw. Accordingly, the same mechanical model, an inverted pendulum with lateral walls, was considered (Fig. 1).

The governing equations are

$$\ddot{x} + 2\delta \dot{x} - x = f(t), \quad |x|<1; \quad \dot{x}(t^+) = -r\,\dot{x}(t^-), \quad |x|=1, \quad (1)$$

where $x=\vartheta/\vartheta_{max}$, ϑ is the deflection angle measured from the vertical, $\delta \in [0,1]$ is the damping coefficient of free motion, $r \in [0,1]$ is the coefficient of restitution for the instantaneous impacts and $f(t)=\Sigma_i \gamma_i \sin(i\omega t+\psi_i)$ is the T-periodic excitation force ($T=2\pi/\omega$).

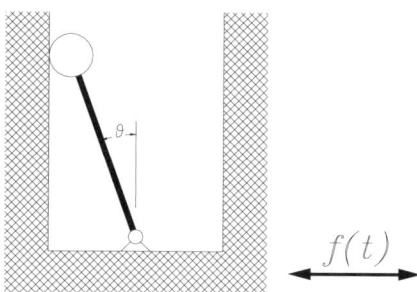

Fig. 1. The inverted pendulum with rigid unilateral constraints.

One of the main contributions was the *formulation* of the general procedure for *applying* the control method, thereafter used in subsequent papers in different contexts. It can be summarized as follows.

(1) Identifying an homo/heteroclinic bifurcation responsible for unwanted dynamical phenomena. This bifurcation can be arbitrary, but only for homo/heteroclinic bifurcations of hilltop saddles it is possible to develop the procedure analytically by means of the Melnikov method. For other bifurcations, a numerical analysis is required[135].

(2) Detecting the dependence of the global bifurcations on the excitation, in particular on certain parameters of the excitation which can be modified for control purposes. Again, this can be done analytically, by the Melnikov method, for bifurcations of hilltop saddles, while it has to be done numerically in the other cases.

(3) Formulating and solving the mathematical problem of optimization, which consists in determining the (optimal) theoretical excitation which maximizes the distance between stable and unstable manifolds for a fixed excitation amplitude or, equivalently, the critical excitation amplitude for occurrence of homoclinic or heteroclinic bifurcation.

How these points have been implemented in the inverted pendulum is summarized in the sequel. In the phase space $(x, \dot{x})$, there exist two homoclinic symmetric loops for the trivial saddle $x=0$ of the unperturbed ($\delta=0$, $r=1$, $f(t)=0$) system. Thus, as far as point **(1)** above is concerned, it is quite natural to consider the homoclinic bifurcation of the hilltop saddle. This is the same as previously done by Shaw[113], and permits analytical computations.

When the perturbed system is considered, on the right Poincarè section $\Sigma = \{(x,y,t) \in [-1,1] \times \mathbb{R} \times S \mid x = +1, y \geq 0\}$ the distance between the invariant manifolds is

$$d(\tau) = C_0/2 + \gamma_1 C_1 h(\tau). \qquad (2)$$

Here $\tau = \omega t \in [0, 2\pi]$, and C_0, C_1 and the function $h(\tau)$ depend on $f(t)$[132]. There is homoclinic intersection if there exists some τ for which $d(\tau) = 0$. Let $M = \max_\tau\{h(\tau)\}$. We have homoclinic bifurcation whenever the excitation amplitude γ_1 reaches the critical threshold

$$\gamma_{1,cr} = - C_0/(2C_1 M). \qquad (3)$$

So far, we have accomplished point **(2)** above, i.e., we have determined how the homoclinic bifurcation threshold depends on the excitation parameters. In particular, the dependence on the excitation shape is summarized in the number M.

The third point of the general "program" consists in determining the optimal excitation. It was realized that this can be obtained by maximizing – through variation of the excitation shape coefficients γ_i

and ψ_i – the parameter $G=1/M$, which is called the *gain* and represents the ratio between the critical amplitude $\gamma_{1,cr}$ of the actual periodic excitation and the critical amplitude $\gamma^h_{1,cr}$ of the harmonic excitation, to be considered as a reference to measure the relative improvement of the proposed optimal solution. Thus, the objective consists in increasing, as far as possible, the value of the excitation amplitude for which homoclinic intersection occurs. This is obtained by varying the shape of the excitation.

Another important result[132] was that of solving exactly the underlying optimization problem by means of an abstract mathematical tool, the Ghizzetti's theorem. The solution was found to be

$$f(t)=f^*(t)+\frac{\gamma_1 C_1 r\pi}{2(1+r)\sin[\pi/(1-c)]}\left[-\delta\left(t-\frac{\pi}{2\omega}\frac{1-3c}{1-c}\right)+\delta\left(t-\frac{\pi}{2\omega}\frac{5-3c}{1-c}\right)\right], \quad (4)$$

where $f^*(t)$ is a bounded function, $\delta(t)$ is the Dirac delta function and $c\leq-1$ is a parameter introduced for mathematical reasons. From a physical point of view, Eq. (4) means that, apart from $f^*(t)$, the optimal excitation consists in a couple of impulses with amplitude proportional to |c|. It corresponds to the best gain $G=2(1-c)/\pi \sin[\pi/(1-c)]$, which can theoretically grow up to 2.

Also some preliminary numerical simulations aimed at checking the practical performances of control were made[132]. In fact, it was realized quite early that it is necessary to check the practical effectiveness of the method, and that this point is far from being trivial because it is not aimed at stabilizing a single orbit, but rather the whole system dynamics, so that the check of the effectiveness is more involved. Thus, the previous list must be updated with the following fourth point.

(4) Practically implementing the optimal excitation, which is required to confirm theoretical predictions and to check the feasibility and the actual performances of the technique. While the first three steps are mainly theoretic, even in the case of a completely numerical approach, this last point is essentially practical, and requires numerical and/or physical experimentations. It depends on the kind of dynamical phenomenon the control method is applied to (reduction of safe basin, single-well to cross-well chaos transition, *etc.*).

While this fourth point was initially addressed only in a preliminary way[132], it was successively investigated systematically in a double paper[133] entirely devoted to numerically check the practical performances of control. Before summarizing the main results of the numerical simulations, we note that the important concept of "global" or "one-side" control was first introduced in this paper. The latter refers to the control of a single homoclinic bifurcation, while the former is concerned with the simultaneous control of both homoclinic bifurcations in a two-well potential system.

According to this distinction, the numerical simulations were grouped in three parts: (i) harmonic excitation, which was investigated to check, by comparisons, the improvements due to controlling excitation; (ii) global control, which is actually a symmetric excitation, so that it is studied together with the harmonic excitation in the first part of the double paper; (iii) one-side control, which is asymmetric and constitutes the object of its second part.

The investigation has been performed through extensive and combined use of bifurcation diagrams and attractor-basin phase portraits, aimed at capturing and describing the bifurcation and attractors scenario. This involves a number of meaningful classical (SN, PD and PF) and non-classical local bifurcations, and a rich variety of sudden changes of chaotic attractors (subduction and different kinds of boundary and interior crises). A simplified synthetic behavior chart summarizing the dynamical responses under the three different excitations is reported in Fig. 2[133].

By comparing harmonic and global control excitations, which are symmetric, we can draw the following conclusions. The global control:

(i) increases the homoclinic bifurcation threshold, as predicted by the theoretical treatment;
(ii) keeps the symmetry of the system response and therefore controls the whole phase space;
(iii) increases the threshold for appearance of scattered steady dynamics, which occur approximately in correspondence of the homoclinic bifurcations;

(iv) regularizes the dynamics, as highlighted by the fact that there are windows with only periodic attractors within the bifurcation diagram.

A drawback of the global optimal control is that it furnishes a relatively low gain. If one wishes to increase the gain, then the (unsymmetrical) one-side optimal excitation must be used. The unsymmetric excitation controlling one-side (the upper bifurcation diagram in Fig. 2(c)) possesses several interesting properties as regards to both dynamics and control. For example, the transition from confined to scattered attractors is very involved: while occurring nominally in "correspondence" with the homoclinic bifurcation, it is actually spread over a large interval with a complicated attractors and bifurcation scenario. Furthermore, we have observed strong coexistence and competition of scattered and confined (chaotic and/or periodic) attractors, a circumstance not so evident with symmetric excitations.

The main features of the response scenarios with the optimal excitations are:

(i) The considerable increase of the excitation amplitude corresponding to the onset of scattered dynamics with respect to reference excitation. In the case of global control, this occurs in "correspondence" with the homoclinic bifurcations; in the case of one-side control, due to the presence of an uncontrolled side (the lower part of the diagram in Fig. 2(c)), it precedes the homoclinic bifurcation on the controlled side which, however, occurs at a notably larger parameter value.

(ii) The overall regularization of the dynamics, witnessed by the presence of large periodic windows in the case of global control, and by the predominance of periodic attractors distributed all over the considered excitation range in the case of one-side control.

All previous points show the effectiveness – in average sense – of the proposed procedure for controlling steady nonlinear dynamics, though it is not possible to conclude that all solutions are regularized.

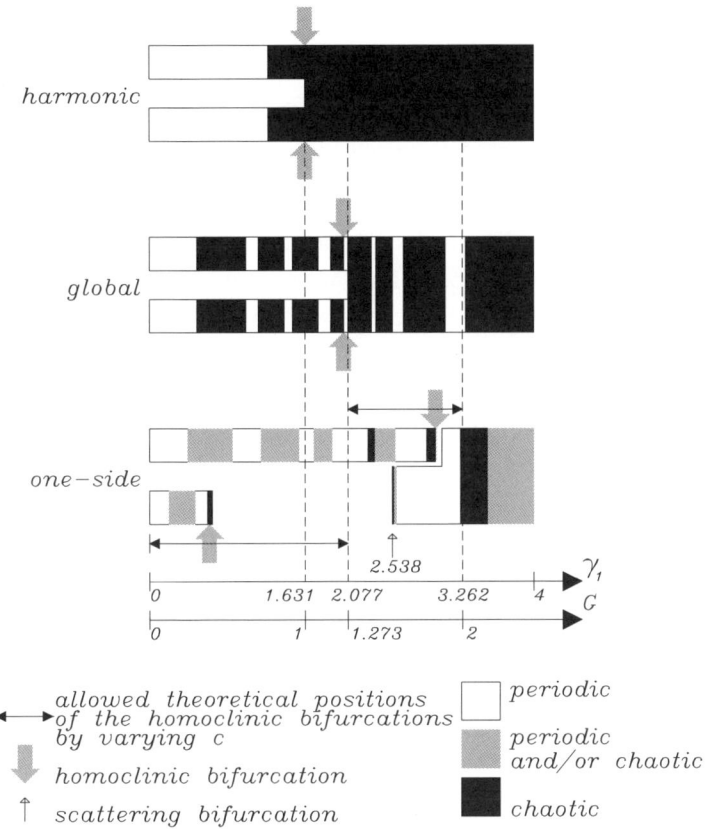

Fig. 2. Simplified behavior chart of the different excitations.

After checking the control performances for the inverted pendulum, and before considering oscillators representative of more involved mechanical systems, we followed two different developments, which are actually complementary for orderly being mainly theoretical (the first) and very practical (the second).

In the first development[131], we addressed the optimization problem by considering also the constraint of boundedness of the optimal excitation. This is not required from a theoretical point of view, but from a practical point of view it is much more interesting to have optimal bounded excitations instead of optimal impulses, as in (4). By using a different mathematical tool, the Neyman-Pearson lemma instead of the Ghizzetti's theorem, we were able to obtain the optimal bounded excitation.

In the second development[134], we considered two feed-back implementations of the optimal excitations able to fulfill two specific technical requirements. The first one consists in alternating application of right optimal, as far as the pendulum mass is in the corresponding well, and harmonic excitations elsewhere. This approach was called "Optimal-Harmonic"[134], and permits to increase the percentage of time spent by the mass in the desired potential well (Fig. 3).

The second feedback implementation consists in applying always the one-side optimal control, in particular, left control as long as the mass is in the left potential well, and right control when it is in the right potential well. This approach was named "Optimal-Optimal" control, and succeeds in reducing the number of jumps between wells (Fig. 4), which is another important technical requirement from, e.g., the system integrity viewpoint.

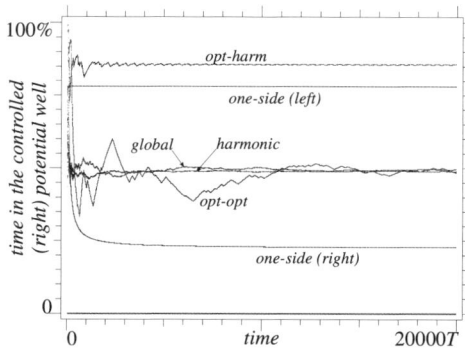

Fig. 3. The percentage of time spent in the controlled (right) potential well.

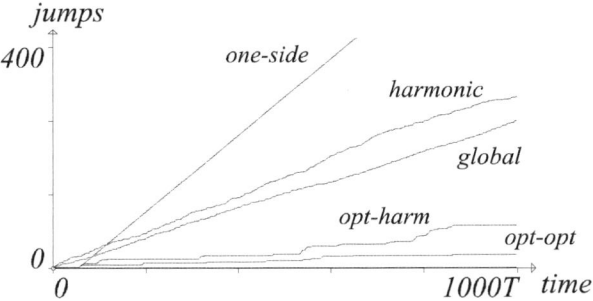

Fig. 4. The number of jumps versus time.

3.2. Smooth classical oscillators

After investigating in detail the behavior of the inverted pendulum, more realistic mechanical models have been considered to check the performances of control. In particular, smooth systems were addressed. This passage has deep consequences. In fact, from one side it permits to consider models representative of different classes of mechanical/structural systems, but from the other side it requires the Melnikov perturbative method, whereas in the case of the inverted pendulum the homoclinic bifurcation was computed exactly, without using asymptotic arguments.

Together with the passage from non-smooth to smooth systems, we also introduced different mechanical characteristics by considering both bounded (or hardening) and unbounded (or softening) oscillators, as well as different methodological approaches, by considering also an homoclinic bifurcation of a non-hilltop saddle, which requires an entirely numerical treatment.

The first investigated smooth system was the Helmholtz oscillator[136], which is the archetypal single-degree-of-freedom nonlinear oscillator with one escape direction also describing some mechanical phenomenon (e.g., ship capsizing), to a first approximation. The governing equation is

$$\ddot{x} + \varepsilon \delta \dot{x} - [1 + \varepsilon \alpha(\omega t)] x + [1 + \varepsilon \beta(\omega t)] x^2 = \varepsilon \chi(\omega t),$$
$$\alpha(s) = \Sigma_i \alpha_i \sin(is + a_i), \quad \beta(s) = \Sigma_i \beta_i \sin(is + b_i), \quad \chi(s) = \Sigma_i \chi_i \sin(is + c_i). \quad (5)$$

The proposed method was again (as in the inverted pendulum) applied to the homoclinic bifurcation of the hilltop saddle. From a mathematical point of view, this permits to detect analytically, though now in an approximate way, the critical amplitude for homoclinic bifurcation. From a physical point of view, on the other hand, this homoclinic intersection is seen to directly or indirectly play a fundamental role in the organization of the regular and, moreover, non regular and chaotic dynamics of the oscillator. Although the main homoclinic bifurcation of the hilltop saddle does not correspond exactly to the escape boundary, it initiates the penetration of the escaping tongues into the safe basin, and therefore it constitutes the first event of the complex sequence of topological phenomena, called erosion, finally leading to escape.

The previous Eq. (5) shows the presence of simultaneous parametric ($\alpha(\omega t)$ and $\beta(\omega t)$) and external ($\chi(\omega t)$) excitations, while in the inverted pendulum we were concerned with only external excitations. This generality does not entail special difficulties in the analytical treatment. In fact, the general Melnikov function for the homoclinic bifurcation of the hilltop saddle

$$M(m) = -\frac{6}{5}\delta - 3\pi \sum_i \alpha_i \frac{(i\omega)^2 + (i\omega)^4}{\sinh(i\omega\pi)} \cos(im + a_i)$$

$$+ 3\pi \sum_i \beta_i \frac{4(i\omega)^2 + 5(i\omega)^4 + (i\omega)^6}{5\sinh(i\omega\pi)} \cos(im + b_i)$$

$$- 6\pi \sum_i \chi_i \frac{(i\omega)^2}{\sinh(i\omega\pi)} \cos(im + c_i), \qquad (6)$$

can be rewritten in the simpler form[136]

$$M(m) = -\frac{6}{5}\delta - \sum_i \gamma_i \frac{6\pi(i\omega)^2}{\sinh(i\omega\pi)} \cos(im + \Psi_i) \qquad (7)$$

by introducing an equivalent external excitation $\chi(s)=\Sigma_i\gamma_i\sin(is+\psi_i)$. Note that both (6) and (7) are the first order ε-terms of the distance between stable and unstable-manifolds. Thus, in this case it is crucial that excitations and damping are small (they are multiplied by ε in (5)), while this is not required for the inverted pendulum.

The main achievements[136] are concerned with both theoretical and practical results. From a theoretical point of view, it was realized that the mathematical solution of the underlying optimization problem, considering an infinite number of controlling superharmonics added to the basic harmonic excitation, leads to a physically inadmissible solution in terms of optimal excitation (as for the inverted pendulum where it leads to a couple of impulses). Thus, the mathematical optimization problem was reconsidered. Three further solutions were obtained.

(i) Optimal solutions with a finite number of superharmonics. These solutions overcome the unfeasibility problem, which comes from the divergence of the related series and automatically disappears by considering only a finite number of superharmonics.
(ii) Constrained problem with an infinite number of superharmonics. This consists in introducing an adequate constraint capable of guaranteeing the physical admissibility of the solution. The modified optimal problem was then addressed, and the solution was found for large values of the excitation frequency ω, while for lower values of ω an upper bound was obtained.
(iii) Constrained problems with finite but bounded number of superharmonics.

Note that for the Helmholtz oscillator there is only one homoclinic loop of the hilltop saddle, so that "global" control does not apply.

The second main achievement[136] was the extensive numerical simulation aimed at checking the performances of optimal control excitations. In particular, the following points, which are actually sub-points of point **(4)** above, were addressed.

(4.1) Numerical verification of theoretical prediction of homoclinic bifurcations. This is required because in this case the Melnikov method only provides an asymptotic estimation of the bifurcation threshold, thus giving accurate results only for small values of damping and excitation amplitude.

(4.2) Analysis of regularization of fractal basin boundaries, which is the first predicted effect of the homoclinic bifurcation elimination and provides first hints on the practical effectiveness of optimal excitations. An example is reported in Fig. 5.

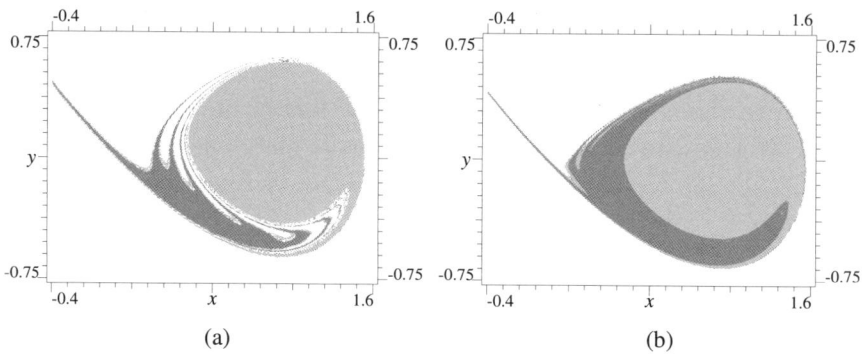

Fig. 5. Basins of attraction for different excitations. (a) harmonic (reference) and (b) optimal excitation with two controlling superharmonics.

(4.3) Analysis of reduction of fractal erosion of basins of attractions. This has been done by the construction of the so-called erosion profiles (i.e., a measure of the integrity of the safe basins[109] versus a varying parameter, in particular the excitation amplitude), and by determining how these profiles are modified by the optimal excitation. An example is reported in Fig. 6. The shift toward higher excitation amplitudes of the erosion profiles shows the effectiveness of the method.

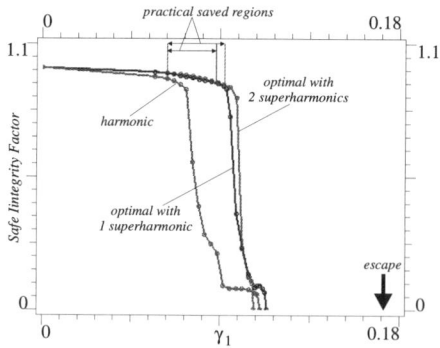

Fig. 6. The erosion profile with harmonic and optimal excitations.

It was realized that the erosion profiles, and how they are modified by the optimal excitation, are a reliable tool for checking the practical performances of the control as applied to the homoclinic bifurcation of the hilltop saddle. When applicable, they have been used in all successive works.

Recently, avoiding the escape from a potential well has been also investigated by, e.g., Seoane et al.[112], in a different context.

After the Helmholtz oscillator, we studied the Duffing oscillators[137] due to several reasons, the most important being the following.

(i) The Duffing oscillator describes many mechanical models and the single d.o.f. dynamics of many structural systems, so it is important to see how the control works on this oscillator.
(ii) The Duffing equation is an archetype for smooth and hardening oscillators, that were not investigated before.
(iii) In the Duffing oscillator it is possible to deal with both "global" and "one-side" controls, the former not applying for the Helmholtz oscillator and being only mentioned in the inverted pendulum.

The attention was focused on the homoclinic bifurcations of the hilltop saddle, again because this constitutes a very important dynamical event even if it is not directly related to any immediately observable dynamical phenomenon, and because it permits an analytical treatment.

The theoretical analysis went along the same line as of the Helmholtz oscillator, apart from the fact that now there are two homoclinic loops and thus also "global" control is possible. Only solutions of the optimization problem with a finite number of controlling superharmonics were considered, in the case of either "one-side" or "global" control, because only these solutions are appealing from a practical point of view.

Detailed numerical investigations aimed at checking the practical performances of control were performed, by organizing them along the same points **(4.1)**, **(4.2)** and **(4.3)** previously discussed for the Helmholtz oscillator. It was checked that the theoretical Melnikov predictions of the homoclinic bifurcation are reliable, along with the predicted regularization of the basin of attractions with both "global" and "one-side" control (see Fig. 7), and with the ensuing shift (different in the two cases) of the erosion profiles. Furthermore, it was numerically checked, by comparison of the relevant bifurcation diagrams, whether the control is able to delay the onset of cross-well attractors, although there are no theoretical expectations on the occurrence of this welcome phenomenon.

So far the control method was applied always to the homoclinic bifurcation of the hilltop saddle. Thus, we decided to consider also the homoclinic bifurcations of a non-hilltop saddle, which is the object of a subsequent paper[135]. While requiring a complete numerical treatment, although along the same guidelines as of the other cases, this approach permits to focus on an homoclinic bifurcation directly responsible for an unwanted dynamical event. Thus, while in this case the application of control is more onerous and less general (because it is done numerically), it is expected to be much more performant.

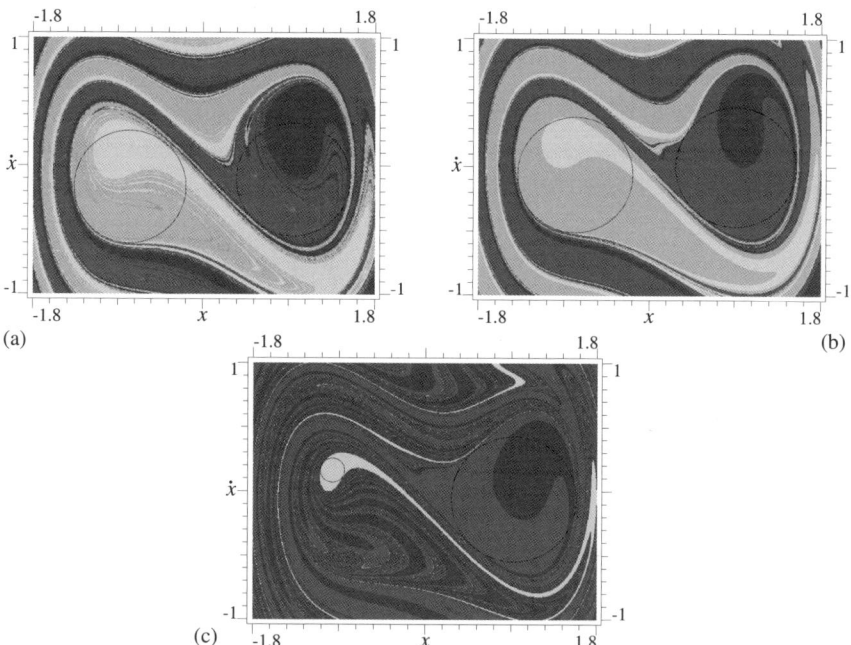

Fig. 7. Basins of attractions for (a) harmonic excitation, (b) global optimal excitation with one superharmonic, and (c) right one-side optimal excitation with one superharmonic.

We considered the Duffing equation

$$\ddot{x} + 0.164\dot{x} - 0.2x + x^3 = A[\sin(t) + c_1\sin(nt+c_2)]. \quad (8)$$

The numbers c_1 and c_2 are the control parameters we added to the basic harmonic excitation, with c_1 measuring the relative amplitude of the superharmonic, and c_2 measuring the phase difference. $n=2$ was used for "one-side" control (asymmetric excitation), and $n=3$ for "global" control (symmetric excitation).

We focused the attention on the transition from single-well to cross-well chaos illustrated in Fig. 8, which is due to a homoclinic bifurcation of a period 3 direct saddle D^3. It was identified by Katz and Dowell[83], and is illustrated in Fig. 9 together with the underlying boundary crisis of the confined attractors. Our aim was to eliminate or, better, to shift in parameters space, the homoclinic bifurcation responsible for the appearance of the scattered chaotic attractor, by properly choosing the control parameters c_1 and c_2.

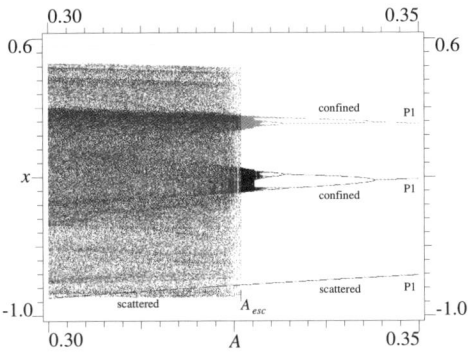

Fig. 8. Bifurcation diagrams of Eq. (8) in the case of harmonic excitation, $c_1 = 0$.

The control method was applied by computing numerically the distance between the stable and the unstable manifold of D^3 as a function of the control parameters c_1 and c_2, while taking fixed the excitation amplitude at $A=0.3252$ corresponding to the homoclinic bifurcation for the harmonic excitation. This is the "heavy" part of the procedure, and requires a lot of CPU time, although being conceptually similar to the analytical computation of the distance by the Melnikov method.

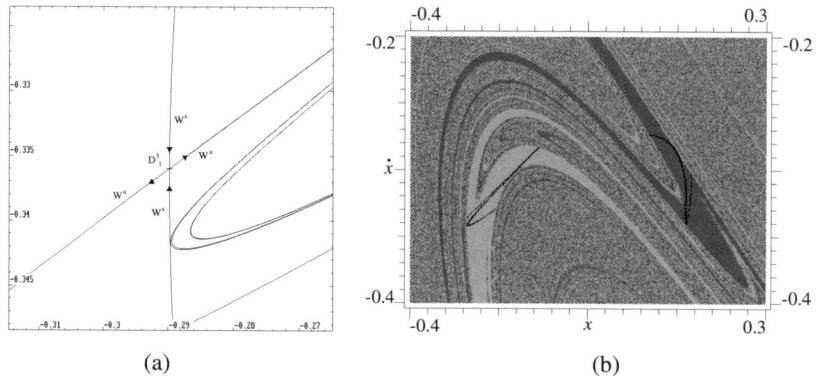

Fig. 9. (a) Stable and unstable manifolds of the saddle D^3 at the homoclinic bifurcation, $A = 0.3252$, and (b) the underlying boundary crisis of the confined attractors ($A = 0.326$). Harmonic excitation, $c_1 = 0$.

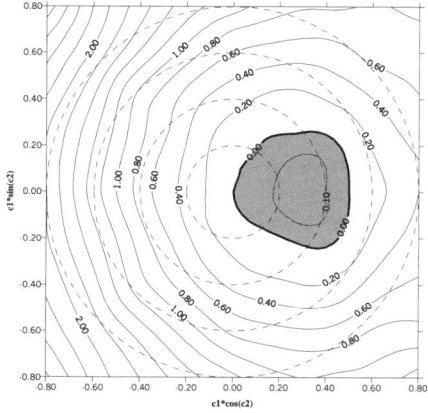

Fig. 10. The contour plot of the function $d(c_1, c_2)$ in polar coordinates for $A_{esc} = 0.3252$ and $n = 3$.

For the case of symmetric excitation ($n=3$), corresponding to "global" control, the polar coordinates contour plot of the distance between stable and unstable manifolds for $A=0.3252$ is reported in Fig. 10. In the white region we have detachment of the manifolds. We conclude that with symmetric excitation the control is theoretically effective in a very large region of the parameter space. In particular, the maximum distance is attained for $c_2 \approx 1.08\pi$, while the influence of c_1 is simpler, because the distance is an increasing function of c_1. We conclude that the optimal excitation has $c_2 \approx 1.08\pi$ and the largest possible c_1.

The practical performances of the optimal excitation are illustrated in Fig. 11, where it is seen able to shift (toward the left) the onset of the scattered chaotic attractors (compare this bifurcation diagram with that of Fig. 8). Furthermore, the larger is c_1, the larger is the distance between the stable and the unstable manifold, and the larger is the shift of the critical threshold for cross-well chaos (Fig. 11).

The case of asymmetric excitation ($n=2$) is analogous, and is not illustrated.

In implementing the control of the homoclinic bifurcation of the hilltop saddle of the oscillators encountered so far (inverted pendulum, Helmholtz and Duffing), we noted that the associated mathematical problems of optimization have the same structure. So, we investigated in detail this point[138,139]. The gains obtained by increasing the number of superharmonics were also evaluated in the various cases.

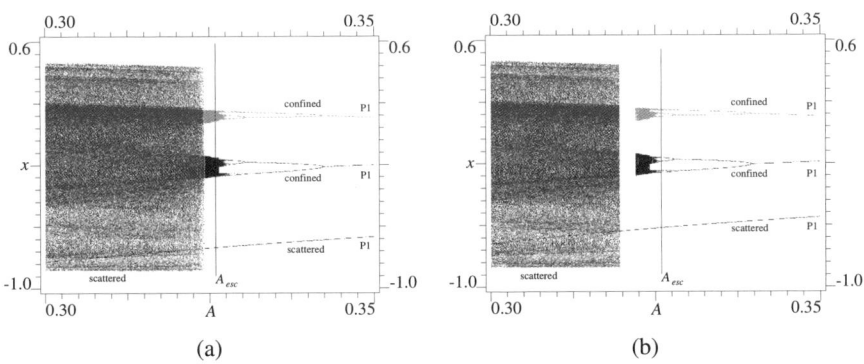

Fig. 11. Bifurcation diagrams in the case of symmetric optimal excitation $n=3$, $c_2=1.08\pi$ and a) $c_1=0.4$, b) $c_1=0.8$.

Four softening (Helmholtz, Helmholtz-Duffing with negative cubic term, Duffing with negative cubic term and mathematical pendulum) and two hardening (Duffing with positive cubic term and a nonlinear oscillator with large odd stiffness) smooth oscillators were considered within a *unified framework*[139]. After an extensive theoretical analysis, it was shown that the optimization problem for "one-side" control,

$$\text{Maximize } G \text{ by varying the shape of the excitation} \qquad (9)$$

(here G is one between the left G^l and the right G^r gain), is "universal," i.e., it is system independent, so that it can be solved once and for all. In fact, (9) can be recast into the mathematical form[139]

$$\text{Maximize } \min_{m\in[0,2\pi]}\{h(m)\} \text{ by varying the Fourier coefficients}$$

$$h_i \text{ and } \Psi_i \text{ of } h(m)=\cos(m+\Psi_1)+\Sigma_{i\geq 2}\, h_i \cos(im+\Psi_i), \qquad (10)$$

which no longer contains the system parameters.

The same result holds for the "global" control of the two-well symmetric oscillators, and the optimization problem

$$\text{Maximize } G=\min\{G^r,G^l\} \text{ by varying the shape of the excitation} \qquad (11)$$

becomes "universal" in the form

$$\text{Maximize } \min_{m\in[0,2\pi]}\{h(m)\} \text{ by varying the Fourier coefficients}$$

h_i and Ψ_i of $h(m)=\cos(m+\Psi_1)+\Sigma_{i\geq 2}\, h_i \cos(im+\Psi_i)$ under the

$$\text{constraint } -\min_{m\in[0,2\pi]}\{h(m)\}=\max_{m\in[0,2\pi]}\{h(m)\}. \qquad (12)$$

It has been shown[139] that the constraint is automatically satisfied by assuming $h_i=0$ for even values of i.

A different behavior is encountered for the "global" control of two-well non-symmetric systems[138] or for oscillators having more than two potential wells. Here the universality of the solution is lost, and a much more involved situation occurs. We focus attention[138] to the "global" control of the Helmholtz-Duffing oscillator

$$\ddot{x}+\varepsilon\delta\, \dot{x}-\sigma x-\frac{3}{2}(\sigma-1)x^2+2x^3=\varepsilon\chi(\omega t), \quad \chi(s)=\gamma_1 \sum_i \frac{\gamma_i}{\gamma_1}\sin(is+\psi_i), \qquad (13)$$

which, due to the asymmetry parameter σ, has two different homoclinic loops of the hilltop saddle and, consequently, two different homoclinic bifurcation thresholds $\gamma^r_{1,cr}$ and $\gamma^l_{1,cr}$, as shown in the (normalized) excitation parameter space of Fig. 12.

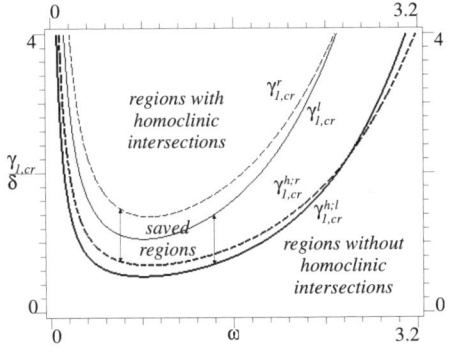

Fig. 12. The curves $\gamma^{h;r}_{1,cr}$, $\gamma^{h;l}_{1,cr}$ and $\gamma^{r}_{1,cr}$, $\gamma^{l}_{1,cr}$ for $\sigma = 1.2$.

To control simultaneously the right and left homoclinic tangencies ("global" control "without symmetrization"), we have to increase simultaneously the right and left gains G^r and G^l, which are strictly correlated. Mathematically, this entails increasing the minimum value of G^r and G^l, namely, the optimization problem (11).

In this way, we increase the critical threshold of both right and left homoclinic bifurcations, even if the proportional increments (i.e., the gains) with respect to the harmonic case might be different in the two cases. However, the solutions are seen to actually satisfy the condition $G^r=G^l=G$. The main novelty of this mathematical problem of optimization is that the solution depends on σ and ω, and thus is no longer system-independent and no longer universal. In fact, problem (11) cannot be recast in the form (12) in this case.

In the previous strategy the right and left homoclinic bifurcation thresholds are $\gamma^{r}_{1,cr} = G\gamma^{h;r}_{1,cr}$ and $\gamma^{l}_{1,cr} = G\gamma^{h;l}_{1,cr}$, respectively, and they are distinct ($\gamma^{h;r}_{1,cr}$ and $\gamma^{h;l}_{1,cr}$ correspond to the reference harmonic excitation). The difference due to the asymmetry of the system is thus maintained by the controlling excitation. Another strategy ("global" control "with symmetrization") can be followed, by choosing to increase the lowest of the two values $\gamma^{r}_{1,cr} = G\gamma^{h;r}_{1,cr}$ and $\gamma^{l}_{1,cr} = G\gamma^{h;l}_{1,cr}$ corresponding to a generic excitation, instead of increasing the lowest gain. In this case, the aim is to increase the threshold of the first critical event, thus improving the worst performance of the system while still keeping an overall control. From this viewpoint, the critical threshold is $\gamma_{1,cr} = \min\{\gamma^r_{1,cr}, \gamma^l_{1,cr}\}$. The associated optimization problem is:

$$\text{Maximize } \gamma_{1,cr} = \min\{G^r \gamma^{h;r}_{1,cr}; G^l \gamma^{h;l}_{1,cr}\} \text{ by varying the excitation shape.} \quad (14)$$

As in the case of "global" control "without symmetrization", the solution depends on the system. Here, however, the things are even more involved, since the solution behaves differently according to whether $\gamma^{h;r}_{1,cr}$ is close to $\gamma^{h;l}_{1,cr}$ or away from it. The former case, in which the optimal solution is characterized by the condition $G^r \gamma^{h;r}_{1,cr} = G^l \gamma^{h;l}_{1,cr}$ and thus the right and left homoclinic bifurcations occur simultaneously, is called "global" control "with symmetrization".

In the latter case, the procedure exhausts all of its resources in increasing the lower threshold between $\gamma^r_{1,cr}$ and $\gamma^l_{1,cr}$, and it is not able to symmetrize the system. Thus, strictly speaking, this is still a case of global control without symmetrization. However, this case is similar to the previous one because the same control strategy is adopted, irrespective of the solution of the problem which is an a posteriori result. In this respect, one can indeed consider the solution as the one realizing the maximum symmetrization allowed for by the specific characteristics of the system and of the excitation. We named the former case ($\gamma^{h;r}_{1,cr} \approx \gamma^{h;l}_{1,cr}$) as "global" control with "achieved symmetrization" and the latter case as "global" control with "pursued symmetrization".

3.3. *Special applications*

The next step of the research consisted in applying the control method to further mechanical systems, from the practical point of view, and in also considering the anti-control problem, from a theoretical point of view. The latter issue, which consists in enhancing instead of removing chaos, can also be of practical interest.

We considered two somehow complementary new systems: the nonlinear thermoelastic electrically actuated microbeam[142] depicted in Fig. 13, which is a very modern application in the area of MEMS (Micro Electro Mechanical Systems), and the rocking rigid block[140,141], which is an old but evergreen mechanical model.

The application of the control method to MEMS[142] is standard, and does not require any special care from a control point of view. Here the deep motivation of the paper was to investigate how the control method works in the coupled thermoelectromechanical context involving a physical event – the dynamic pull-in phenomenon – different from the purely mechanical ones (e.g., hitting, capsizing, structural instability) implicitly considered in former oscillators. Yet, the pull-in phenomenon is related to the softening behaviour of the system, thus somehow resembling the behaviour of the Helmholtz oscillator.

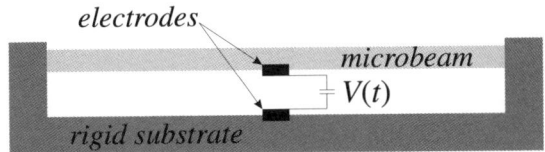

Fig. 13. A schematic picture of the electrically actuated microbeam.

According to the previous observation, herein we pay attention to checking the effectiveness of the control method, and not to its application features. In particular, we focus on the safe basin erosion by control. We report in Fig. 14 the basins of attraction for various relative amplitudes η_2/η_1 of the (single) controlling superharmonic[142]. The case of harmonic excitation with the associated eroded basin boundary is also reported for comparison (Fig. 14(b). The addition of the superharmonic term is indeed able to significantly reduce the erosion, as shown in Fig. 14(c). Then, by increasing η_2/η_1 up to the optimal value (Fig. 14(d)), the erosion is further reduced. The crucial role of the added superharmonic is underlined by Fig. 14(a), which shows the dramatic effects obtained if the superharmonic is not properly chosen, in particular if considering a wrong sign corresponding to a half period phase shift.

Representative erosion profiles with harmonic ($\eta_2/\eta_1=0$), control ($\eta_2/\eta_1=0.5$), and optimal control ($\eta_2/\eta_1=1.6591$) excitations are compared in Fig. 15, from which it is clear that the previous observations on control effectiveness, made for a fixed amplitude value, do extend and generalize.

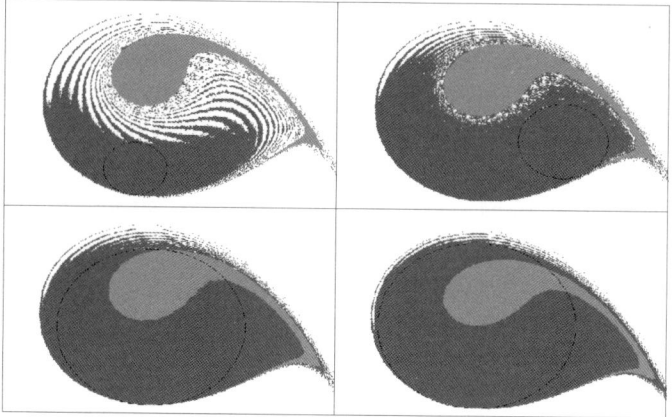

Fig. 14. Safe basins for (a) $\eta_2/\eta_1=-1.5$ (upper left), (b) $\eta_2/\eta_1=0$ (harmonic), (c) $\eta_2/\eta_1=0.5$ and (d) $\eta_2/\eta_1=1.6591$ (optimal, bottom right).

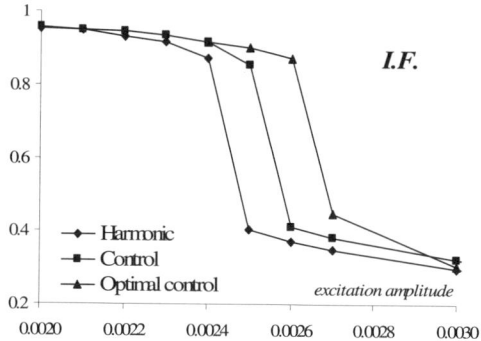

Fig. 15. Erosion profiles for harmonic ($\eta_2/\eta_1=0$), control ($\eta_2/\eta_1=0.5$) and optimal control ($\eta_2/\eta_1=1.6591$) excitations for $\Omega=0.7$.

The research concerning the rocking rigid block dynamics had a double motivation. From the one-side, in this case heteroclinic instead of homoclinic bifurcations are the unwanted dynamical phenomena to be eliminated by control[140]. From the other side, for the first time interest is also in implementing *anti-control* of chaos, so that we need a system as simple as possible in order not to hide the anti-control features behind any mechanical complexity[141].

We initially focused on the control-induced differences between slender and non-slender blocks[140]. There are differences in the computation of the heteroclinic bifurcation threshold, which can be done exactly in the first case and asymptotically (by the Melnikov method) in the second case. However, this difference has no effects in terms of the underlying mathematical optimization problem, which is again the same encountered in previous works, thus confirming the "universality" of the previously discussed method[139].

The work[141] contains two main parts. In the first one, the control problem is addressed. Here the element of novelty is that, in addition to the application of the control method to the heteroclinic bifurcation, the control ideas are applied to another critical threshold of mechanical interest, i.e. the immediate overturning threshold, where the block topples down instantaneously without preliminarily rocking.

However, the main element of novelty is in the second part of the paper, where the anti-control problem is addressed. Changing the point of view, from control to anti-control, implies an important change in the approach to the problem. In fact, there is no longer a given harmonic excitation, of amplitude γ_1, to be controlled by adding superharmonics, but rather a "completely" free periodic excitation. Accordingly, we need an overall amplitude measure which takes into account the amplitudes of each of the superharmonics, and we must look for the *worst* excitation in the class of shapes sharing the same magnitude, even accepting that the amplitude γ_1 of the first harmonic, which has no more special meaning as in the control case, is reduced. Practically, we use the input energy as a measure of the amplitude of the excitation. Physical considerations support this choice, since it is not difficult to trigger undesired dynamical events, in particular overturning, by increasing the input energy. The interesting question is instead decreasing the critical threshold by keeping fixed the input energy.

We considered the anti-control of the heteroclinic and of the immediate overturning thresholds, and in both cases "one-side" and "global" anti-control strategies are proposed, similarly to what happens for control. The underlying optimization problems were stated and found to be mathematically identical for the heteroclinic and the immediate overturning thresholds. They have been solved by the Lagrange multipliers theorem, and the solutions have been compared with each other. The anti-controlled gains measuring the *reduction* (while in control there is an *increment*) of the critical threshold are also computed, and an example is reported in Fig. 16.

The theoretical results have been compared with numerical simulations aimed, as usual, at checking the practical performances. Here, attention has been focused on the charts of overturning, which is the most important dynamical event for rocking rigid blocks, and on checking how the control is able to increase the stability region, while the anti-control is able to enlarge the toppling region.

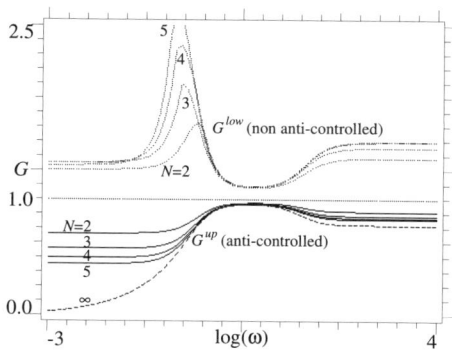

Fig. 16. Optimal one-side anti-control gains, for different values of the number N of employed superharmonics.

3.4. *The infinite dimension*

The most recent research step[143] is concerned with a decisive point, the infinite dimension of most of the controlled mechanical/structural systems. In fact, because of focusing the attention on control/anti-control and on their theoretical developments and practical consequences, all of the previous papers dealt with single d.o.f. systems, so that it is deemed necessary to also consider actually continuous structures, which are of course more realistic.

From a mathematical point of view this means that partial differential equations are considered instead of ordinary differential equations. In fact, we considered the buckled beam, whose dynamics are governed by

$$\ddot{w}+w''''+\Gamma w''-kw''\int_0^1 (w')^2 d\zeta =\varepsilon[F(z,t)-\delta \dot{w}],\qquad(15)$$

where $w(z,t)$ is the time dependent transversal displacement of the point at abscissa $z\in[0,1]$, $\Gamma>\Gamma_{cr}$ is the dimensionless first-order axial load (positive=compression, Γ_{cr}=buckling load), k is the dimensionless stiffness due to the membrane effects, δ is the coefficient of viscous damping, and

$$F(z,t)=\Sigma_n f_n(z)\sin(n\omega t+\psi_n)\qquad(16)$$

is the external $T=2\pi/\omega$ time periodic, spatially distributed, excitation.

To focus on the infinite dimension of the problem, we applied the control method to the homoclinic bifurcations of the hilltop saddle, i.e., the simplest case considered so far. However, we can now vary not only the *temporal* but also the *spatial* shape of the excitation, which is assumed to be periodic in time.

Detecting the homoclinic bifurcations of the hilltop saddle is, as usual, the starting point for the application of control. This has been done by means of the Holmes and Marsden[77] theorem, which is a quite involved extension of the classical Melnikov method to infinite dimension. This theorem requires that the homoclinic orbits of the unperturbed (no damping and no excitation) system lie on a two-dimensional invariant manifold embedded in the infinite-dimensional phase space. Looking for this manifold, we realized that two different families of boundary conditions have to be considered. For the first one, which is the simplest but the less general, the manifold is flat, while for the second one, which is more difficult but more general, the manifold bends in the phase space.

We explicitly applied the control to one representative case of each family. The hinged-hinged beam, which belongs to the first family of boundary conditions, is considered first. In the application of the method we exploit the possibility of modifying both the spatial and temporal shape of the excitation, and look for the best shape which permits to increase the homoclinic bifurcation thresholds as much as possible, both in "global" and "one-side" sense.

Optimizing the spatial shape of the excitation is easy for the first family. In fact, the flatness of the manifold implies that all spatial modes perpendicular to the two modes spanning the manifold do not modify the homoclinic bifurcation threshold. Thus, they are useless in view of modifying the chaotic region, and accordingly they are not considered, $f_n(z)=0$, $n=2,3,\ldots$. This entails two meaningful circumstances:

(i) it permits to minimize the cost of control, because we only use what is effective;
(ii) it permits to automatically satisfy some non-resonance conditions involved in the Holmes and Marsden theorem.

We consider practically "optimal" this assumption although, contrary to what happens with the temporal shape, such a choice does not ensue from a mathematical optimization problem, but just from mainly qualitative considerations.

With this assumption, we then passed to the optimization of the temporal shape of the excitation. Fortunately enough, we found that it leads to the same mathematical problem of optimization as that obtained for the reduced-order (single-mode) Duffing model, so we take advantage from our previous results.

The fixed-fixed case is considered as a representative of the second family of boundary conditions. Here the main issue is the detection of the invariant non-planar manifold, which has been done by the nonlinear normal mode technique, in particular by following Rand's approach[107]. Since this involves extensive computations, the axial load is fixed slightly above the buckling load. This choice, which is of course restrictive, is supported by the interest in obtaining some first results in this direction.

Upon detecting the invariant manifolds, we computed the homoclinic bifurcation threshold by the Holmes and Marsden theorem. To have an expression of the unperturbed homoclinic solutions, we considered reduced order models governing the dynamics on the invariant non-planar manifold[87]. Then, we passed to the optimization problem, i.e. we looked for the best spatial and temporal shape permitting the best shift of the critical thresholds. As in the case of the first family, the spatial shape of the excitation is "optimized" by qualitative considerations, while the temporal shape is optimized mathematically. This is easy because we encounter again the same optimization problems as of single d.o.f. systems, and again take advantage of previous results.

The general conclusion[143] is that the non-flatness of the manifold, while being theoretically important, has minor consequences from a practical point of view. This result is due to the closeness of Γ to the buckling value, and cannot be generalized.

4. Conclusions

Upon establishing an overall framework within which addressing control of system complex dynamics, an overview of recent relevant achievements in the area of solid and structural mechanics – along with some open questions – has been provided, by referring to some main classes of mechanical/structural systems and to a few paradigmatic problems.

Thereafter, a method for controlling nonlinear dynamics and chaos via elimination of homo/heteroclinic bifurcations, previously developed by the authors, has been systematically reviewed. Contrary to previous review papers[27,144], where the topic has been addressed from a methodological point of view, in the present work the control technique, and its applications, are revisited in view of their chronological developments. Apart from better illustrating the main lines along which the research has been driven, this approach also permits to highlight the sequence of theoretical problems addressed and solved, along with the increasingly wider range of technical problems considered.

Several mechanical models and systems have been addresses within an organized framework, by explaining the motivations behind the choices made in various works and by mostly referring to the application aspects of the method, whose relevant details are contained in the original papers.

We feel that this chapter, providing a comprehensive and unified overview on the control method and on its practical features for various mechanical/structural systems, contributes a better understanding of the former, while representing a bridge towards further applications. Among them, we mention the extension to other, more involved, systems, the combined use of parametric and external excitations, of quasi-periodic or random excitations, the role played by the reduced order modeling in complex dynamics, and the physical experimentation.

References

Books on control of chaos

1. G.R. Chen and X. Dong, *From Chaos to Order: Methodologies, Perspectives and Applications* (World Scientific, Singapore, 1998).
2. G.R. Chen, *Controlling Chaos and Bifurcations in Engineering Systems* (CRC Press, Boca Raton, 1999).
3. K. Judd, A.I. Mees, K.L. Teo and T. Vincent, *Control of Chaos* (Birkhauser, Boston, 1997).
4. A.L. Fradkov and A.Yu. Pogromsky, *Introduction to Control of Oscillations and Chaos* (World Scientific, Singapore, 1998).
5. T. Kapitaniak, *Controlling Chaos* (Academic Press, Lodz, 1996).
6. M. Lakshmanan and K. Murali, *Chaos in Nonlinear Oscillators: Controlling and Synchronization* (World Scientific, Singapore, 1996).
7. H.-G. Schuster, Ed., *Handbook of Chaos Control: Foundations and Applications* (Wiley & Sons, 1999).
8. R. Chacon, *Control of Homoclinic Chaos by Weak Periodic Perturbations*, Series on Nonlinear Science, Series A, Vol. 55 (World Scientific, 2005).

Special Issues

9. T. Arecchi, S. Boccaletti, M. Ciofini, C. Grebogi and R. Meucci, Eds., Control of Chaos: New Perspectives in Experimental and Theoretical Nonlinear Science, *Int. J. Bif. Chaos* **8** (1998).
10. G.R. Chen and M.J. Ogorzalek, Eds., Control and Synchronization of Chaos, *Int. J. Bif. Chaos* **10** (2000).
11. G.R. Chen, Ed., *Int. J. Bif. Chaos* **12** (2002).
12. W.L. Ditto and K. Showalter, Eds., Control and Synchronization of Chaos, *Chaos* **7** (1997).
13. M. Hasler and J. Vandevalle, Eds., Communications, Information Processing and Control Using Chaos, *Int. J. Circuit Theor. Appl.* **27** (1999).
14. T. Kapitaniak, Ed., Controlling Chaos, *Chaos, Solit. & Fract.* **8** (1997).
15. M. Kennedy and M. Ogorzalek, Eds., Chaos Control and Synchronization, *IEEE Trans. Circuits and Systems* **44** (1997).
16. H. Nijmeijer, Ed., Control of Chaos and Synchronization, *Systems & Control Letters* **31** (1997).

17. S. Lenci and G. Rega, Eds., Exploiting Chaotic Properties of Dynamical Systems for Their Control: Suppressing, Enhancing, Using Chaos, *Phil. Trans. Roy. Soc. A* **364** (2006).

Reviews or surveys

18. T. Arecchi, S. Boccaletti, M. Ciofini, R. Meucci and C. Grebogi, The Control of Chaos: Theoretical Schemes and Experimental Realizations, *Int. J. Bif. Chaos* **8**(8), 1643-1655 (1998).
19. B. Blazejczyk, T. Kapitaniak, J. Wojewoda and J. Brindley, Controlling Chaos in Mechanical Systems, *Applied Mechanics Review* **46**(7), 385-391 (1993).
20. S. Boccaletti, C. Grebogi, Y.C. Lai, H. Mancini and D. Maza, The Control of Chaos: Theory and Applications, *Physics Reports* **329**, 103-197 (2000).
21. G.R. Chen and X. Dong, From Chaos to Order - Perspectives and Methodologies in Controlling Chaotic Nonlinear Dynamical Systems, *Int. J. Bifur. Chaos* **3**(6), 1363-1409 (1993).
22. M.Z. Ding, E.J. Ding, W.L. Ditto, B. Gluckman, V. In, J.H. Peng, M.L. Spano and W.M. Yang, Control and Synchronization of Chaos in High Dimensional Systems: Review of Some Recent Results, *Chaos* **7**(4), 644-652 (1997).
23. A.L. Fradkov, Chaos Control Bibliography (1997-2000), Russian Systems and Control Archive (RUSYCON), www.rusycon.ru/chaos-control.html (2000).
24. A.L. Fradkov and R.L. Evans, Control of Chaos: Methods and Applications in Engineering, *Ann. Rev. Contr.* **29**, 33-56 (2005).
25. A.L. Fradkov, R.L. Evans and B.R. Andrievski, Control of Chaos: Methods and Applications in Mechanics, *Phil. Trans. Roy. Soc. A* **364**, 2279-2307 (2006).
26. J.F. Lindner and W.L. Ditto, Removal, Suppression, and Control of Chaos by Nonlinear Design, *Applied Mechanics Review* **48**(12), 795-807 (1995).
27. G. Rega and S. Lenci, Bifurcations and Chaos in Single-d.o.f. Mechanical Systems: Exploiting Nonlinear Dynamics Properties for their Control, in *Recent Research Developments in Structural Dynamics*, Ed. A. Luongo (Transworld Research Network, 2003), pp. 331-369.
28. G. Rega and S. Lenci, Introduction to Theme Issue "Exploiting Chaotic Properties of Dynamical Systems for Their Control: Suppressing, Enhancing, Using Chaos," *Phil. Trans. Roy. Soc. A* **364**, 2269-2277 (2006).
29. M.L. Spano and E. Ott, Controlling Chaos, *Physics Today* **48**(6), 34-40 (1995).

Research papers

30. H.N. Agiza, Controlling Chaos for the Dynamical System of Coupled Dynamos, *Chaos Solit. Fract.* **13**, 341-352 (2002).
31. A.K. Agrawal, J.N. Yang and J.C. Wu, Non-linear Control Strategies for Duffing Systems, *Int. J. Non-Linear Mech.* **33**(5), 829-841 (1998).
32. L.A. Aguirre and S.A. Billings, Closed-loop Suppression of Chaos in Nonlinear Driven Oscillators, *J. Nonlin. Sci.* **5**, 189-206 (1995).
33. A. Alasty and H. Salarieh, Nonlinear Feedback Control of Chaotic Pendulum in Presence of Saturation Effect, *Chaos Solit. Fract.* **31**, 292-304 (2007).

34. J. Alvarez-Ramirez, G. Espinosa-Paredes and H. Puebla, Chaos Control Using Small-amplitude Damping Signals, *Phys. Lett. A* **316**, 196-205 (2003).
35. E. Barreto, F. Casas, C. Grebogi and E.J. Kostelich, Control of Chaos: Impact Oscillators and Targeting, in *IUTAM Symp. Interaction Between Dyn. Contr. in Adv. Mech. Syst.*, Ed. D.H. van Campen (Kluwer, Eindhoven, 1997), pp. 17-26.
36. G.L. Baker, Control of the Chaotic Driven Pendulum, *Am. J. Phys.* **63**, 832-838 (1995).
37. M. Belhaq and M. Houssni, Suppression of Chaos in Averaged Oscillation Driven by External and Parametric Excitation, *Chaos Solit. Fract.* **11**, 1237-1246 (2000).
38. C.M. Bird and P.J. Aston, Targeting in the Presence of Noise, *Chaos, Solit. & Fract.* **9**(1/2), 251-259 (1998).
39. S.R. Bishop and D. Xu, Control of the Parametrically Excited Pendulum, in *IUTAM Symp. Interaction Between Dyn. Contr. in Adv. Mech. Syst.*, Ed. D.H. van Campen (Kluwer, Eindhoven, 1997), pp. 43-50.
40. S.R. Bishop and D. Xu, The Use of Control to Eliminate Subharmonic and Chaotic Impacting Motions of a Driven Beam, *J. Sound Vibr.* **205**(2), 223-234 (1997).
41. S.R. Bishop, D.J. Wagg and D. Xu, Use of Control to Maintain Period-1 Motions During Wind-up or Wind-down Operations of an Impacting Driven Beam, *Chaos, Solit. & Fract.* **9**(1/2), 261-269 (1998).
42. S. Boccaletti, A. Farini, E.J. Kostelich and T. Arecchi, Adaptive Targeting of Chaos, *Physical Review E* **55**(5), 4845-4848 (1997).
43. S. Bowong and F.M.M. Kakmeni, Chaos Control and Duration Time of a Class of Uncertain Chaotic Systems, *Physics Letters A* **316**(3-4), 206-217 (2003).
44. Y. Braiman and I. Goldhirsch, Taming Chaotic Dynamics with Weak Periodic Perturbation, *Phys. Rev. Lett.* **66**, 2545-2548 (1991).
45. G. Cai and J. Huang, Optimal Control Method with Time Delay in Control, *J. Sound Vibr.* **251**(3), 383-394 (2002).
46. H.J. Cao, X.B. Chi and G.R. Chen, Suppressing or Inducing Chaos by Weak Resonant Excitations in an Externally-forced Froude Pendulum, *Int. J. Bif. Chaos* **14**(3), 1115-1120 (2004).
47. H.J. Cao, X.B. Chi and G.R. Chen, Suppressing or Inducing Chaos in a Model of Robot Arms and Mechanical Manipulators, *J. Sound Vibr.* **271**(3-5), 705-724 (2004).
48. H.J. Cao, Primary Resonant Optimal Control for Homoclinic Bifurcations in Single-Degree-of-Freedom Nonlinear Oscillators, *Chaos Solit. Fract.* **24**, 1387-1398 (2005).
49. H.J. Cao, G.R. Chen, Global and Local Control of Homoclinic and Heteroclinic Bifurcations, *Int. J. Bif. Chaos* **15**(8), 2411-2432 (2005).
50. H.J. Cao, G.R. Chen, A Simplified Optimal Control Method for Homoclinic Bifurcations, *Nonlin. Dyn.* **42**, 43-61 (2005).
51. F. Casas and C. Grebogi, Control of Chaotic Impacts, *Int. J. Bif. Chaos* **7**(4), 951-955 (1997).
52. H.K. Chen and C.I. Lee, Anti-control of Chaos in Rigid Body Motion, *Chaos, Solit. & Fract.* **21**, 957-965 (2004).
53. N.J. Corron, S.D. Pethel and B.A. Hopper, Controlling Chaos with Simple Limiters, *Phys Rev. Lett.* **84**(17), 3835-3838 (2000).

54. S.L.T. de Souza and I.L. Caldas, Controlling Chaotic Orbits in Mechanical Systems with Impacts, *Chaos Solit. Fract.* **19**, 171-178 (2004).
55. S.L.T. de Souza, I.L. Caldas, R.L. Viana, J.M. Balthazar and R.M.L.R.F. Brasil, Impact Dampers for Controlling Chaos in Systems with Limited Power Supply, *J. Sound Vibr.* **279**, 955-967 (2005).
56. S.L.T. de Souza, I.L. Caldas and R.L. Viana, Damping Control Law for a Chaotic Impact Oscillator, *Chaos Solit. Fract.* **32**, 745-750 (2007).
57. S.L.T. de Souza, I.L. Caldas, R.L. Viana, J.M. Balthazar and R.M.L.R.F. Brasil, A Simple Feedback Control for a Chaotic Oscillator with Limited Power Supply, *J. Sound Vibr.* **299**, 664-671 (2007).
58. S.L.T. de Souza, I.L. Caldas, R.L. Viana and J.M. Balthazar, Control and Chaos for Vibro-impact and Non-ideal Oscillators, to appear on *J. Theor. Appl. Mech.* (2008).
59. M. di Bernardo and D.P. Stoten, MCS Adaptive Control of Nonlinear Systems: Utilizing the Properties of Chaos, *Phil. Trans. Roy. Soc. A* **364**, 2397-2415 (2006).
60. M. Ding, E. Ott and C. Grebogi, Controlling Chaos in a Temporally Irregular Environment, *Physica D* **74**, 386-394 (1994).
61. M. Ding, Controlling Chaos in a Temporally Irregular Environment and Its Application to Engineering Systems, in *IUTAM Symp. Interaction Between Dyn. Contr. in Adv. Mech. Syst.*, Ed. D.H. van Campen (Kluwer, Eindhoven, 1997), pp. 109-117.
62. A.R. Dzhanoev, A. Loskutov, H.J. Cao and M.A.F. Sanjuán, A New Mechanism of the Chaos Suppression, *Discrete and Contin. Dyn. Systems B* **7**(2), 275-284 (2007).
63. B. Epureanu, S.T. Trickey and E. Dowell, Stabilization of Unstable Limit Cycles in Systems with Limited Controllability: Expanding the Basin of Convergence of OGY-type Controllers, *Nonlin. Dyn.* **15**, 191-205 (1998).
64. B. Epureanu and E. Dowell, On the Optimality of the Ott-Grebogi-Yorke Control Scheme, *Physica D* **116**, 1-7 (1998).
65. G. Filatrella, G. Rotoli and M. Salerno, Suppression of Chaos in the Perturbed Sine-Gordon System by Weak Periodic Signals, *Physics Letters A* **178**, 81-84 (1993).
66. Z.M. Ge, C.M. Chang and Y.S. Chen, Anti-control of Chaos of Single Time Scale Brushless DC Motor, *Phil. Trans. Roy. Soc. A* **364**, 2449-2462 (2006).
67. Z.M. Ge, J.W. Cheng and Y.S. Chen, Chaos Anti-control and Synchronization of Three Time Scales Brushless DC Motor System, *Chaos Solit. Fract.* **22**, 1165-1182 (2004).
68. Z.M. Ge and C.I. Lee, Non-linear Dynamics and Control of Chaos for a Rotational Machine with Hexagonal Centrifugal Governor with a Spring, *J. Sound Vibr.* **262**, 845-864 (2003).
69. Z.M. Ge and C.I. Lee, Control, Anti-control and Synchronization of Chaos for an Autonomous Rotational Machine System with Time Delay, *Chaos Solit. Fract.* **23**, 1855-1864 (2005).
70. Z.M. Ge and W.Y. Leu, Anti-control of Chaos of Two-degree-of-freedom Louder Speaker System and Chaos Synchronization of Different Order System, *Chaos Solit. Fract.* **20**, 503-521 (2004).
71. Z.M. Ge and T.M. Lin, Chaos, Chaos Control and Synchronization of Electro-mechanical Gyrostat System, *J. Sound Vibr.* **259**(3), 585-603 (2003).

72. Z.M. Ge and J.S. Shiue, Nonlinear Dynamics and Control of Chaos for Tachometer, *J. Sound Vibr.* **253** (4), 773-793 (2002).
73. Z.M. Ge and H.W. Wu, Chaos Synchronization and Chaos Anti-control of a Suspended Track with Moving Load, *J. Sound Vibr.* **270**, 685-712 (2004).
74. J. Guémez, J.M. Gutiérrez, A. Iglesias and M.A. Matias, Suppression of Chaos through Changes in the System Variables: Transient Chaos and Crises, *Physica D* **79**, 164-173 (1994).
75. D.L. Hill, On the Control of Chaotic Dynamical Systems Using Nonlinear Approximations, *Int. J. Bif. Chaos* **11**(1), 207-213 (2001).
76. D.L. Hill, On the Control of High Dimensional Chaotic Dynamical Systems Using Nonlinear Approximations, *Int. J. Bif. Chaos* **11**(6), 1753-1760 (2001).
77. P. Holmes and J. Marsden, A Partial Differential Equation with Infinitely Many Periodic Orbits: Chaotic Oscillations of a Forced Beam, *Arch. Rat. Mech. Anal.* **76**, 135-165 (1981).
78. H.Y. Hu, Controlling Chaotic Motion of a Mechanical System with a Set-up Elastic Stop, in *IUTAM Symp. Interaction Between Dyn. Contr. in Adv. Mech. Syst.*, Ed. D.H. van Campen (Kluwer, Eindhoven, 1997), pp. 151-158.
79. B. Hübinger, R. Doerner, W. Martienssen, M. Herdering, R. Pitka and U. Dressler, Controlling Chaos Experimentally in Systems Exhibiting Large Effective Lyapunov Exponents, *Phys. Rev. E* **50**, 932-948 (1994).
80. E.R. Hunt, Stabilizing High-periodic Orbits in a Chaotic System: The Diode resonator, *Phys. Rev. Lett.* **67**(15), 1953-1955 (1991).
81. E.R. Hunt, Keeping Chaos at Bay, *IEEE Spectrum* **30**, 32-36 (1993).
82. A. Isidori, *Nonlinear Control Systems* (Springer-Verlag, Rome, 1995).
83. A. Katz and E.H. Dowell, From Single Well Chaos to Cross Well Chaos: A Detailed Explanation in Terms of Manifold Intersections, *Int. J. Bif. Chaos* **4**, 933-941(1994).
84. K. Kecik and J. Warminski, Control of Regular and Chaotic motions of an Autoparametric System with Pendulum by Using MR Damper, in 9th *Conf. Dyn. Syst. Theory Appl.*, Eds. J. Awrejcewicz, P. Olejnik and J. Mrozowski, Lodz, **2**, (2007), pp. 649-656.
85. G. Kovacic and S. Wiggins, Orbits Homoclinic to Resonance, with an Application to Chaos in a Model of the Forced and Damped Sine-Gordon Equation, *Physica D* **57**, 185-225 (1992).
86. J.Y. Lee and J.J. Yan, Control of Impact Oscillator, *Chaos Solit. Fract.* **28**, 136-142 (2006).
87. S. Lenci and G. Rega, Dimension Reduction of Homoclinic Orbits of Buckled Beams via Nonlinear Normal Modes Technique, *Int. J. Non-Lin. Mech.* **42**, 515-528 (2007).
88. R. Lima and M. Pettini, Suppression of Chaos by Resonant Parametric Perturbations, *Phys. Rev. A* **41**, 726-733 (1990).
89. G. Litak, A. Syta and M. Borowiec, Suppression of Chaos by Weak Resonant Excitations in a Nonlinear Oscillator with a Non-symmetric Potential, *Chaos Solit. Fract.* **32**, 694-701 (2007).
90. E.N. Macau and C. Grebogi, Control of Chaos and its Relevance to Spacecraft Steering, *Phil. Trans. Roy. Soc. A* **364**, 2462-2475 (2006).

91. G.M. Mahmoud, A.A. Mohamed and A.A. Shaban, Strange Attractors and Chaos Control in Periodically Forced Complex Duffing's Oscillators, *Physica A: Statistical Mechanics and its Applications* **292**(1-4), 193-206 (2001).
92. F.C. Moon, J. Cusumano and P.J. Holmes, Evidence for Homoclinic Orbits as a Precursor to Chaos in a Magnetic Pendulum, *Physica D* **24**, 383-390 (1987).
93. F.C. Moon, *Chaotic and Fractal Dynamics* (J. Wiley & Sons, New York, 1992).
94. F.C. Moon, M.A. Johnson and W.T. Holmes, Controlling Chaos in a Two-Well Oscillator, *Int. J. Bif. Chaos* **6**(2), 337-347 (1996).
95. F.C. Moon, A.J. Reddy and W.T. Holmes, Experiments in Control and Anti-Control of Chaos in a Dry Friction Oscillator, *J. Vibr. Control* **9**, 387-397 (2003).
96. B.R. Nana Nbendjo, Y. Salissou and P. Woafo, Active Control with Delay of Catastrophic Motion and Horseshoes Chaos in a Single Well Duffing Oscillator, *Chaos Solit. Fract.* **23**, 809-816 (2005).
97. B.R. Nana Nbendjo, R. Tchoukuegno and P. Woafo, Active Control with Delay of Vibration and Chaos in a Double-well Duffing Oscillator, *Chaos Solit. Fract.* **18**, 345-353 (2003).
98. B.R. Nana Nbendjo and P. Woafo, Active Control with Delay of Horseshoes Chaos Using Piezoelectric Absorber on a Buckled Beam Under Parametric Excitation, *Chaos Solit. Fract.* **32**, 73-79 (2007).
99. H. Nijmeijer, Adaptive/robust Control of Chaotic Systems, in *IUTAM Symp. Interaction Between Dyn. Contr. in Adv. Mech. Syst.*, Ed. D.H. van Campen (Kluwer, Eindhoven, 1997) pp. 255-262.
100. A.B. Nordmark, Non-Periodic Motion Caused by Grazing Incidence in an Impact Oscillator, *J. Sound Vibr.* **145**, 279-297 (1991).
101. E. Ott, C. Grebogi and J.A. Yorke, Controlling Chaos, *Phys. Rev. Lett. E* **64**, 1196-1199 (1990).
102. G. Osipov, L. Glatz and H. Troger, Suppressing Chaos in the Duffing Oscillator by Impulsive Actions, *Chaos Solit. Fract.* **9**, 307-321 (1998).
103. F.H.I. Pereira-Pinto, A.M. Ferreira and M.A. Savi, Chaos Control in a Nonlinear Pendulum Using a Semi-continuous Method, *Chaos Solit. Fract.* **22**, 653-668(2004).
104. O.C. Pinto and P. Gonçalves, Non-linear Control of Buckled Beams Under Step Loading, *Mech. Syst. Sign. Proc.* **14**, 967-985 (2000).
105. O.C. Pinto and P. Goncalves, Active Non-linear Control of Bucking and Vibrations of a Flexible Buckled Beam, *Chaos Solit. Fract.* **14**, 227-239 (2002).
106. K. Pyragas, Continuous Control of Chaos by Self-controlling Feedback, *Physics Letters A* **170**, 421-428 (1992).
107. H.R. Rand, Lecture Notes on Nonlinear Vibrations, Cornell University, available on line at www.tam.cornell.edu/randdocs/ (2003).
108. B. Ravindra and A.K. Mallik, Dissipative Control of Chaos in Non-linear Vibrating Systems, *J. Sound Vibr.* **211**(4), 709-715 (1998).
109. G. Rega and S. Lenci, Identifying, Evaluating, and Controlling Dynamical Integrity Measures in Nonlinear Mechanical Oscillators, *Nonlin. Analysis* **63**, 902-914 (2005).
110. M.G. Rozman, M. Urbakh and J. Klafter, Controlling Chaotic Frictional Forces, *Phys. Rev. E* **57**, 7340-7343 (1998).

111. M.A.F. Sanjuán, Remarks on Transitions Order-chaos Induced by the Shape of the Periodic Excitation in a Parametric Pendulum, *Chaos Solit. Fract.* **7**, 435-440(1996).
112. J.M. Seoane, S. Zambrano, S. Euzzor, R. Meucci, F.T. Arecchi and M.A.F. Sanjuán, Avoiding escapes in open dynamical systems using phase control, in press on *Phys. Rev. E* (2008).
113. S.W. Shaw, The Suppression of Chaos in Periodically Forced Oscillators, in *Nonlinear Dynamics in Engineering Systems*, Ed. W. Schiehlen, Proc. of IUTAM Symposium, Stuttgart, August 21-25, 1989 (Springer-Verlag, Berlin, 1990).
114. T. Shinbrot, W. Ditto, C. Grebogi, E. Ott, M. Spano and J.A. Yorke, Using the Sensitive Dependence of Chaos (the "Butterfly Effect") to Direct Trajectories in an Experimental Chaotic System, *Phys. Rev. Lett.* **68**(19), 2863-2866 (1992).
115. T. Shinbrot, E. Ott, C. Grebogi and J.A. Yorke, Using Chaos to Direct Trajectories to Targets, *Phys. Rev. Lett.* **65**(26), 3215-3218 (1990).
116. T. Shinbrot, E. Ott, C. Grebogi and J.A. Yorke, Using Small Perturbations to Control Chaos, *Nature* **363**, 411-417 (1993).
117. M.K. Sifakis and S.J. Elliott, Strategies for the Control of Chaos in a Duffing-Holmes Oscillator, *Mech. Syst. Sign. Process.* **14**, 987-1002 (2000).
118. J. Singer, Y.Z. Wang and H.H. Bau, Controlling a Chaotic System, *Phys. Rev. Lett.* **66**(9), 1123-1125 (1991).
119. J.C. Sommerer and C. Grebogi, Determination of Crisis Parameter Values by Direct Observation of Manifold Tangencies, *Int. J. Bif. Chaos* **2**(2), 383-396 (1992).
120. J. Starrett, Control of Chaos by Occasional Bang-Bang, *Phys.Rev. E* **67**, 036203-1-4 (2003).
121. J. Starrett and R. Tagg, Control of a Chaotic Parametrically Driven Pendulum, *Phys. Rev. Lett.* **74**(11), 1974-1977 (1995).
122. V. Tereshko, R. Chacon and V. Preciado, Controlling Chaotic Oscillators by Altering Their Energy, *Phys. Lett. A* **320**, 408-416 (2004).
123. K. Yagasaki and M. Kumagai, External Feedback Control of Chaos Using Approximate Periodic Orbits, *Phys. Rev. E* **65**, 026204-1-7 (2002).
124. K. Yagasaki and Y. Tochio, Experimental Control of Chaos by Modifications of Delayed Fedback, *Int. J. Bif. Chaos* **11**, 3125-3132 (2001).
125. K. Yagasaki and T. Uozumi, Controlling Chaos in a Pendulum Subjected to Feedforward and Feedback Control, *Int. J. Bif. Chaos* **7**, 2827-2835 (1997).
126. K. Yagasaki and T. Uozumi, A New Approach for Controlling Chaotic Dynamical Systems, *Phys. Lett. A* **238**, 349-357 (1998).
127. K. Yagasaki and T. Uozumi, Controlling Chaos Using Nonlinear Approximations and Delay Coordinate Embedding, *Phys. Lett. A* **247**, 129-139 (1998).
128. J. Yang and Z. Jing, Inhibition of Chaos in a Pendulum Equation, *Chaos Solit. Fract.* **35**, 726-737 (2006).
129. R. Wang and Z. Jing, Chaos Control of Chaotic Pendulum System, *Chaos Solit. Fract.* **21**, 201-207 (2004).
130. W. Zhang, Chaotic Motion and Its Control for Nonlinear Nonplanar Oscillations of a Parametrically Excited Cantilever Beam, *Chaos Solit. Fract.* **26**, 731-745 (2005).

Reviewed authors' contributions

131. S. Lenci, On the Suppression of Chaos by Means of Bounded Excitations in an Inverted Pendulum, *SIAM J. Appl. Math.* **58**(4), 1116-1127 (1998).
132. S. Lenci and G. Rega, A Procedure for Reducing the Chaotic Response Region in an Impact Mechanical System, *Nonlin. Dyn.* **15**(4), 391-409 (1998).
133. S. Lenci and G. Rega, Controlling Nonlinear Dynamics in a Two-well Impact System. Parts I&II, *Int. J. Bif. Chaos* **8**(12), 2387-2424 (1998).
134. S. Lenci and G. Rega, Numerical Control of Impact Dynamics of Inverted Pendulum Through Optimal Feedback Strategies, *J. Sound Vibr.* **236**(3), 505-527 (2000).
135. S. Lenci and G. Rega, Optimal Numerical Control of Single-well to Cross-well Chaos Transition in Mechanical Systems, *Chaos, Solit. Fract.* **15**, 173-186 (2003).
136. S. Lenci and G. Rega, Optimal Control of Homoclinic Bifurcation: Theoretical Treatment and Practical Reduction of Safe Basin Erosion in the Helmholtz Oscillator, *J. Vibr. Control* **9**(3), 281-315 (2003).
137. S. Lenci and G. Rega, Optimal Control of Nonregular Dynamics in a Duffing Oscillator, *Nonlin. Dyn.* **33**(1), 71-86 (2003).
138. S. Lenci and G. Rega, Global Optimal Control and System-dependent Solutions in the Hardening Helmholtz-Duffing Oscillator, *Chaos, Solit. & Fract.* **21**(5), 1031-1046 (2004).
139. S. Lenci and G. Rega, A Unified Control Framework of the Nonregular Dynamics of Mechanical Oscillators, *J. Sound Vibr.* **278**, 1051-1080 (2004).
140. S. Lenci and G. Rega, Heteroclinic Bifurcations and Optimal Control in the Nonlinear Rocking Dynamics of Generic and Slender Rigid Blocks, *Int. J. Bif. Chaos* **15**(6), 1901-1918 (2005).
141. S. Lenci and G. Rega, Optimal Control and Anti-control of the Nonlinear Dynamics of a Rigid Block, *Phil. Trans. R. Soc. A* **364**, 2353-2381 (2006).
142. S. Lenci and G. Rega, Control of Pull-in Dynamics in a Nonlinear Thermoelastic Electrically Actuated Microbeam, *J. Micromech. Microeng.* **16**, 390-401 (2006).
143. S. Lenci and G. Rega, Control of the Homoclinic Bifurcation in Buckled Beams: Infinite Dimensional vs Reduced Order Modeling, *Int. J. Non-Linear Mech.* **43**, 474-489 (2008).
144. G. Rega and S. Lenci, Nonsmooth Dynamics, Bifurcation and Control in an Impact System, *Syst. Anal. Mod. Simul.* **43**(3), 343-360 (2003).

Chapter 8

CLIPPING CHAOS TO CYCLES

Sudeshna Sinha

Institute of Mathematical Sciences, C.I.T. Campus, Chennai 600 113, India
and
Indian Institute of Science Education and Research (IISER) Mohali,
Transit Campus: MGSIPAP Complex, Sector 26 Chandigarh 160 019, India
sudeshna@imsc.res.in
sudeshna@iisermohali.ac.in

1. Introduction

In this chapter we will review a powerful control strategy based on the simple and easily implementable *threshold* mechanism [1]. The central idea is as follows: consider a general N-dimensional dynamical system, described by the evolution equation $\frac{d\mathbf{x}}{dt} = F(\mathbf{x}; t)$ where $\mathbf{x} \equiv (x_1, x_2, \ldots x_N)$ are the state variables. Say variable x_T is chosen to be monitored and threshold controlled. The prescription for thresholding in this system is as follows: control will be triggered whenever the value of the monitored variable exceeds the critical threshold x^* (i.e. when $x_T > x^*$) and the variable x_T will then be re-set to x^*. The dynamics continues till the next occurence of x_T exceeding the threshold, when control resets its value to x^* again. So in this method no parameters are adjusted, and only one state variable is occasionally reset.

This method is not based on stabilising unstable periodic orbits [2]. Rather, the threshold mechanism *clips* the chaotic orbit to periodic time sequences of desired lengths. So the effect of this scheme is to *limit the dynamic range* slightly, i.e. "snip" off small portions of the available phase space, and this action effectively yields a very wide range of stable regular dynamics [3].

In the sections below we will analyse the control achieved by thresholding in different prototypical systems, with varying levels of complexity,

Table 1. Controlling a 1-dimensional chaotic map by thresholding.

Threshold	Nature of Controlled Orbit
$x^* < 0.75$	Period 1 (fixed point)
$0.75 < x^* < 0.905$	Period 2 Cycle
$x^* \sim 0.965$	Period 3 Cycle
$0.905 < x^* < 0.925$	Period 4 Cycle
$x^* \sim 0.979$	Period 5 Cycle
$x^* \sim 0.93$	Period 6 Cycle
$x^* \sim 0.9355$	Period 7 Cycle
$x^* \sim 0.932$	Period 8 Cycle
$x^* \sim 0.981$	Period 9 Cycle
$x^* \sim 0.95$	Period 10 Cycle

Threshold values vs. periodicity, of a few representative controlled cycles, for the chaotic logistic map $x_{n+1} = 4x_n(1 - x_n)$. Note that cycles of the same period, but different geometries, can be obtained in different threshold windows.

including the challenging task of controlling hyperchaos. We will also discuss the experimental implementation of the scheme on a range of strongly nonlinear electronic circuits.

2. Application to One-dimensional Maps

When the dynamics of the uncontrolled system is given by $x_{n+1} = f(x_n)$ where f is a nonlinear function, the threshold mechanism is simply implemented as the following condition: if variable $x_{n+1} > x^*$ then the variable is adjusted back to x^*, namely the threshold x^* is the critical value the state variable is not allowed to exceed, and control is triggered whenever the variable grows larger than this threshold. The effect of this simple thresholding is dramatic: it yields stable periodic orbits of all orders. See Table 1 for a illustrative list of controlled orbits obtained for a range of threshold values.

The threshold controlled chaotic map is effectively a beheaded map (or the "flat top" map), i.e. the unimodal map cut off by the $x_{n+1} = x^*$ line [1, 4]. The level at which the map is chopped off depends on the threshold x^* (see Fig. 1). This map can yield periodic orbits of various orders under different threshold values. Note also that these orbits are *stable* and the low order cycles have fairly large windows of stability in threshold parameter space. Control latency is very short, and once the system exceeds the critical value it is trapped immediately in a stable cycle whose order is determined by the value of the threshold.

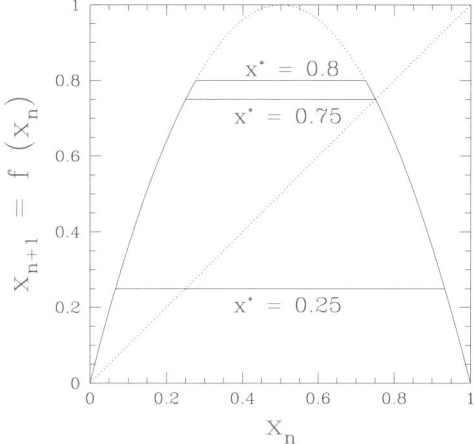

Fig. 1. The chaotic logistic map under threshold control, with threshold values 0.25, 0.75 and 0.8. The figure shows the first iterate x_{n+1} (—) of the effective thresholded map, as well as the $x_{n+1} = x_n$ line (i.e. the 45° line (...)). It is clear that the intersection of the flat portion of the map x_{n+1} with the 45° line yields a superstable fixed point of period 1.

The basis of the marked success of this method is clear for one dimensional maps. It is best rationalised through the fixed points of the effective map obtained from the chaotic unimodal map under threshold mechanism, namely the beheaded map mentioned above. The fixed points of this map under varying heights of truncation (i.e. different thresholds) give the different periods. For instance, Fig. 1 shows the first iterate x_{n+1} of the thresholded chaotic logistic map. One obtains a fixed point from the intersection of this map and the $x_{n+1} = x_n$ line, namely the 45° line. When the threshold is lower than 0.75 one gets an intersection of the flat portion of the map (i.e. $x = x^*$) and the $x_{n+1} = x_n$ line. This fixed point solution at $x = x^*$ has slope zero and thus is *superstable*. This happens for all thresholds $x^* < 3/4$, and one obtains a stable fixed point at x^* for those thresholds. Thus one can control the chaos to a *continuous set of fixed points* in the range $[0, 3/4]$ by thresholding.

It is also evident from Fig. 1 that when the threshold exceeds $3/4$ (for example $x^* = 0.8$) the thresholded map does not yield an intersection of the flat portion and the $x_{n+1} = x_n$ line. Rather, it only has the usual fixed point at $x = 0.75$, which is the same as that obtained for the (un-thresholded) chaotic map. This has a slope greater than one and is thus unstable. So for thresholds greater than $x^* = 0.75$, the thresholding action does not yield

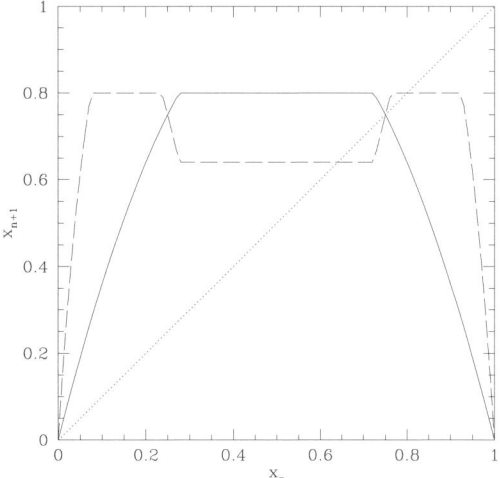

Fig. 2. The chaotic logistic map under threshold control, with threshold value 0.8. The figure shows the first iterate x_{n+1} (—) and second iterate x_{n+2} (- - -) of the effective thresholded map, as well as the 45° line (...). It is clear that the intersection of the flat portion of the map x_{n+2} with the 45° line yields a superstable fixed point of period 2.

a stable period 1 solution. However *higher order periods* can be obtained. For example Fig. 2 shows the second iterate of the thresholded map, with threshold $x^* = 0.8$, yielding a fixed point from the intersection of the flat portion of the map and the $x_{n+2} = x_n$ line. This point has slope zero and thus yields a *superstable period 2 orbit*.

Similarly, in Fig. 3 one sees the first four iterates of the threshold map and the fourth iterate yields a superstable fixed point at the intersection of the flat portion of the effective map and $x_{n+4} = x_n$ line. Thus one obtains a stable period 4 orbit at this value of threshold.

In terms of probability densities, the chaotic map under threshold mechanism will map large intervals onto a severely contracting region. Essentially large intervals will get mapped onto a point. This is the reason why the transient control period is so small, and the method is so powerful and stable.

2.1. *Analysis*

For one dimensional maps, the *correspondence of the periodicity of the controlled orbit and the threshold can be obtained exactly*. So one can directly calculate what periodicity will emerge when a certain threshold is set.

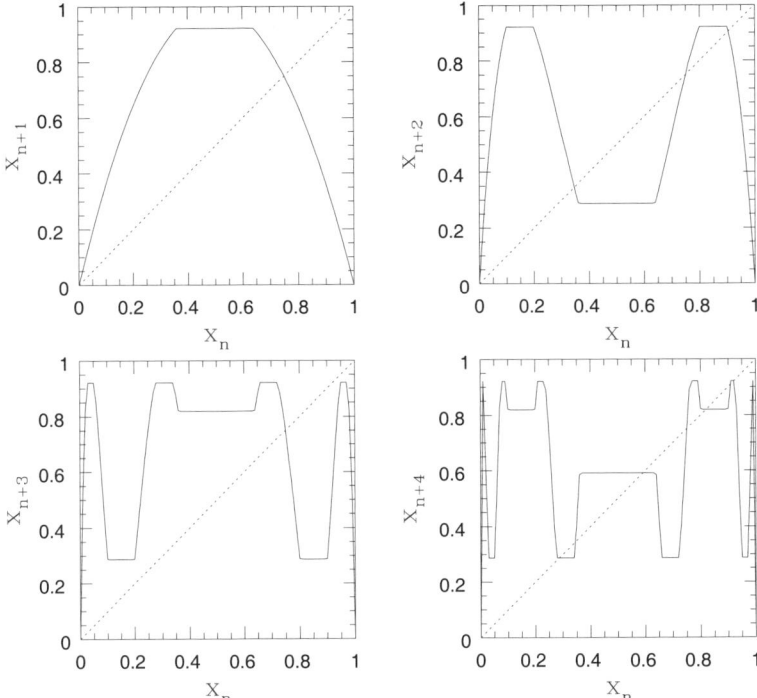

Fig. 3. The chaotic logistic map under threshold control, with threshold value 0.922. The figure shows the first four iterates of the effective thresholded map: x_{n+1}, x_{n+2}, x_{n+3}, x_{n+4}, as well as the 45° line (...). It is clear that the intersection of the flat portion of the map x_{n+4} with the 45° line yields a superstable fixed point of period 4. Note that only for the fourth iterate x_{n+4} does one obtain a solution at the intersection of the flat portion of the effective thresholded map, where slope is zero. So only the period 4 orbit is stable at this threshold.

Further one can obtain the answer to the reverse (important) question as well: what threshold do we need to set in order to obtain a certain period [1].

Now the starting point of the analysis is the fact that the chaotic system is ergodic and thus it is guaranteed to exceed threshold at some point in time. At that point its state is re-set to x^*. One then studies the forward iterations of the map, starting from this state $x = x^*$, i.e. $f_0(x^*), f_1(x^*) \ldots$, where $f_k(x^*)$ is the kth iterate of the map. That is:

1. $k = 0$; $f_0(x^*) = x^*$
2. $k = 1$; $f_1(x^*) = 4x^*(1 - x^*)$
3. $k = 2$; $f_2(x^*) = 4(4x^*(1 - x^*))(1 - 4x^*(1 - x^*))$

and so on. In general

$$f_k(x^*) = f \circ f_{n-1}(x^*) = f \circ f \circ \ldots f \circ (x^*)$$

Whenever the $f_k(x^*)$ vs. x^* curve crosses above the $f_0 = x^*$ line (i.e. the 45° line) we have a k cycle, as this implies that the kth iterate exceeds the critical value x^* and thus is adapted back to x^* ($= f_0$, which is the first point in the cycle). For instance, in the range $0 \le x^* \le \frac{3}{4}$ the $f_1(x^*)$ curve lie above the f_0 curve (i.e. $f_1(x^*) > x^*$). So the chaotic element is adapted back to x^* at every iterate, yielding a period 1 fixed point. In the range $\frac{3}{4} < x* < 0.9$ the $f_1(x^*)$ curve dips below the 45° line, but the $f_2(x^*)$ curve lies above the 45° line. This imples that the second iterate of the map (starting from $x = x^*$) exceeds threshold and is adapted back to x^*, thus giving rise to a period 2 cycle. Thus the cycle at each value of threshold is the smallest k such that the kth iterate of the map (starting from $x_0 = x^*$) is greater than x^*, i.e. $f_k(x^*) > x^*$. In this manner the threshold mechanism leads to regular cyclic evolution, whose period depends on the threshold. The chaotic element can then yield a wide variety of dynamical behavior determined by the threshold.

Thus in threshold parameter space we can find "windows" of various cycles. These are intervals where the following equation is satisfied: Period $P(x^*) = k$ iff $f_k(x^*) \ge x^*$ and $f_l(x^*) < x^*$ for all $l < k$. $P(x^*)$ is a piecewise continuous function of x^*.

For every cycle of periodicity k there will be several windows (with an upper bound of 2^{k-1} windows for period k). The "middle" of the period k windows lies approximately where the curve $f_k(x^*)$ touches 1 (since if it touches 1 it has to have exceeded x^*, as the value of x^* is bounded by 1). Then the solutions of the equation $f_k(x^*) = 1$ gives the x^* values corresponding to a period k. The solutions can be formulated as: $f^{-1} \circ f^{-1} \circ f^{-1} \circ f^{-1}(1)$ where f^{-1} is the (double valued) invere map. The inverse map: $f^{-1}(y) = \frac{1}{2} \pm \frac{\sqrt{1-y}}{2}$ has two values – one on the right and one on the left of the centre of the interval ($x = \frac{1}{2}$). We will denote these as R and L respectively. Note that for $f^{-1}(1)$ the value of $L(1) = R(1) = \frac{1}{2}$. So the number of distinct values arising from the expression $f^{-1} \circ f^{-1} \ldots f^{-1}(1)$ is 2^{k-1} (arising from the 2^{k-1} different possible combinations of R and L).

The evaluation of this algebraic expression for various values of k is simple and direct. Now the existence of a window of period k ($k > 1$) is dependent on the pervious iterates as well, i.e. a solution for period k may be masked by the fact that some iterate l, $l < k$, may have $f_l(x^*) > x^*$. For instance for $k > 1$ all combinations starting with symbol L are masked by period 1 (as the period 1 window extends from 0 to $\frac{3}{4}$ and $L(x) \leq \frac{1}{2}$). So half of the combinations of $f^{-1} \circ f^{-1} \ldots f^{-1}(1)$ are swallowed by period 1. One has to examine the remaining 2^{k-2} combinations to check which ones survive swallowing by lower order windows.

However note that one family of windows is guaranteed to exist, namely $RL^{k-1}(1)$, as all iterates leading up to 1 here (i.e. all the subsequences $L(1), L^2(1), \ldots L^{k-1}(1)$) have value less than $\frac{1}{2}$ (as they are all composed of L). Since all relevant thresholds for $k > 1$ are greater than $\frac{3}{4}$ it implies that all the iterates leading up to $f_k(x^*)$ have value less than x^* and so this sequence will always yield period k (not any other lower period). So *all possible periods k have atleast one stable window in threshold space.* That is, threshold control for a one dimensional map yields periods of all orders.

Now the analytical results based on symbolic dynamics [1] outlined above are exactly corroborated in circuit realizations of one-dimensional discrete time systems [5] (see Fig. 4 for traces of representative controlled orbits). Note that chaos is advantageous here, as it possesses a rich range of temporal patterns which can be clipped to wide ranging stable behaviors. This immense variety is not available from thresholding regular systems.

Also note that arrays of thresholded nonlinear elements have been designed, fabricated, and tested in CMOS [6] and such arrays show a wide range of controlled spatiotemporal cycles, consistent with the analytical results.

In marked contrast to many control methods where chaotic trajectories in the vicinity of unstable fixed points are controlled onto these points, in threshold control the system *does not have to be close to any particular unstable fixed point before implementing the control.* Once the trajectory exceeds the threshold it is caught immediately in a stable orbit. So there is no significant interval between the onset of control and the achievement of control, as a wide interval is open to targetting.

Caveat: While threshold control will always yield some regular orbit, it is not clear at the outset exactly what kind of dynamic behavior will result from a given threshold value [7]. This limitation is overcome easily however through one initial exploratory run over threshold parameter space to map out the dynamic behaviors obtained for different thresholds. Such a "calibration run" once done, makes the scope of the threshold mechanism apparent at the outset, and yields a *look-up table* for the system (relating threshold value to controlled period) for all further applications.

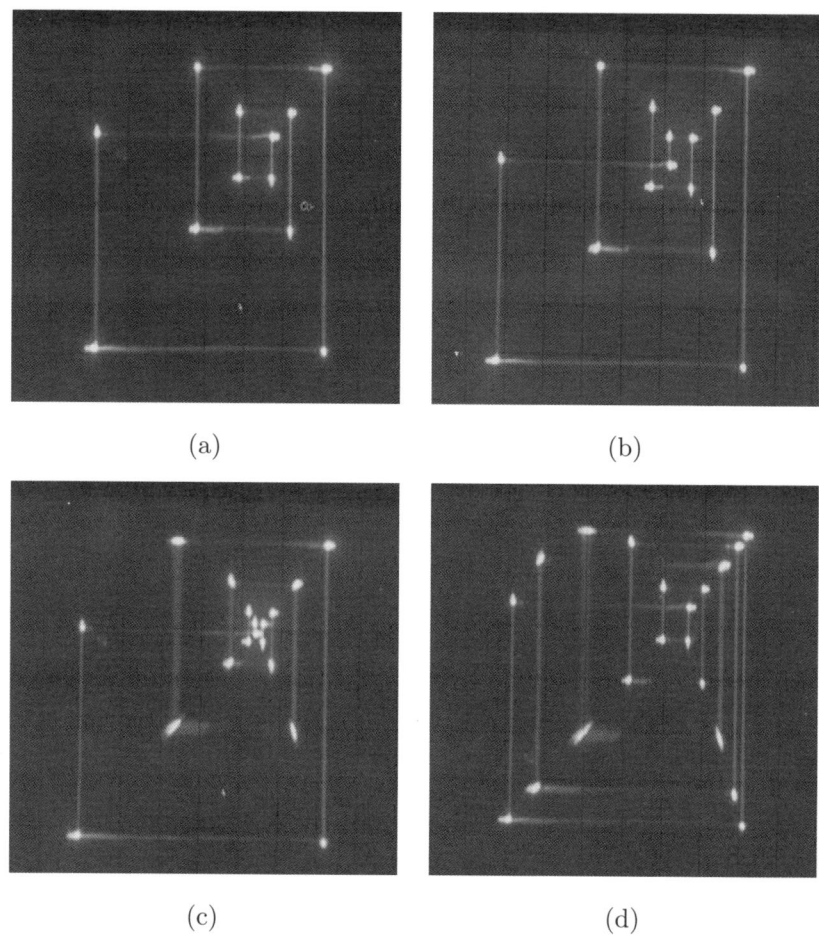

Fig. 4. Experimental verification of a range of controlled periods in a circuit realization of the logistic map. The ordinate and abscissa represent traces of x_{n+1} and x_n for (a) period 6 cycle (b) period 7 cycle (c) period 9 cycle with threshold and (d) period 10 cycle. The threshold levels at which these cycles were obtained coincide exactly with those predicted theoretically.

The possibility of having this kind of a look-up table to effect control to a "library" of patterns is one of the most powerful features of this method.

2.2. *Application to encoding and information storage*

Information storage is a fundamental function of computing devices. Computer memory is implemented by computer components that retain data for some interval of time. Storage devices have progressed from punch cards and paper tape to magnetic, semiconductor and optical disc storage by exploiting different natural physical phenomena to achieve information storage. For instance, the most prevalent memory element in electronics and digital circuits is the flip-flop or bistable multivibrator which is a pulsed digital circuit capable of serving as a one-bit memory, namely storing value 0 or 1. More meaningful information is obtained by combining consecutive bits into larger units. Here we briefly review a different direction in designing information storage devices: namely we describe schemes to store data using the vast variety of patterns and distinct behaviors that can be extracted by thresholding nonlinear maps.

The aim is to utilize arrays of nonlinear elements to stably encode and store various items of information (such as patterns and strings) to create a database [8]. Further we indicate how this storage method also allows one to utilize the nonlinear dynamics of the array elements in order to determine the number of matches (if any) to specified items of information in the database.

Encoding information: We consider encoding N data elements, each comprised of one of M distinct items. N can be arbitrarily large and M is determined by the kind of data being stored. For instance for storing English text one can consider the letters of the alphabet to be the natural distinct items building the database, namely $M = 26$. Or, for the case of data stored in decimal representation $M = 10$, and for databases in bioinformatics comprised typically of symbols A, T, C, G, one has $M = 4$. One can also consider strings and patterns as the items. For instance for English text one can also consider the keywords as the items, and this will necessitate larger M as the set of keywords is large.

Now we demonstrate a method which stores data by utilizing the abundance of distinct stable behaviors obtained by thresholding a chaotic system. This ability of the thresholded chaotic map to be in a variety of fixed states gives it the capacity to represent a large set of items.

A database of size N is stored by N thresholded chaotic elements. Each dynamical element stores one element of the database, encoding any one of the M items comprising our data. Now in order to hold information one must confine the dynamical system to a fixed point behavior, i.e. a state that is stable and constant throughout the dynamical evolution of the system.

As analyzed above, typically a large window of threshold values can be found where the system is confined on fixed points, namely, the state of the element under thresholding is stable at T i.e. $x = T$, where T is the threshold, for all times. So one can choose a large set of distinct thresholds $T^1, T^2, \ldots, T^M$, within the fixed point window, with each threshold having a one-to-one correspondence with a distinct item of our data. Thus the number of distinct items that can be stored in a single dynamical element is typically large, with the size of M limited only by the precision of the threshold setting.

In particular, consider a collection of storage elements that evolve in discrete time according to the tent map, $f(x) = 2\min(x, 1-x)$, with each element storing one element of the given database. Each element can hold any one of the M distinct items. As described above, a threshold will be applied to each dynamical element to confine it to the fixed point corresponding to the item to be stored.

For the tent map, thresholds ranging from 0 to 2/3 yield fixed points, namely $x = T$, for all time, when threshold $0 < T < 2/3$. This can be obtained exactly from the fact that $f(x) > T$ for all x in the interval $(0, 2/3)$, implying that the subsequent iteration of a state at T will always exceed T, and thus get reset to T. So x will always be held at value T.

In our encoding, the thresholds are chosen from the interval $(0, 1/2)$, namely a sub-set of the fixed point window $(0, 2/3)$. For specific illustration, with no loss of generality, consider each item to be represented by an integer i, in the range $[1, M]$. Defining a resolution r between each integer as $r = \frac{1}{2}\frac{1}{M+1}$ gives a lookup map from the encoded number to the threshold, namely relating the integers i in the range $[1, M]$, to threshold in the range $[r, 1/2 - r]$, by: $T = ir$.

Therefore we obtain a direct correspondence between a set of integers ranging from 1 to M, where each integer represents an item, and a set of M threshold values. So we can store N database elements by setting appropriate thresholds on N chaotic maps. Clearly, if the threshold setting has more resolution, namely smaller r, then a larger range of values can be encoded. Note however that precision is not a restrictive issue here, as different representations of data can always be chosen in order to suit a given precision of the threshold mechanism.

Processing Information: Now we briefly indicate how we can search an arbitrarily large unsorted database set up as above, for the existence of a specific item, by performing just one global operation on the whole array. The basic principle here is that one can construct a single suitable global operation that acts on the nonlinear elements encoding the database such that only elements encoding the matching items yield a prescribed, easily measurable property. This enables the occurrence(s) of matches to be identified easily. We give some details below.

Given a database stored by setting appropriate thresholds on N dynamical elements, we can query for the existence of a specific item in the database by globally shifting the state of all elements of the database up by the amount that represents the item searched for. Specifically the state of all the elements is raised to $x + \mathcal{Q}$, where $\mathcal{Q}$ is a search key given by: $\mathcal{Q}^k = 1/2 - \mathcal{T}^k$, where k is the number being queried for. So the value of the search key is simply $1/2$ minus the threshold value corresponding to the item being searched for. This addition shifts the interval that the database elements can span, from $[r, 1/2 - r]$ to $[r + \mathcal{Q}^k, 1/2 - r + \mathcal{Q}^k]$, where $\mathcal{Q}^k$ is the globally applied shift.

Notice that the information item being searched for, is coded in a manner "complimentary" to the encoding of the items in the database (much like a key that fits a particular lock), namely $\mathcal{Q}^k + \mathcal{T}^k$ adds up to $1/2$. This guarantees that *only* the element matching the item being queried for will have its state shifted to $1/2$. The value of $1/2$ is special in that it is the only state value that on the subsequent update will reach the value of 1.0, which is the maximum state value for this system. So only the elements holding an item matching the queried item will reach extremal value 1.0 on the dynamical update following a search query. Note that the important feature here is the nonlinear dynamics that maps the state $1/2$ to 1, while all other states (both higher and lower than $1/2$) get mapped to values lower than 1.

Basically in unimodal maps, the maximal point can act as a "pivot" for the "folding". This provides us with a single global monitoring operation to push the state of all the elements matching the queried item to the unique maximal point, in parallel.

To complete the search we now must detect the extremal state. This can be accomplished in a variety of ways. For example, one can simply employ a level detector to register all elements at the maximal state. This will directly give the total number of matches, if any. So the total search process is rendered simpler as the state with the matching pattern is selected out and mapped to the maximal value, allowing easy detection. Further, by relaxing the detection level by a prescribed "tolerance", we can check for the existence within our database of numbers or patterns that are close to the number or pattern being searched for. So nonlinear dynamics works as a powerful *preprocessing tool*, reducing the determination of matching patterns to the detection of maximal states, an operation that can be accomplished by simple means, in parallel.

3. Application to Multi-dimensional Systems

The action of thresholding on 1-dimensional chaotic maps can be analyzed exactly, as outlined in the section above. However for multi-dimensional systems such an exact calculation is not possible. So one has to rely on numerics and experiments to gauge the scope of threshold control on such systems.

The central issue in multi-dimensional systems is whether or not the thresholded state variable can enslave the rest of the variables to some regular dynamical behavior. With this in mind we present several examples below, of controlling highly coupled strongly nonlinear high dimensional systems, by thresholding just *one* variable.

3.1. *Controlling the Lorenz system*

First we demonstrate the action of the threshold mechanism on a system of 3 coupled ODE's: the chaotic Lorenz attractor (a system known to be relevant to lasers [9]). It is given by

$$\dot{x} = \sigma(y - x),$$
$$\dot{y} = rx - y - xz,$$
$$\dot{z} = xy - bz, \qquad (1)$$

with parameters $\sigma = 10$, $r = 28$, $b = 8/3$.

In order to check whether or not *one* thresholded state variable can drag the rest of the multidimensional system to some fixed dynamical behavior, we impose threshold control on any one of the three variables of the Lorenz system, i.e. one demands that either variable x or y or z must not exceed a prescribed threshold value x^*. Figure 5 show some representative results of this thresholding action for a range of threshold values. It is clear that the mechanism successfully controls limit cycles of varying sizes and geometries.

3.2. *Controlling neuronal spikes*

In neuronal systems, a wealth of complex patterns have been experimentally observed in a variety of cases [10]. However the mechanisms by which such complex spiking patterns can be manipulated are not well understood. It is thus of considerable interest and potential utility to devise control algorithms capable of achieving the desired type of behavior in such complex systems.

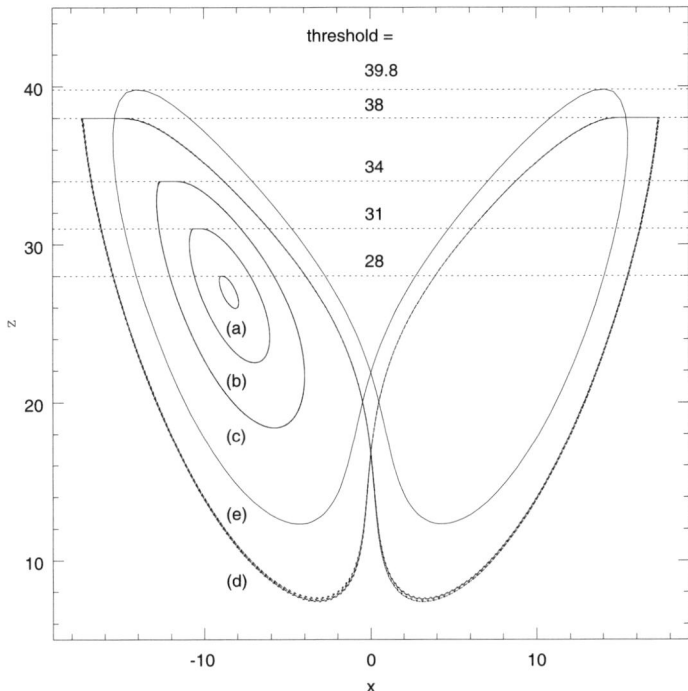

Fig. 5. The chaotic Lorenz attractor (with $\sigma = 10, r = 28, b = 8/3$), under threshold control of variable z. The dotted lines indicate the value of the threshold.

Here we use threshold control to target desired firing patterns in a prototypical model of a Hippocampal neuron: the Pinsky-Rinzel model [11]. This model neuron consists of somatic and dendritic compartments resistively coupled at different potentials. A patch of the cell membrane is modeled as an equivalent electrical circuit consisting of a resistor and a capacitor in parallel. The current balance equations for the two compartments follow from differentiating the capacitance definition.

The model has 8 variables: the 5 gating variables: h, n, s, c and q, the Ca level and the soma voltage V_s and dendrite voltage V_d. The parameters include the coupling conductance between soma and dendrite g_c, reversal potentials V_{Na}, V_{Ca}, V_K, V_l, V_{syn}, ionic conductances g_l, g_{Na}, g_{KDR}, g_{Ca}, g_{KAHP}, g_{KCa}, synaptic conductances g_{MMDA}, g_{AMPA}, relative area of soma to dendrite p, membrane capacitance c_m, and the applied soma current i_s. The details of the dynamical equations and parameters are given in [11].

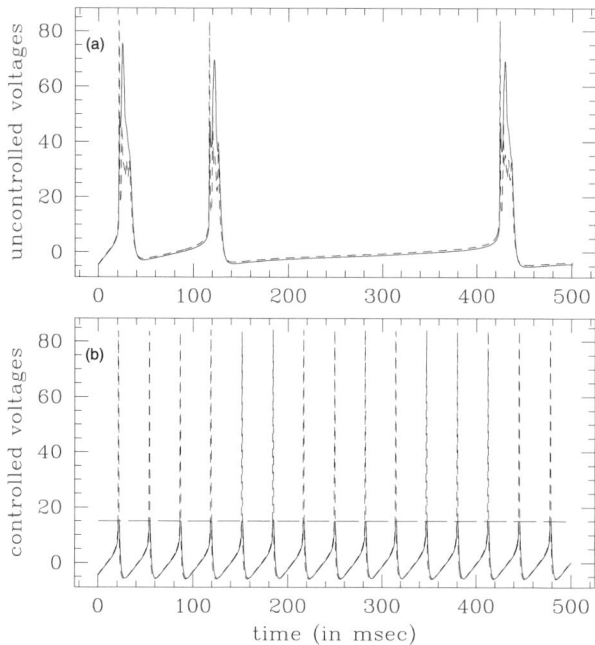

Fig. 6. The time evolution of the voltages V_s and V_d (in mV) for Pinsky-Rinzel neuron for the cases of: (a) uncontrolled neuron showing infrequent and irregular spiking behavior, (b) the same neuron, with voltage V_d under threshold control, with threshold $V^* = 15$ mV. (Here $i_s = 1$ nA) Clearly the controlled neuron spikes at very regular intervals. The solid lines show V_d (—) and the dashed lines show V_s (- - -). The threshold voltage of $V^* = 15$ mV is shown by a dashed line (- -).

In the context of neuronal systems, it is unrealistic to implement the threshold mechanism on the gating variables or the Ca levels as it is unlikely that one can manipulate these externally with ease. On the other hand, it is natural to try and implement the threshold action on the somatic or dendritic voltages V_s or V_d as they are much more accessible to measurement and monitoring. Thus we demand that variable V_s (or V_d) must not exceed a prescribed threshold value V_* (1 mV $< V^* < 20$ mV), and examine the scope of this mechanism to yield regular firing behavior. The noteworthy feature here is that *only one variable is thresholded in this strongly nonlinear, highly coupled, 8-dimensional system.*

Figure 6 shows a representative result of thresholding the neuronal system. It is clearly evident that the mechanism manages to yield complete regularity, as compared with the very irregular and infrequent firing behavior of the neuron with no thresholding. So the thresholded variable has the ability to drag the rest of this high-dimensional excitable system to regular dynamical behavior (see Fig. 6). The threshold mechanism typically yields two types of behavior: fixed states and states with regular spiking (with interspike intervals ranging from about 14 msec to 60 msec). Low threshold leads to the first dynamics and higher thresholds leads to regular firing states.

Similar control is achieved by thresholding the somatic voltage V_s. We also checked that the method works for slightly delayed threshold action, that is a scenario where the variable is brought down to the threshold value after a small delay (as is conceivable in real set-ups where there may be a small delay between the detection of the crossing of the threshold condition and the re-setting of the state variable). We find that the method is still as effective [12].

3.3. *Smart matter application*

Here we present an application of threshold control to the interesting problem of controlling an unstable elastic array, which has been used as a prototypical model for "smart matter" [13]. It is clear that in such a context, where the system contains many elements, the effectiveness of control algorithms which rely on access to the full state of the system and detailed knowledge of its behavior, is limited. Hence the present approach, which needs *local information from very few sites* (and no knowledge of the dynamics) in order to implement the necessary control, can prove to be of considerable utility.

For example, consider the general extended system:

$$\frac{d^2\mathbf{x}}{dt^2} = A\mathbf{x} - G\dot{\mathbf{x}}, \qquad (2)$$

where the vector $\mathbf{x}$ gives the displacements of the elements in the array, matrix A contains the system's coupling parameters and G is the damping. In this array we now implement threshold control on a few selected sites.

Specifically, one can consider a model of buckling instability of beams [13]: an elastic array of N elements coupled to nearest neighbors by springs with spring constants α, and a destabilizing force coefficient f. The dynamics of the beam is given by Eq. (2) where the N-dimensional vector $\mathbf{x} = x_1, x_2, \ldots x_N$ gives the displacements of the elements, damping matrix G is gI (where I is the identity matrix) and coupling matrix A has elements $A_{mn} = -2\alpha + f$ for $m = n$, $A_{mn} = \alpha$ for $m = n \pm 1$, and $A_{mn} = 0$ otherwise [13].

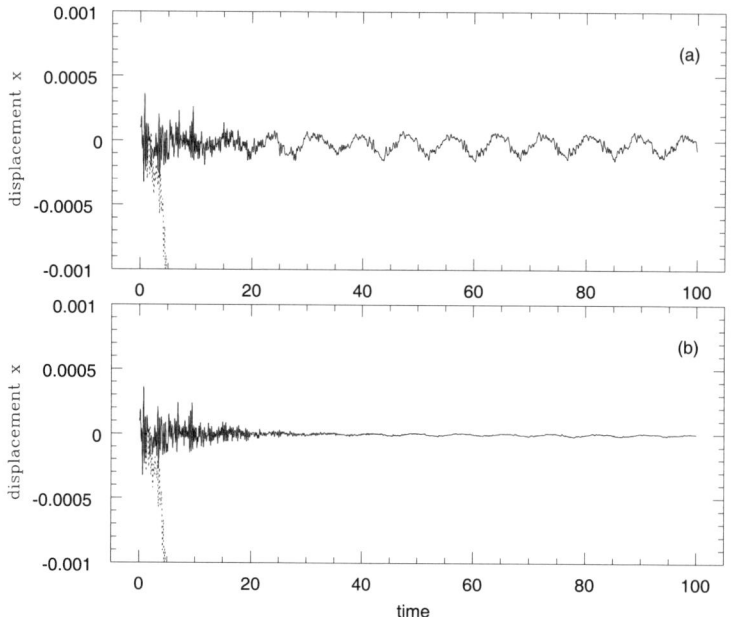

Fig. 7. Displacement of a representative element ($i = 20$) in an unstable elastic array w.r.t. time in the controlled (—) and uncontrolled case (....). The number of elements N in the chain is 100. Here we have placed 2 threshold controllers in the array at sites $i = 33$ and $i = 66$. The threshold value T is equal to (a) 0.0001 and (b) 0.00001. In the absence of control the system moves exponentially away from the steady state, while under control it manages to maintain the buckling within prescribed limits (namely, within 0.0001 and 0.00001 respectively).

In this array we now choose a few sites for threshold control. For effective control of the entire beam one needs to control a minimum of *two* sites [14]. *The amount of buckling tolerated determines the threshold T.* In order to effect control in this system we must implement a slight variation of the threshold method, with control action being triggered whenever $|x| > T$. At the controlled sites here, one then imposes the following control condition: if $x < -T$, the value of x is re-set to T, and if $x > T$, the value of x is re-set to $-T$. This control action has the effect of "bending" the beam in the direction opposite to that of the buckling, thus having a local "straightening" effect on the array. Note that the value of the threshold T can be made very small indeed, leading to arrays which are only slightly deformed.

Figure 7 shows the displacements of a representative element in an array of size 100, threshold controlled at only 2 sites. This displacement is compared to the uncontrolled situation, where the array deforms exponentially fast. Two cases are presented here: one with allowed buckling tolerance equal to 0.0001, i.e. the threshold $\mathcal{T} = 0.0001$, and the other with lower tolerance, $\mathcal{T} = 0.00001$. It is evident from the figure, that in the absence of control very weak environmental perturbations drive the system exponentially away from the desired configuration, while threshold control manages to achieve the goal, typical in smart matter applications, of preventing the beam from buckling more than the pre-assigned tolerance. Further note, that at the controlled sites the instances of control were quite infrequent. Typically, for an array with 2 controlled sites, with threshold $\mathcal{T} = 0.0001$, the control was 'triggered on' approximately one-tenth the number of times the variables were monitored for control.

Now the positions of the controlling sites are crucial for the success of this control. The controlling sites must span the array at roughly equidistant points. For instance, when 2 controllers are used in an array of size 100, one of these should be placed at some site between $i = 30$ and 40 and another between $i = 60$ and 70. Here we are in fact exploiting the natural coupling of the system to influence a large neighborhood with only a few sites. Thus very weak and infrequent control at very few sites manages to control the entire array.

It is thus evident that the threshold scheme can successfully control this extended nonlinear system. The present approach needs to implement the control on very few sites (minimum two). It utilizes no knowledge of the dynamics or system parameters, and also does not entail any computation in implementing the control. Further, there is no communication involved, as no site needs to know the state of its neighbors. One only needs to check if the value of the controlled variable at the controlled sites exceeds the user defined critical value or not. Since no communication or computation is involved, the control is very easily and simply implemented.

3.4. *Thresholding at varying intervals to obtain different temporal patterns*

Now we discuss how stroboscopic threshold mechanisms can be effectively employed to obtain a wide range of stable cyclic behavior from chaotic systems, by simply *varying the frequency of control* [15]. For instance, consider the action of infrequent threshold control on the chaotic Lorenz-like attractor given by Eq. (1), using the three parameters corresponding to the coherently pumped far-infrared ammonia laser system, obtained by detailed comparison with experiments [9]: $\sigma = 2$, $r = 15$ and $b = 0.25$.

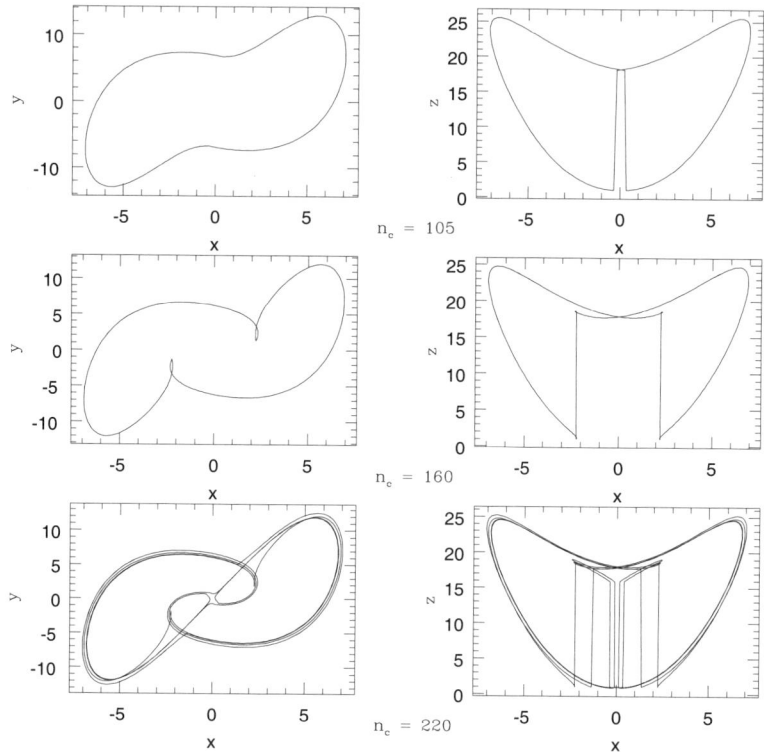

Fig. 8. The chaotic laser-Lorenz system under threshold control of variable z, with threshold value $z^* = 1$. The control acts at intervals of $n_c \times \delta t$, with $\delta t = 0.01$. Here $n_c = 105, 160, 220$. The controlled cycles in x-y space and x-z space are displayed. Notice the period doubling of the cycles as control interval n_c is increased.

Consider the particular case of the threshold mechanism imposed on the z variable. The stroboscopic threshold action occurs at an interval of $n_c \delta t$. Figure 8 shows the different temporal patterns obtained when the threshold is fixed at $z^* = 1$ and the n_c is increased.

In [15], a large range of numerical evidence is reported, with the frequency of control spanning three orders of magnitude, to show that stroboscopic threshold action of any variable in this multi-dimensional chaotic system successfully yields regular temporal patterns, displaying a wide variety of periods and geometries. In fact, the interval of control may be very large in many cases and still lead to very effective control onto simple limit cycles. So varying the interval of threshold control offers flexibility and cost-effectiveness in regulating chaotic systems onto different cyclic patterns.

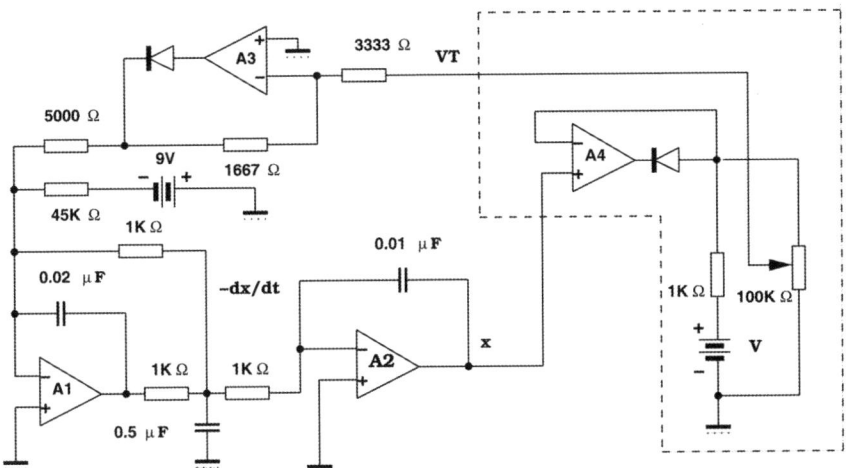

Fig. 9. A general circuit for solving Eq. (3) using a nonlinear feedback element $G(x) = B|x| - C$. The precision clipping control circuit is shown in the dotted box. Here V_T corresponds to the threshold controlled signal.

4. Experimental Verification

Here we review the experimental verification of thresholding as a versatile tool for efficient and flexible chaos control. We demonstrate the success of the technique in rapidly controlling different chaotic electrical circuits, including a hyperchaotic circuit, onto stable fixed points and limit cycles of different periods, by thresholding just one variable. The simplicity of this controller entailing no run-time computation, and the ease and rapidity of switching between different targets it offers, suggests a potent tool for chaos based applications.

4.1. Controlling a circuit realization of Jerk equations

The first experimental set-up is a realization of nonlinear third-order ordinary differential equations (ODE), a form known in literature as Jerk equations:

$$\frac{d^3x}{dt^3} + A\frac{d^2x}{dt^2} + \frac{dx}{dt} = G(x), \qquad (3)$$

where $G(x)$ is a piecewise linear function: $G(x) = B|x| - C$ with $B = 1.0, C = 2.0$ and $A = 0.6$ [16]. The circuit realization of the above uses resistors, capacitors, diodes and operational amplifiers as shown in Fig. 9.

Table 2. Thresholding a third-order nonlinear system.

Threshold Value	Nature of Controlled Orbit
$x^* < -2.00$	fixed point
$-2.00 < x^* < 1.477$	Period 1 Cycle
$1.477 < x^* < 2.242$	Period 2 Cycle
$2.242 < x^* < 2.321$	Period 4 Cycle
$2.321 < x^* < 2.325$	Period 8 Cycle
$2.325 < x^* < 2.331$	Period 16 Cycle

Threshold ranges (in V) vs. periodicity of the controlled cycle, for the chaotic circuit described by Eq. (3).

The implementation involves three successive active integrators to generate $\frac{d^2x}{dt^2}$, $\frac{dx}{dt}$ and x from $\frac{d^3x}{dt^3}$, coupled with a nonlinear element that generates $G(x)$ and feeds it back to $\frac{d^3x}{dt^3}$.

Now we implement the threshold mechanism on variable x, i.e. whenever $x > x^*$, x is clipped to x^*. A precision clipping circuit [17] as depicted in the dotted box in Fig. 9 is employed for threshold control. We have chosen component values for the control circuit to be: [op-amp = μ A 741, diode = IN4148, load resistor = $1k\Omega$ and threshold reference voltage = V, which sets the x^*].

Figure 10(a) displays the uncontrolled attractor and Fig. 10(b) to 10(d) shows some representative results of the threshold action on this chaotic system for a range of threshold values x^* ($x^* < 2.4$). It is clear that the mechanism manages to yield cycles of varying periodicities. Further, detailed comparison shows *complete agreement between our experimental results and our numerical simulation results.*

So the single thresholded variable x has the ability to drag the rest of this 3-dimensional system to regular dynamical behavior. The characteristics of the controlled states can be easily varied by just changing the threshold x^* (see Table 2). Also note that simply setting the threshold beyond the bounds of the attractor gives back the original dynamics.

The control transience is very short here (typically of the order of 10^{-3} times the controlled cycle). This makes the control practically instantaneous. The underlying reason for this is that the system does not have to be close to any particular unstable fixed point, as in OGY based schemes, before implementing control. Once a specified state variable exceeds the threshold it is caught immediately in a stable orbit.

The changes in state effected by thresholding, namely $(x - x^*)$ when $x > x^*$, are typically small (as adjustments are made just after x crosses x^*). Further for higher periods the controlling action is infrequent and occurs

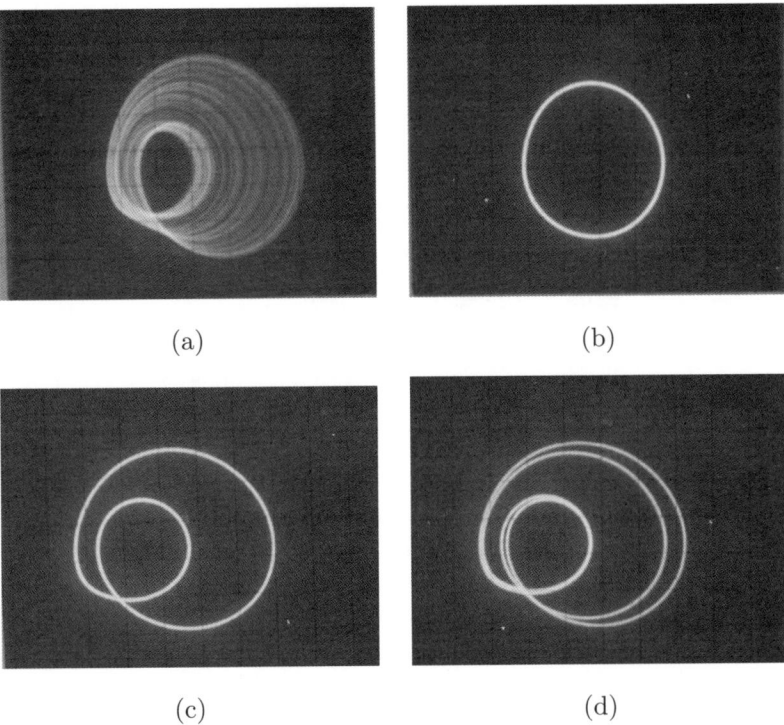

Fig. 10. Attractors in the x–$\dot{x}$ plane: (a) the uncontrolled chaotic system obtained from the circuit realization of Eq. (3), (b) period 1 cycle obtained when $x^* = 1\ V$, (c) period 2 cycle obtained when $x^* = 2\ V$ and (d) period 4 cycle obtained when $x^* = 2.1\ V$.

for short intervals in every controlled cycle. For instance to control to a 16-cycle with $x^* = 2.327$, the thresholding is operational for only ~ 0.22 msec in an interval of 50 msec.

4.2. Controlling Chua's circuit

Now we consider a realization of the double scroll chaotic Chua's attractor given by the following set of (rescaled) 3 coupled ODEs [18]

$$\dot{x} = \alpha(y - x - g(x)), \quad (4)$$
$$\dot{y} = x - y + z, \quad (5)$$
$$\dot{z} = -\beta y, \quad (6)$$

where $\alpha = 10.$ and $\beta = 14.87$ and the piecewise linear function $g(x) = bx + \frac{1}{2}(a-b)(|x+1| - |x-1|)$ with $a = -1.27$ and $b = -0.68$. The corresponding

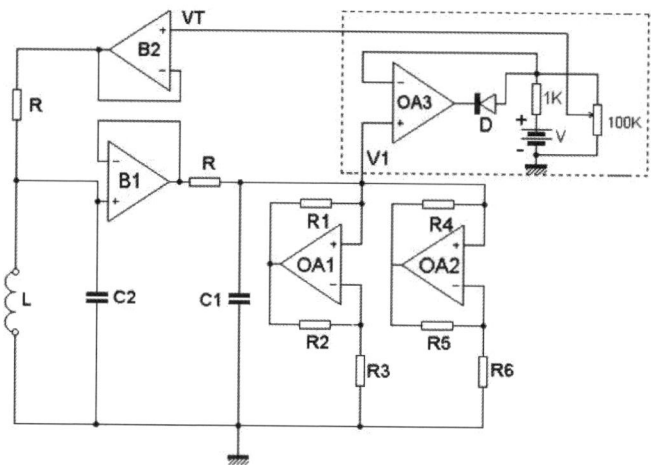

Fig. 11. Chua's chaotic circuit with threshold level controlling circuit (shown in the dotted box). Here V_T is the threshold controlled signal.

circuit component values are: [$L = 18\ mH$, $R = 1710\ \Omega$, $C_1 = 10\ nF$, $C_2 = 100\ nF$, $R_1 = 220\ \Omega$, $R_2 = 220\ \Omega$, $R_3 = 2.2\ k\Omega$, $R_4 = 22\ k\Omega$, $R_5 = 22\ k\Omega$, $R_3 = 3.3\ k\Omega$, $D = $ IN4148, B_1, $B_2 = $ Buffers, OA1 - OA3: opamp μA741]. Note that the circuit of Fig. 11 is the ring structure configuration of the classic Chua's circuit [18, 19]. The uncontrolled attractor from this system is displayed in Fig. 12(a).

Now we implement an even more minimal thresholding. Instead of demanding that the x variable be reset to x^* if it exceeds x^* we only demand this in Eq. (5). This has very easy implementation, as it avoids modifying the value of x in the nonlinear element $g(x)$, which is harder to do. So then all we do is to implement $\dot{y} = x^* - y + z$ instead of Eq. (5), when $x > x^*$, and there is no controlling action if $x \leq x^*$. In the circuit the voltage V_T

Table 3. Thresholding the Chua's circuit.

Threshold Value	Nature of Controlled Orbit
$x^* < 1.84375$	fixed point
$1.84375 < x^* < 2.235$	Period 1 Cycle
$2.235 < x^* < 2.258$	Period 2 Cycle
$2.258 < x^* < 2.264$	Period 4 Cycle
$2.264 < x^* < 2.265$	Period 8 Cycle
$2.265 < x^* < 2.2653$	Period 16 Cycle

Threshold ranges (in V) vs. periodicity of the controlled cycle, for the chaotic system given by Eqs. (4) to 6.

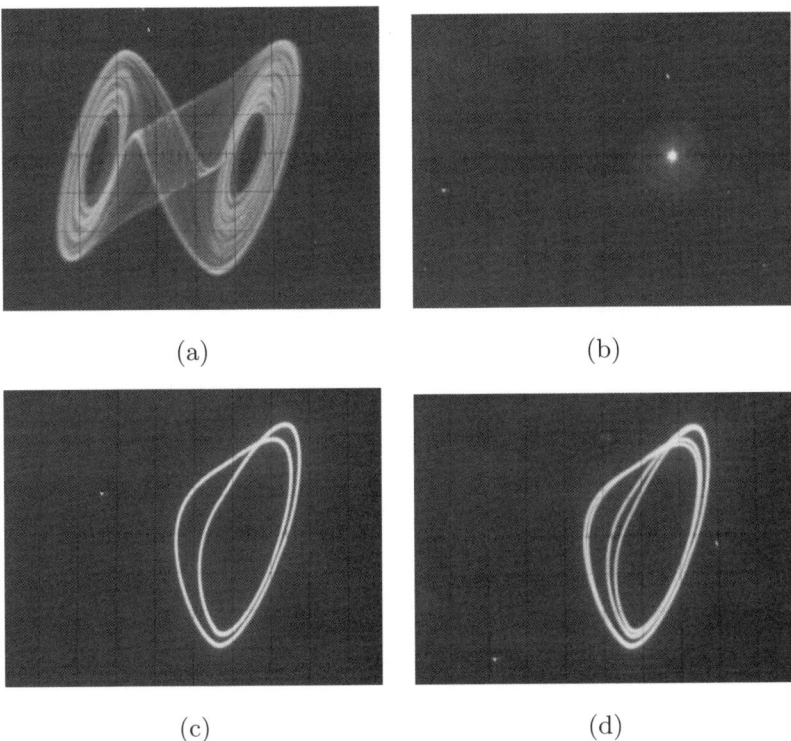

Fig. 12. Attractors in the $V_1 - V_2$ plane, corresponding to the $x - y$ plane of Eqs. (4) to (6): (a) uncontrolled chaotic attractor, (b) fixed point obtained when $x^* = 1.8\ V$, (c) period 2 cycle obtained when $x^* = 2.7\ V$ and (d) period 4 cycle obtained when $x^* = 2.71\ V$.

corresponds to x^* (see Fig. 12). The resulting controlled orbits with respect to threshold x^* is given in Figs. 12(b) to 12(d) ($x^* < 2.7$). So the threshold control works on the system rapidly and can control to a wide range of temporal behaviors (see Table 3).

4.3. *Controlling hyperchaos*

Now we demonstrate the method on a hyperchaotic electrical circuit. This constitutes a stringent test of the control method [20] since the system posseses more than one positive lyapunov exponent, and so more than one unstable eigendirection has to be reigned in by thresholding a single variable. In particular we consider the realization of four coupled nonlinear

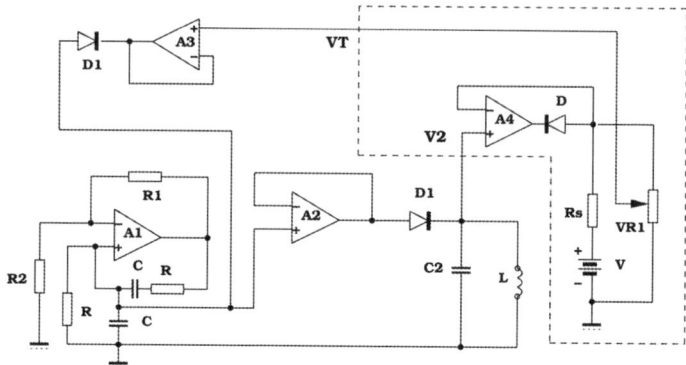

Fig. 13. Circuit implementation of Eqs. (7) to (10), with the precision clipping control circuit in the dotted box. V_T is the threshold controlled signal.

(rescaled) ordinary differential equations of the form

$$\dot{x}_1 = (k-2)x_1 - x_2 - G(x_1 - x_3), \quad (7)$$
$$\dot{x}_2 = (k-1)x_1 - x_2, \quad (8)$$
$$\dot{x}_3 = -x_4 + G(x_1 - x_3), \quad (9)$$
$$\dot{x}_4 = \beta x_3, \quad (10)$$

where

$$G(x_1 - x_3) = \frac{1}{2}b[|x_1 - x_3 - 1| + (x_1 - x_3 - 1)]$$

with $k = 3.85$, $b = 88$ and $\beta = 18$ [21]. The circuit realization of the above is displayed in Fig. 13, with component values: [$L = 18$ mH, $C_2 = 68$ nF, $R = 1.8 k\Omega$, $C = 68$ nF, $R_1 = 2.8$ $k\Omega$, $R_2 = 1$ $k\Omega$, $D_1 = $ IN4148]. Figure 14(a) displays the (uncontrolled) hyperchaotic attractor resulting from this circuit, and it is characterised by two maximal positive lyapunov exponents: $\lambda_1 = 0.13$, $\lambda_2 = 0.05$.

Again we implement a *partial* thresholding on variable x_3: whenever $x_3 > x^*$ in the system, $G(x_1 - x_3)$ in Eq. (7) becomes $G(x_1 - x^*)$, i.e. we have $\dot{x}_1 = (k-2)x_1 - x_2 - G(x_1 - x^*)$, while Eqs. (8) to (10) are unchanged. When $x_3 \leq x^*$ there is no action at all. A precision clipping circuit [17] as depicted in the dotted box in Fig. 13 is employed for the above scheme, which is even simpler to implement than thresholding x_3 throughout the system. We have chosen component values for the control circuit to be: [op-amp = μA741, diode (D) = IN4148 or IN34A, series resistor $Rs = 1$ $k\Omega$ and threshold reference voltage = V, which sets the x^*].

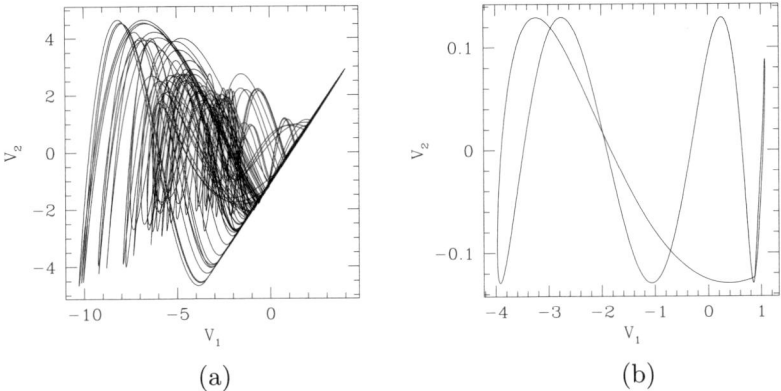

Fig. 14. (a) Uncontrolled hyperchaotic attractor, (b) controlled attractor for threshold = 0 V, in the $V_1 - V_2$ plane, corresponding to the $x_1 - x_3$ plane of Eqs. (7) to (10).

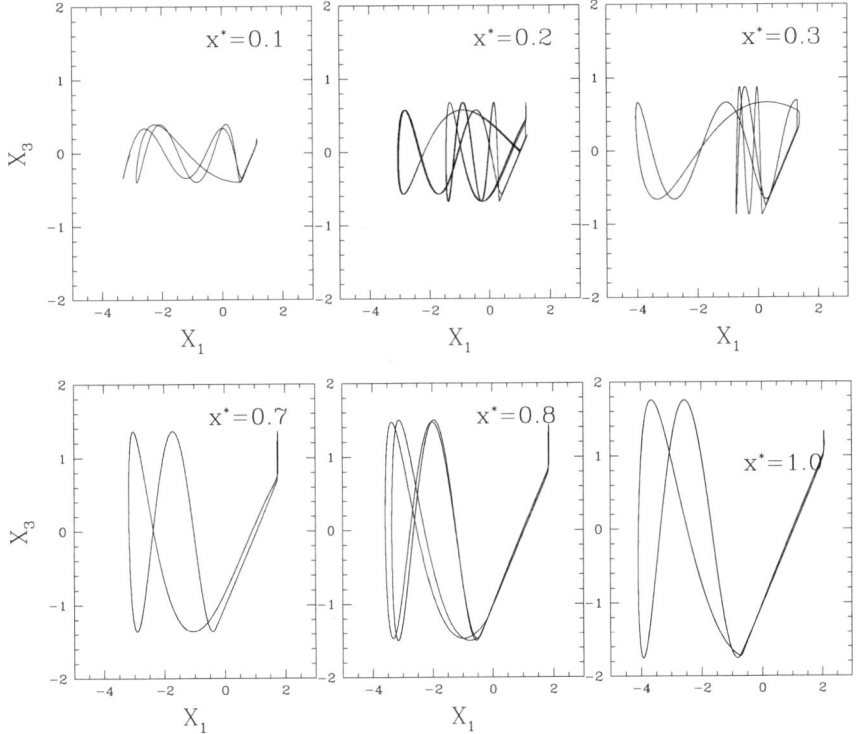

Fig. 15. Controlled attractors in the $x_1 - x_3$ plane, obtained from the hyperchaotic system by thresholding the x_3 variable in Eq. (7), with threshold values: (i) $x^* = 0.1$, (ii) $x^* = 0.2$, (iii) $x^* = 0.3$, (iv) $x^* = 0.7$, (v) $x^* = 0.8$, and (vi) $x^* = 1.0$,

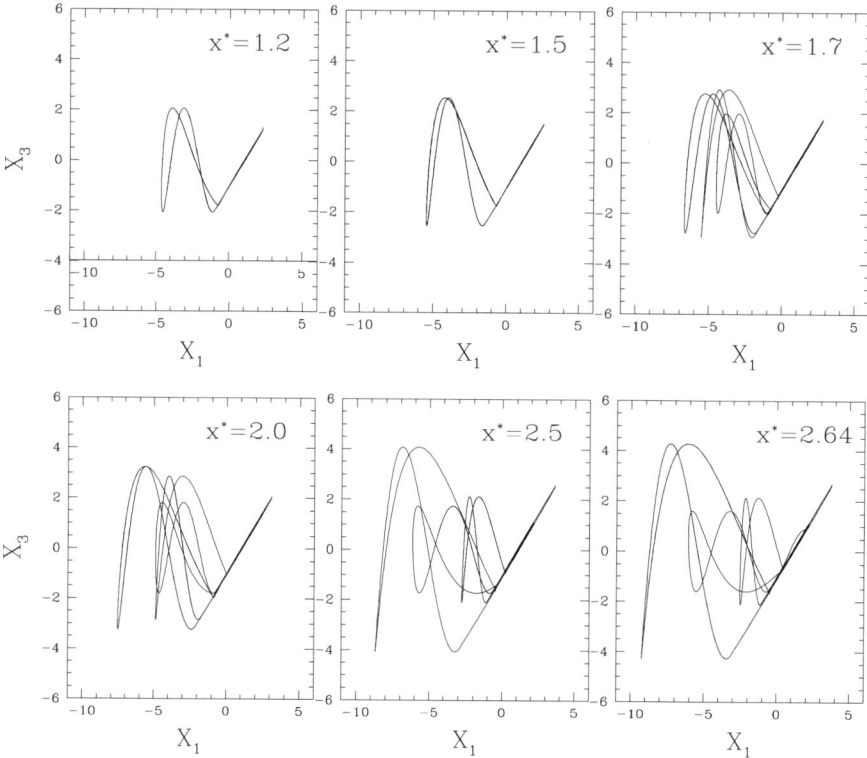

Fig. 16. Controlled attractors in the $x_1 - x_3$ plane, obtained from the hyperchaotic system by thresholding the x_3 variable in Eqs. (7) to (10), with threshold values: (i) $x^* = 1.2$, (ii) $x^* = 1.5$, (iii) $x^* = 1.7$, (iv) $x^* = 2.0$, (v) $x^* = 2.5$ and (vi) $x^* = 2.84$.

Both our experiments and our numerical simulations (which are in complete agreement) show that this scheme successfully yields regular *stable* cycles under a very wide range of thresholds. A representative example with threshold set at 0 V is displayed in Fig. 14(b), which shows the controlled cycle in the $V_1 - V_2$ plane, which corresponds to the rescaled $x_1 - x_3$ plane of Eqs. (7) to (10).

So it is evident that *a single thresholded variable has the ability to clip the full 4-dimensional hyperchaotic system to regular dynamical behavior* (see Figs. 15 and 16 for some examples of the geometries of the controlled orbits). Thus the period and geometry of the controlled states can be easily varied by setting x^* in different windows. For instance, thresholding at 0 V yields a 1 T attractor (with respect to the x_1 variable), while thresholding at 0.3 V yields period 3 T, 0.32 V yields period 8 T, 0.33 V yields period 5 T and 0.35 V yields period 13 T.

5. Conclusions

It is clearly evident then through analytical results, numerics and experiments that threshold control is a powerful, efficient and robust technique to extract a wide range of regular behaviors from a chaotic system. The method involves no adjustment of parameters, but merely the manipulation and re-setting of *one* state variable, even in hyperchaotic systems possesing more than one unstable eigendirections. The richness of chaotic dynamics is an useful feature here, as it ensures that the dynamics can be "clipped" or "truncated" to many different kinds of patterns, i.e. under threshold mechanism the chaos yields a very wide variety of stable dynamical behavior. In fact for one-dimensional systems one can obtain exact results for the effect of thresholding, and these theoretical results have been completely verified in electronic circuit experiments, as well as in arrays of 1-d systems implemented in CMOS based VLSI design [5, 6].

Threshold control is especially useful in the situation where one wishes to design components with the ability to *switch between patterns*, operating as potential *pattern generators*. Basically it can provide a *look up table* for a *"library of patterns"* for very swift control. So the simplicity and versatility of the threshold controller has much potential utility for chaos-based applications, such as chaos computing [22, 23] and communications [24].

References

[1] S. Sinha and D. Biswas, Phys. Rev. Letts. **71** (1993) 2010; S. Sinha, Phys. Rev. E, **49** (1994) 4832; Phys. Letts. A, **199** (1995) 365; Int. Jour. Mod. Phys. B **9** (1995) 875; S. Sinha, in *Recent Developments in Nonlinear Systems* (Narosa, 2002).

[2] E. Ott, C. Grebogi and J. Yorke, Phys. Rev. Letts. **66** (1990) 1196.

[3] Note that a similar method was developed independently by Wagner and Stoop, and goes by the name *Limiter Control*: see for instance, C. Wagner, R. Stoop, *Phys. Rev. E* **63** 017201 (2001); C. Wagner, R. Stoop, *J. Stat. Phys.*, **106** 97-107 (2002).

[4] L. Glass and W. Zheng, Int. J. Bif. and Chaos, **4** (1994) 1061.

[5] K. Murali and S. Sinha, *Phys. Rev. E* **68** (2003) 016210.

[6] D. Atuti, T. Morie, and K. Aihara, *IEEE Int. Workshop on Nonlinear Dynamics of Electronic Systems* (NDES 2007), pp. 145-148 (2007).

[7] This method may be better suited for optical and electrical systems which lend themselves easily to limiters, while it may be harder to implement limiters in chemical and mechanical systems.
[8] A. Miliotis, S. Sinha and W.L. Ditto, Proceedings of *IEEE Asia-Pacific Conference on Circuits and Systems* (APCCAS06), Singapore (2006) pp. 1843-1846; A. Miliotis, S. Sinha and W.L. Ditto, *Int. J. of Bif. and Chaos* (to appear).
[9] U. Hubner, N.B. Abraham and C.O. Weiss, Phys. Rev. A, **40** (1989) 6354 (and references therein).
[10] *Patterns, Information and Chaos in Neuronal Systems*, ed. B.J. West (World Scientific, 1993) and references therein.
[11] Pinsky, P. and Rinzel, J., Journ. Comp.Neuroscience, **1** (1994) 39; Lindner, J.F. and Ditto, W.L., Proceedings of the 3rd Technical Conference on Nonlinear Dynamics (Chaos) and Full-Spectrum Processing, Mystic, Connecticut, (AIP Conference Proceedings, **375** (1996) 709.
[12] S. Sinha and W. Ditto, Phys. Rev. E **63** (2001) 056209.
[13] T. Hogg and B. Huberman, Smart Materials and Structures, **7** (1998) R1, and references therein.
[14] The very successful smart matter controls based on markets, i.e. multiagent methods of control [13], depend on the information from many sites, with sensors located at various points.
[15] S. Sinha, Phys. Rev. E, **63** (2001) 036212.
[16] J.C. Sprott, Phys. Letts. A **266** (2000) 19.
[17] Maddock, R.J. and Calcutt, D.M., *Electronics: A Course for Engineers*, Addison Wesley Longman Ltd., (1997) p. 542.
[18] Dimitriev, A.S. *et al*, J. Comm. Tech. Electronics, **43** (1998) 1038.
[19] Lakshmanan, M. and Murali, K., *Chaos in Nonlinear Oscillators: Controlling and Synchronisation*, World Scientific, Singapore, 1996.
[20] M. Ding, *et al*, Chaos **7** (1997) 644; L. Yang *et al*, Phys. Rev. Letts. **84** (2000) 67.
[21] Murali, K. *et al*, *Proc. 6th Experimental Chaos Conference*, Potsdam, July 17-22, 2001.
[22] Sinha, S. and Ditto, W.L. *Phys. Rev. Lett.* **81** (1998) 2156; Sinha, S. and Ditto, W.L. *Phys. Rev. E* **59** (1999) 363; T. Munakata *et al*, IEEE *Trans. Circ. and Systems* **49** (2002) 1629; Sinha, S. *et al*, Phys. Rev. E **65** (2002) 036214; Sinha, S. *et al Phys. Rev. E* **65** (2002) 036216.
[23] K. Murali, Sudeshna Sinha and W.L. Ditto, *Phys. Rev. E* **68** (2003) 016205; K. Murali, S. Sinha and W.L. Ditto, *Int. J. Bif. and Chaos* (Letts) **13** (2003) 2669; M.R. Jahed-Motlagh, B. Kia, W.L. Ditto and S. Sinha, *Int. J. Bif. and Chaos* **17** (2007) 1955; W.L. Ditto and S. Sinha, *Phil. Trans. of the Royal Soc. of London* **364 A** (2006) 2483.
[24] Hayes, S. *et al*, Phys. Rev. Lett. **70** (1993) 3031.

Chapter 9

A MINIMAL MODEL OF CITY TRAFFIC: CHAOS, CRITICAL BEHAVIOR AND CONTROL

J. A. Valdivia, B. A. Toledo*, V. Muñoz and J. Rogan

Departamento de Física, Facultad de Ciencias,
Universidad de Chile, Santiago, Chile
*btoledo@macul.ciencias.uchile.cl

1. Introduction

The complex behavior displayed in traffic patterns is an interesting field of physics that has been attracting some attention for several decades [7], in particular for their statistical [10, 19] and dynamical [9, 24] properties. There are a number references on traffic jams, chaotic traffic flows, bus-route problems, pedestrian flows, etc. [6, 12–14, 16, 17]. In particular, the development of complex behavior in traffic flows determines, in a certain way, the efficiency of the transportation infrastructure of a city, region, or country. In this context, traffic flows, with and without passing, have been studied extensively in the literature [1, 15], e.g., cellular automaton models, mean-field theories which test the microscopic evolution, hydrodynamic models which approach collective behavior, etc. [11, 20].

In this chapter, we will formulate "a minimal model of city traffic", where we will follow the behavior of cars moving through a sequence of street light signals, and discuss different control schemes. In this model, the street lengths can be fixed or variable and the control is applied to the frequency and relative phase of the traffic lights.

It is worth noticing that the timing of traffic lights must be close to the characteristic traveling time e.g., including car interaction and so on between signals, since longer or shorter timing will slow down the car mean speed, and may contribute to jam the road [18]. This suggests that resonant conditions may lead to efficient traffic systems, but more importantly, resonance and control are related. Moreover, it will be shown that around resonance, for our model, dynamical variables follow certain power laws. Such power laws resemble scaling relations near second order phase transitions, and in view of this analogy we refer to them as critical behavior. We plan to characterize this criticality and derive the critical behavior close to the resonance in terms of traveling time, velocity, and fuel consumption. In particular, we will discuss in detail a common control strategy used in cities, the "green wave" [3], in which a green signal is made to propagate with velocity v_{wave}. The applicability to other synchronization strategies will be discussed below.

This particular control method tends to produce clusters of vehicles, and due to this high correlation, a precise knowledge of the leading car can provide us with information about the cluster itself. Therefore, as long as the leading car represents the behavior of the cluster to which it belongs (this occurs for low noise conditions), we can describe with a single car model some common states in traffic behaviors involving clusters of vehicles [11]. Because of this, we will limit ourselves in this chapter to study a single car moving through a sequence of traffic lights [22, 23].

Hence, within this model, we will analyze three control strategies: (a) the zero phase strategy, (b) the green wave strategy, and (c) a Parrondo-like strategy that considers the transients.

2. The Microscopic Model

The aim of our approach is to follow the details of one vehicle moving through a sequence of traffic lights in one dimension. The separation between the nth and $(n+1)^{th}$ traffic light is L_n. The nth light is green if $\sin(\omega_n t + \phi_n) > 0$ and red otherwise, where ω_n is the frequency of the nth traffic light, and ϕ_n is the phase shift. Note that these two parameters are important if we were trying to control the traffic flow.

A car in this sequence of traffic lights can have (a) an acceleration a_+ until its velocity reaches the cruising speed v_{max}, (b) a constant speed v_{max} with zero acceleration, or (c) a negative acceleration $-a_-$ until it stops, hence

$$\frac{dv}{dt} = \begin{cases} a_+ \theta(v_{max} - v), & \text{accelerate,} \\ -a_- \theta(v), & \text{brakes,} \end{cases} \quad (1)$$

where $\theta(x)$ is the Heaviside step function.

As the car approaches the nth traffic light with velocity v the driver must make a decision, to step on the brakes or not, at the distance (the last

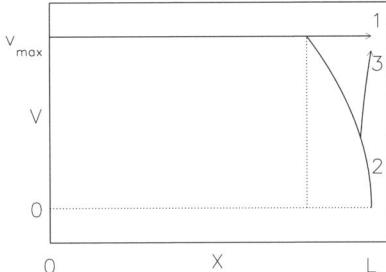

Fig. 1. The possible situations at the decision point, namely, (1) continuing, (2) braking to stop at $x = L$ before the light turns green again, and (3) braking and accelerating again as the light turns green before stopping completely.

stopping point) $v^2/2a_-$ depending on the sign of $\sin(\omega_n t + \phi_n)$. Note that if $(v_{\max}^2/2a_+) + (v_{\max}^2/2a_-) < L_n$, then $v = v_{\max}$ and the car reaches cruising speed before reaching the decision point. Also in general it makes sense that $(2\pi/\omega_n) > (v_{\max}/a_-), (v_{\max}/a_+)$ so that the traffic light does not change too fast from red to green. Of course as the vehicle brakes two things can happen, the car can stop completely and wait until the light turns on again, or it can start accelerating before it stops completely if the light changes. Here we start observing the discontinuous nature of the model.

The type of trajectories between two traffic lights are described in Fig. 1, which clearly shows the typical kinematics associated to this model.

It is interesting to mention that this simplified model may still be relevant in the case of many cars going through the traffic light sequence, but with the effective parameters depending on the density of interacting cars. For example, you may have observed while driving through a city that the effective averaged acceleration seems to depend on the number of cars waiting at the traffic light. Similarly, the averaged effective cruising speed also seems to depend on the density of cars going through the sequence of traffic lights.

We now study the situation of a car traveling through a sequence of N traffic lights, which in essence assumes a city with regular city blocks. We expect that iterating this map may reveal interesting information about the behavior of traffic flow in a city, even with this simplified model. The car enters the sequence of traffic lights with velocity v_0 and time t_0. The set of rules described above determine a 2-D map $M(v_n, t_n)$ that evolves the state (t_n, v_n) at the nth traffic light to state (t_{n+1}, v_{n+1}) at the $(n+1)^{\text{th}}$ traffic light. This map is constructed explicitly in the appendix.

2.1. Fuel consumption

Even though travel time and velocity are good characterizations of the efficiency of a road system, fuel consumption is also of interest to drivers. In general, fuel efficiency will improve if the number of times the car stops is reduced, but it depends on the specific sequence of brakings and accelerations, and thus on the initial conditions. However, general conclusions can be obtained by studying the evolution of the attractor solution.

To account for fuel consumption, we need to study the main sources of dissipation in the car's motion. Fuel consumption is proportional to the mechanical energy produced by the engine, given by $\int_{t_0}^{t_f} Fv\,dt$, where t_0 and t_f are the initial and final times for the complete journey, and F is the forward force or thrust. Besides the engine thrust, we have the rolling friction F_r which opposes the motion, and F_d, where we include other resisting forces such as aerodynamic drag. Therefore, if m is the car mass, the following equation holds:

$$F = ma_+ + F_r + F_d . \qquad (2)$$

An analogous equation for the braking state is not necessary, as we assume that the forward force provided by the engine is zero while braking. Let us consider each term in Eq. (2) separately. The car acceleration is a_+, as given by Eq. (1), and the total injection of energy due to the acceleration from rest to $v_{\max}$ is $mv_{\max}^2/2$. Each time the car stops, this energy is wasted, so this term represents the effect of the driver's behavior on fuel consumption. The rolling friction is estimated as $F_r = \mu mg$, where mg is the weight of the car and μ is the coefficient of rolling friction [5]. Rolling dissipation is thus given by $\int F_r v\,dt \sim F_r L$, which is a function of the distance between traffic lights. Both sources of energy losses can be compared through the dimensionless number $f_r \equiv 2F_r L/mv_{\max}^2 \sim 2\mu gL/v_{\max}^2$ which is $f_r \sim 0.2$ for a car traveling at 50 km/h between lights 200 m apart and a rolling coefficient of $\mu = 0.01$ [21].

Finally, the force F_d is a function of the car velocity. Most of the fuel consumption in a non-stop journey is due to the rolling and drag forces, since accelerations are minimal. However, if the car passes through a sequence of traffic lights, it moves at lower speeds, and then drag is less important than rolling friction. Hence, we neglect drag dissipation in our analysis. We also neglect other dissipative sources such as the energy needed to keep the motor running (in particular, the energy wasted while standing at the traffic light) and the energy lost due to internal frictions in the car mechanisms [4].

Thus, under city traffic conditions, total fuel consumption can be estimated as

$$C = \int_{t_0}^{t_f} Fv\,dt = ma_+ L_+ + F_r\left(L_+ + L_0\right), \qquad (3)$$

where L_+ is the portion of the traveling length in which the driver was accelerating and L_0 is the distance traveled at constant speed.

For now, we will concentrate on studying the dynamics for a given value of $v_{\max}$. Note that this parameter is very relevant in actual city situations since different drivers are willing to reach different values of $v_{\max}$, and traffic light control strategies, achieved through ω_n and ϕ_n, will be very sensitive to its distribution. Furthermore, if we assume that the traffic parameters are, to first order, functions of the density or number of cars, then control strategies must take this into account specially during traffic jams.

3. Zero Phase Control Strategy: Resonant Behavior

In Ref. [23], a specific strategy of traffic light synchronization was considered, namely, all lights have equal phase $\phi_n = 0$. This synchronization, which we consider now, makes sense only if $L_n = L$. Later on we will relax this restriction when we apply other control strategies. Note that we could consider different $L_n = L + \Delta L_n$ values and different frequencies $\omega_n = \omega + \Delta \omega_n$ values as induced phase shifts $\Delta \phi_n = \Delta L_n / v_{\max}$ and $\Delta \phi_n = \Delta \omega_n L / v_{\max}$ respectively. That is why we concentrate for simplicity on the situation $L_n = L$, and $\omega_n = \omega$.

If the period of the signals, $2\pi/\omega$, is equal to the cruising time, T_c, after a short transient (passing a few traffic lights), the car will arrive at each successive decision point when the light's phase is the same. It is important to note that such resonance between the car motion and traffic signals corresponds to a very narrow region of parameters (see the period-1 orbit in Fig. 2). Thus, the interesting regime for controlling traffic situations corresponds to a narrow region around the condition $2\pi/\omega = T_c$. Introducing the dimensionless quantity $\bar{\Omega} = \omega T_c/2\pi$, resonance occurs at $\bar{\Omega} = 1$.

Figure 2 gives the bifurcation diagram of a car starting from rest at the first traffic light. For a given frequency of the traffic lights, characterized by $\bar{\Omega}$, the normalized speed $v_n/v_{\max}$ and time travel between traffic lights $(t_{n+1} - t_n)/T_c$ at the nth light is plotted. A transient of 500 time steps has been removed. This is too large a number of traffic lights to be relevant in real traffic situations, but it is necessary to reach the attractor for all the

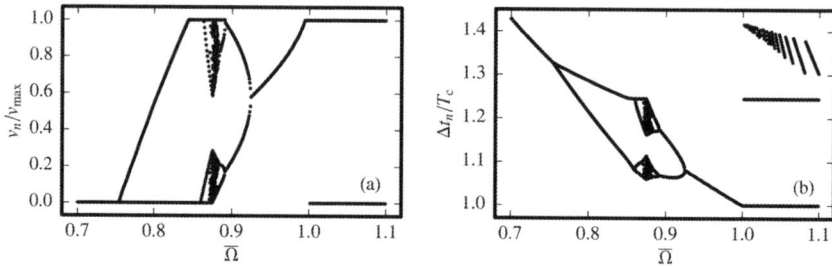

Fig. 2. Bifurcation diagram for the normalized (a) speed at the traffic lights and (b) time travel between traffic lights, versus normalized frequency $\bar{\Omega}$, for $a_+ = 2$ m/s^2, $a_- = 6$ m/s^2, $v_{\max} = 14$ m/s, and $L = 200$ m. A transient of 500 time steps has been removed.

initial conditions plotted (specially in the region very close to the period-doubling bifurcation, where convergence is particularly slow). However, we should point out that most of the initial conditions converge to the attractor in as few as 5–20 traffic lights.

It is important to notice that even in this model there is already an interesting nontrivial behavior in the range $0.75 < \bar{\Omega} < 1$ as displayed in Fig. 2, where a necessary condition for complexity emerges even from the dynamics of a single car. It includes a period doubling bifurcation transition to chaos, where the Lyapunov exponent is estimated in Toledo et al. [2004] for a similar situation. It is interesting to note that this chaotic behavior is produced by the finite accelerating and braking capabilities of the cars, and is thus independent of the interactions between cars. This is one of the reasons for which this model could be an interesting starting point for a first principles approach to traffic in cities.

3.1. Existence of a chaotic regime

The bifurcation diagram of Fig. 2 suggests a period doubling bifurcation to chaos as we increase Ω. A crisis occurs as the chaotic attractor collides with one of the velocity thresholds, producing an inverse period double bifurcation. If we zoom into one of the frequency ranges where the map displays complex behavior, as shown in Fig. 3(a), we find an intricate structure of steady and chaotic behavior.

Estimating the relevance of this chaotic behavior and its sensitivity to perturbation and noise, may be of importance in control strategies. In this sense a finite amplitude Lyapunov exponent can be estimated [2]. Let us take a trajectory in the attractor that starts from (u_0, τ_0) and an initially

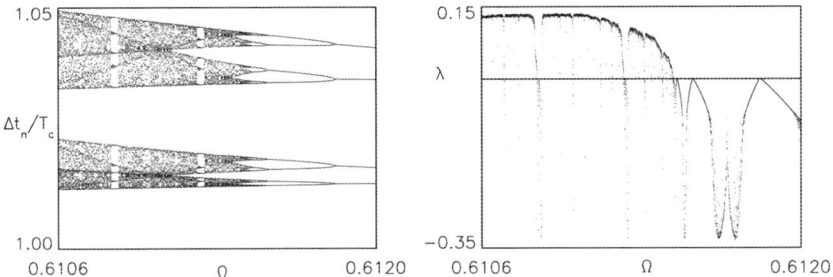

Fig. 3. The bifurcation diagram, (a) zoom for Fig. 4(b), and (b) the associated Lyapunov exponent.

perturbed trajectory that starts from $(u_0, \tau_0 + \delta_0)$, with for example $\delta_0 = 10^{-7}$. The error is iterated n times producing δ_n. Care must be taken to include only the scaling region where

$$\delta_n \sim \delta_0 e^{\lambda n} .$$

Given an initial condition over the attractor an exponent can be estimated by a fitting procedure in the scaling region. Of course, the discontinuous nature of the map complicates this calculation, where for example, both trajectories can reach the same state in one step, yielding $\lambda = -\infty$.

But a final Lyapunov exponent can still be constructed by averaging many initial conditions over the attractor, as shown in Fig. 3(b).

3.2. *Resonance and control*

Intuitively, and from Fig. 2, at $\bar{\Omega} = 1$ the car motion is in resonance with the traffic lights and the traveling time between two given traffic signals is minimized. For $\bar{\Omega} > 1$ (increasing ω), there are a number of resonances, separated by $\Delta \omega = 2\pi/T_c$. Figure 4 displays the average normalized speed $\langle v \rangle / v_{\max}$ (total distance traveled divided by total time elapsed) as a function of frequency. Successive resonant points are found at $\bar{\Omega} = \ell$, where ℓ is a positive integer. We will see below that these resonances display critical behavior. On the other hand for $\bar{\Omega} < 1$ there are situations in which the car covers a distance qL, with q a positive integer, with cruising speed for half the period of the traffic lights, and then is stopped for the other half of the period. In these cases $\bar{\Omega} = 1/q$ and the average normalized speed is $\langle v \rangle / v_{\max} = 1/2$ as shown in Fig. 4. Since for a reasonable city $L \approx 200$ m and $v_{\max} \approx 50$ km/h, the traffic light period of the first resonance

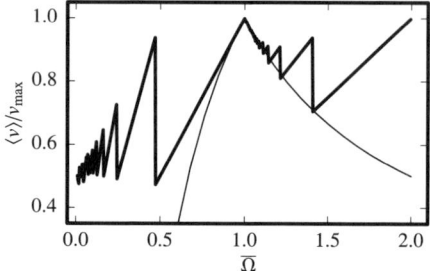

Fig. 4. Resonant tongues showing the average speed (total distance traveled divided by total time elapsed) as a function of the forcing frequency $\bar{\Omega}$. The thin line corresponds to the scaling relation Eq. (23). A transient of 500 time steps has been removed.

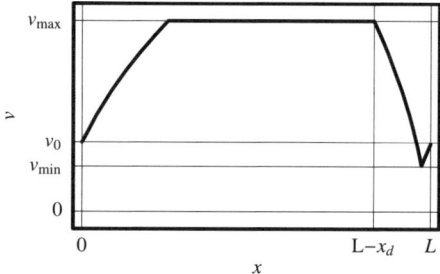

Fig. 5. Speed versus distance for the period-1 attractor below resonance ($\bar{\Omega} < 1$). The car starts at the first traffic light with velocity v_0, accelerates until reaching velocity $v_{\max}$, and arrives at the decision point $L - x_d$ when the next traffic light is red, so it brakes. When the velocity is a certain minimum value $v_{\min}$, the sign turns green, and the car accelerates again, passing the traffic light with the initial speed v_0.

$P = 2\pi/\omega \approx 14$ s is a little unreasonable, and an attempt to control the system using this parameter alone seems impractical, however, exploring this dynamics could allow us to derive more practical control schemes.

In the vicinity of the resonance $\bar{\Omega} \approx 1$, two different dynamics arise depending on the sign of $\bar{\Omega} - 1$. For simplicity, let us consider a car starting at the first traffic light when it changes from red to green, i.e., when the green window begins. If $\bar{\Omega} < 1$, the car will be delayed with respect to the traffic lights, and will reach the second one when it is red, so it will be forced to brake. However, if the delay is small, the traffic light will turn green before the car gets to a full stop, so the car will accelerate again (see Fig. 5), reaching the next traffic light with non-zero velocity. This causes the period-1 orbit below the resonance $\bar{\Omega} = 1$ of Fig. 2.

The situation for $\bar{\Omega} > 1$ is very different. The car reaches the second light a time δt after it has turned green, and this delay increases with each traffic light until it is eventually forced to stop. Thus, for $\bar{\Omega} > 1$, the car moves at maximum speed almost always, except for a stop every p traffic lights, leading to the attractor seen above the resonance in Fig. 2.

To estimate p, we note that the driver arrives at the next signal a small time $\delta t = T_c - 2\pi/\omega > 0$ after the signal turns green, then with a delay $2\delta t$ at the third light, and so on. The journey will continue until the green window is exhausted. The total number of signals, p, that the driver will cross without stopping is given by $p \, \delta t \approx \pi/\omega$, which leads to

$$p \approx \frac{1}{2} \frac{1}{\bar{\Omega} - 1} . \qquad (4)$$

Equation (4) is very interesting, because it also suggests that there is a critical behavior of traffic variables around resonance. However, resonance itself is not a robust feature for $\phi_n = 0$, as it is not independent of the geometry of the road, which is important, because in real situations the distance between traffic lights is not constant, being impossible to maintain resonance while traveling at constant speed.

Fortunately, the opposite is true for another kind of traffic light synchronization strategy, the "green wave", which we now consider.

4. Green Wave Control Strategy

A common strategy for traffic light synchronization is the "green wave", where a green color signal is moved with a speed v_{wave}, so that the color at the nth traffic light, located at a position x_n along the road, is given by $\sin \omega(t - x_n/v_{\text{wave}})$, where ω is the frequency of the traffic light. This implies that $\phi_n = -\sum_{m=1}^{n} L_m \omega / v_{\text{wave}}$. The case $\phi_n = 0$ analyzed in Sec. 3 is equivalent to the green wave case with $v_{\text{wave}} \to \infty$.

In Fig. 6 we plot the bifurcation diagram with $\alpha = v_{\max}/v_{\text{wave}}$ of a car starting from rest for a road with constant distance between traffic signals $L_n = L = 200$ m, constant frequency $\omega = 2\pi/60$ s^{-1}, accelerations $a_+ = 2$ m/s^2 and $a_- = 6$ m/s^2, and $v_{\text{wave}} = 14$ m/s. These parameters are reasonable for an actual road, corresponding to a change of lights every 30 s, and a green wave synchronized with cars moving at 50 km/h. The car will follow a complex path unless the velocity of the car coincides with the wave velocity, *i.e.*, a resonance. Under this condition, the driver will never be stopped. However, resonance is rather fragile, as observed in Fig. 6, hence the dynamics must be observed near the resonant condition $\alpha \sim 1$.

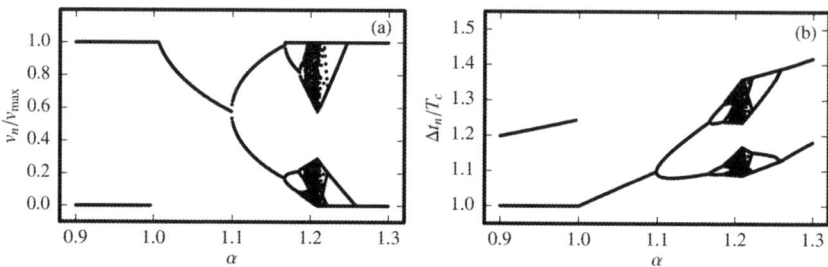

Fig. 6. Bifurcation diagram for (a) normalized speed and (b) normalized time travel between traffic lights, versus α, for $a_+ = 2$ m/s^2, $a_- = 6$ m/s^2, $v_{\text{wave}} = 14$ m/s, $\omega = 2\pi/60$ s^{-1}, $L = 200$ m. The transient has been removed.

The bifurcation diagram in Fig. 6 is very similar to Fig. 2, but reflected horizontally. Thus, it is above resonance, $\alpha > 1$, that a period-1 solution exists, where the car follows a trajectory like Fig. 5, and below resonance the car crosses a certain number p of lights before being stopped. An approximate expression for p can be obtained for the green wave, using similar arguments to those used to derive Eq. (4).

Let us consider the number of traffic lights the car can pass without braking. In the green wave case, close to resonance, we consider a small perturbation $\delta v = v_{\text{wave}} - v_{\text{max}} > 0$. In the optimal case, the driver starts at one extreme of the green semi-period just when the signal changes from green to red, so that at the next signal the driver arrives a time $\delta t = L/v_{\text{max}} - L/v_{\text{wave}}$ before the signal turns red. The journey will continue until the green window is exhausted. The total number of signals, p, that the driver will cross without stopping is given by $p \, \delta t = \pi/\omega$, or

$$p \approx \frac{\lambda/L}{2} \frac{\alpha}{1-\alpha}, \qquad (5)$$

where $\lambda = v_{\text{wave}} \cdot 2\pi/\omega$. Criticality is, once more, explicit. However, unlike the case $\phi_n = 0$, resonance for the green wave holds even if the distance between traffic lights is not constant, in which case $\phi_n = -\sum_{m=1}^{n} L_m \omega / v_{\text{wave}}$. Regarding the quantity p we can do even a little more. If we take advantage of the periodic nature of the solution in the asymptotic regime we can derive the following expression, exact to second order for $\alpha \in (0.67, 1)$,

$$p = \left[(\pi - \xi) \sqrt{\frac{6}{15 - 16 \cos\left(\frac{2\pi L}{\lambda \alpha}\right) + \cos\left(\frac{4\pi L}{\lambda \alpha}\right)}} \right], \qquad (6)$$

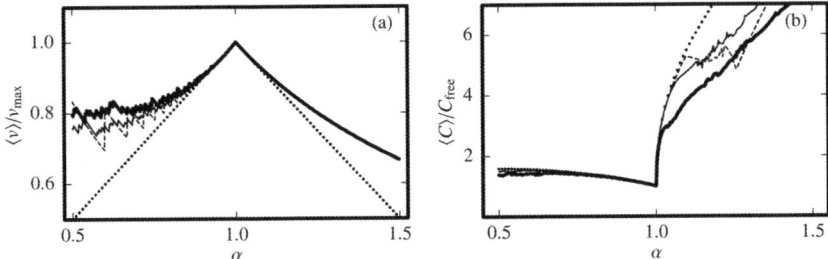

Fig. 7. (a) Resonant tongue showing the average speed (total distance traveled after crossing n signals, divided by time elapsed) as a function of the parameter α. The thin line corresponds to random street length, the thick line corresponds to the Alameda Avenue, the dashed line corresponds to constant street length, and the dotted line corresponds to the scaling laws derived in the text, (b) The corresponding average fuel consumption, normalized to the free consumption $C_{\text{free}} = nF_rL$.

where

$$\xi = \frac{2\pi v_{\max}}{\lambda \alpha} \left\lceil \frac{L}{v_{\max}} + \frac{v_{\max}}{2}\left(\frac{1}{a_+} - \frac{1}{a_-}\right) \right\rceil \mod 2\pi,$$

and the function $\lceil x \rceil$ is the ceiling function, the function that returns the the closest upper integer of x.

An interesting example of this independence of geometry for the behavior near resonance is shown in Fig. 7(a) for the average speed after traveling a large number of traffic lights as a function of $\alpha = v_{\max}/v_{\text{wave}}$. Three cases are compared: (i) a street where distance between traffic lights $L_n = L = 200$ m is constant; (ii) a street with a random distribution of distances $L_n = L + \Delta L_n$, where $\Delta L_n/L$ is a uniform random number in the interval $[-0.5, 0.5]$; and (iii) a real street, namely, the longest city street in Chile (the Avenida del Libertador Bernardo O'Higgins, also known as Alameda Avenue; its precise geometry can be obtained from the Chilean Military Geographic Institute at http://www.igm.cl/). All curves are identical at resonance. The same is true for the average time between traffic lights. This suggests that behavior near resonance for the green wave, at $\alpha = 1$, is indeed universal, regardless of the detailed geometry of the road. Moreover, it will be shown that near resonance, traffic variables behave according to scaling laws. Thus, Fig. 7 shows the universality of this critical behavior. The figure also shows how the efficiency of the strategy degrades as the effective speed of the cars gets away from v_{wave}.

Based on Eq. (5), it is now easy to obtain scaling laws for the traffic variables (time, velocity, fuel consumption). At $\alpha = 1$, the system is at resonance, so that the average travel time $\langle t \rangle$ is equal to the time of "free" travel, when no red lights are found, $T_{\text{free}} \equiv nL/v_{\max}$, where n is the number of passed traffic lights. Average velocity is equal to the corresponding maximum or free velocity $\langle v \rangle = V_{\text{free}} \equiv v_{\max}$. Below resonance these relations change because, if $\alpha < 1$, the car is forced to stop at some point. Since π/ω is the time the red light window lasts, the car is at rest a time $\approx k\pi/\omega$, with k as the number of times the driver brakes. Then the average travel time is

$$\langle t \rangle = T_{\text{free}} + \frac{k\pi}{\omega} . \tag{7}$$

The average velocity in the same run is

$$\langle v \rangle \sim \frac{nL}{\langle t \rangle} . \tag{8}$$

Fuel consumption at resonance, on the other hand, is $\langle C \rangle = C_{\text{free}} \equiv nF_r L$. Below resonance fuel consumption can be estimated by observing that the car stops k times when it covers a distance nL at cruising speed, hence $\langle C \rangle \sim F_r nL + kmV_{\text{free}}^2/2$, which is the total work done by F_r plus the energy wasted in each stop, thus

$$\langle C \rangle \sim C_{\text{free}} \left(1 + \frac{mkV_{\text{free}}^2/2}{nF_r L} \right) . \tag{9}$$

Equations (7) to (9) can be written as

$$\frac{\langle t \rangle - T_{\text{free}}}{T_{\text{free}}} \sim \frac{\lambda}{2L} \frac{k}{n} \alpha ,$$

$$\frac{\langle v \rangle - V_{\text{free}}}{V_{\text{free}}} \sim -\frac{\lambda}{2L} \frac{k}{n} \alpha ,$$

$$\frac{\langle C \rangle - C_{\text{free}}}{C_{\text{free}}} \sim 1 + \frac{1}{f_r} \frac{k}{n} .$$

Since after p traffic signals there is one stop, we can estimate $k/n \sim 1/p$. Then, using (5), yields the following scaling laws:

$$\frac{\langle t \rangle}{T_{\text{free}}} \sim 1 + (1 - \alpha) , \tag{10}$$

$$\frac{\langle v \rangle}{V_{\text{free}}} \sim 1 - (1 - \alpha) , \tag{11}$$

$$\frac{\langle C \rangle}{C_{\text{free}}} \sim 1 + \frac{2L/\lambda}{f_r} \frac{(1 - \alpha)}{\alpha} . \tag{12}$$

Above resonance ($\alpha > 1$), the period-1 solution is possible if the average time to move between two traffic lights is

$$\langle t \rangle = \frac{L}{v_{\text{wave}}} = T_{\text{free}}\, \alpha \approx T_{\text{free}} \left[1 + (\alpha - 1) + \mathcal{O}(\alpha - 1)^2\right], \qquad (13)$$

and the average velocity is

$$\langle v \rangle = v_{\text{wave}} = \frac{v_{\max}}{\alpha} \approx v_{\max} \left[1 - (\alpha - 1) + \mathcal{O}(\alpha - 1)^2\right]. \qquad (14)$$

Equations for $\langle t \rangle$, (10) and (13), and for $\langle v \rangle$, (11) and (14), can be combined as

$$\frac{\langle t \rangle}{T_{\text{free}}} = 1 + |1 - \alpha|, \qquad (15)$$

$$\frac{\langle v \rangle}{V_{\text{free}}} = 1 - |1 - \alpha|, \qquad (16)$$

being symmetrical around resonance.

Symmetric expressions like these cannot be obtained for fuel consumption. In order to estimate fuel consumption above resonance, let us first notice that the trajectory is analogous to Fig. 5. The distance in which rolling friction acts against the engine is

$$x_r = L - \frac{v_{\max}^2 - v_{\min}^2}{2 a_-}, \qquad (17)$$

and the energy lost when breaking is

$$W_a = \frac{m}{2} \left(v_{\max}^2 - v_{\min}^2\right). \qquad (18)$$

Thus, total work between two traffic lights is

$$W = F_r\, x_r + \frac{m}{2} \left(v_{\max}^2 - v_{\min}^2\right) = F_r\, L + \frac{1}{2} \left(v_{\max}^2 - v_{\min}^2\right) \left(m - \frac{F_r}{a_-}\right). \qquad (19)$$

Note that this is equivalent to Eq. (3). In order to obtain $v_{\min}$, we solve the following set of equations:

$$v_0 = v_{\min} \sqrt{1 + \frac{a_+}{a_-}}, \qquad (20)$$

$$T = \left(\frac{v_{\max}}{2} - v_{\min}\right)\left(\frac{1}{a_+} + \frac{1}{a_-}\right) + \frac{v_0^2}{2 v_{\max} a_+} + \frac{L}{v_{\max}}. \qquad (21)$$

These equations follow from Fig. 5. Equation (21) simply states that the time to travel from one light to the next one is equal to $T = L/v_{\text{wave}}$. Thus,

$$\langle C \rangle \sim C_{\text{free}} \left(1 + \frac{2}{f_r}\left[1 - \frac{F_r}{m a_-}\right] \sqrt{\frac{2 a_+ a_-}{a_+ + a_-} \frac{L}{v_{\text{wave}}^2} \frac{(\alpha - 1)^{\frac{1}{2}}}{\alpha}}\right) + \mathcal{O}(\alpha - 1). \qquad (22)$$

Fuel consumption behavior is not symmetrical near resonance. This asymmetry is related to the fact that below resonance the car fully stops only once every p signals, whereas above resonance the car never stops, but brakes at every signal. Since C depends strongly on the detailed pattern of acceleration in the trajectory, scalings are different at each side of the resonance. In Fig. 7(b) numerical results, obtained by iterating the map, are plotted, showing good agreement with the approximated expressions Eqs. (12) and (22) (dotted lines). Let us note that f_r is a function of α if we assume that v_{wave} is constant and we vary v_{max}. For $\alpha > 1$ the scaling law we derived above breaks at the period doubling bifurcation, i.e., $\alpha \approx 1.1$ as seen in Fig. 7(b). The strong asymmetry in this figure also suggests that on average, fuel consumption is higher for "impatient" drivers traveling with velocity above the green wave velocity.

The universality of Eq. (16) is also clearly suggested in Fig. 7(a) for the averaged velocity. This is interesting, as the scaling laws have been obtained for equidistant traffic lights, but also hold for varying street length.

Although this critical behavior has been derived for a single car model, we expect it to have an effect when multiple cars (not too many, otherwise they will form a jam) are in the road as well. Indeed, for a single car, it corresponds to traveling a large number of traffic lights without stopping. Since it would keep its maximum velocity during most of the travel, it would not interact with other cars also in the same situation. Then, the critical behavior, in general, would occur when a bundle of cars passes p lights before being stopped, with $p \gg 1$. This is analogous to a system near a phase transition, when the correlation length goes to infinity. We have obtained analytical results for the critical behavior in our simple model, which could then be compared with more complex simulations and measurements.

It is interesting to notice that the scaling relations for velocity and time traveled derived for the green wave strategy can be mapped to the equivalent scaling laws for the $\phi_n = 0$ strategy by rewriting $\alpha \longrightarrow 1/\bar{\Omega}$. The actual derivation follows along similar arguments as the ones used for the green wave strategy. For instance the velocity scaling is

$$\frac{\langle v \rangle}{V_{\text{max}}} = 1 - \frac{|1 - \bar{\Omega}|}{\bar{\Omega}}, \tag{23}$$

displayed as the thin line in Fig. 4. In the case of fuel consumption for $\alpha > 1$ (and $\bar{\Omega} < 1$), this mapping is even more evident, since we need to carry the same analysis as above, but with $T = L/v_{\text{wave}} \longrightarrow 2\pi/\omega$, i.e., $\alpha \longrightarrow 1/\bar{\Omega}$.

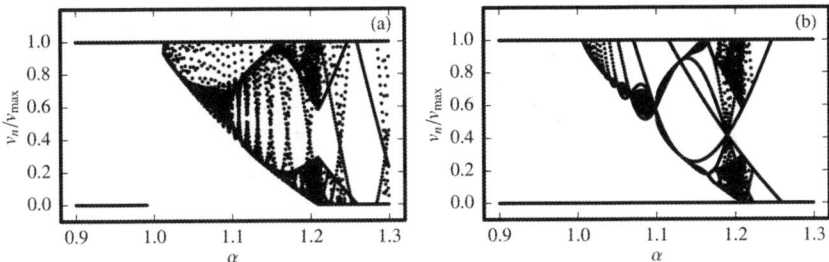

Fig. 8. Bifurcation diagram for the normalized speed v_n/v_{max} as the control parameter $\alpha = v_{max}/v_{wave}$ is varied. Each figure corresponds to a different initial condition: (a) $t_0 = 0$, $v_0 = 0$, and (b) $t_0 = \pi/\omega$, $v_0 = 0$. They contain the transient.

5. Transient Behavior

The results stated in the previous sections regarding resonance and critical behavior for the green wave are valid in the asymptotic regime of the car dynamics. They are valid regardless of the detailed geometry of the system (characterized by the distance L_n between traffic lights). However, trips in cities are typically short, and transient dynamics cannot be neglected in general. In the following sections we intend to describe some features of the transient behavior which may be of interest for city traffic.

Let us consider the green wave strategy. Figure 8 is analogous to Fig. 6(a), but the transient is also shown. In Fig. 8(b) the car starts later. The change in start time is relevant only in the transient part, and of course, both trajectories converge to the same attractor of Fig. 6(a).

Figure 8 shows that, depending on the initial conditions, the evolution can be quite complex, which as mentioned above, may be relevant for city traffic. In particular, strategies for optimizing fuel consumption turn out not to be very obvious even in our simple model. For instance, let us consider the condition $\alpha = 1.3$. The asymptotic solution is a period two orbit with $v_n = 0$ and $v_{n+1} = v_{max}$ (see Fig. 8). This situation represents a simple case with an interesting asymptotic behavior that may be quite annoying for the drivers. The left panel in Fig. 9 shows v_n/v_{max} at traffic lights $n = 3$ and $n = 20$ [Figs. 9(a) and 9(b), respectively] for a range of initial conditions in time and velocity. For the same traffic lights we also compute fuel consumption with Eq. (3). This is plotted in the right panel in Fig. 9. Darker (lighter) color represents lower (higher) fuel consumption. Note that these zones are fairly wide and inhomogeneous. Also, there are points associated to high consumption very near to points of low consumption. This result

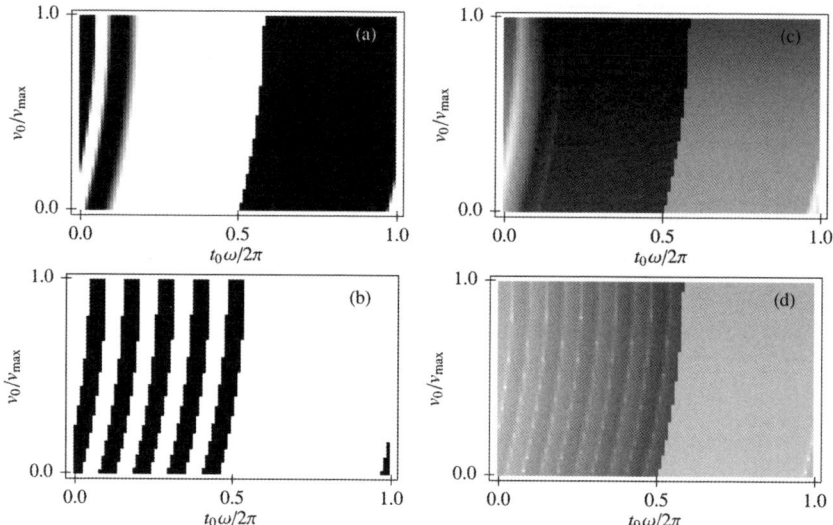

Fig. 9. Transient behavior for $\alpha = 1.3$ according to the initial conditions in the $v_0/v_{\max}$-$\omega t_0/2\pi$ plane. Lighter tones correspond to higher speeds and higher fuel consumption when crossing the traffic light. In Figs. 9(a) and 9(b), we show the distribution of speed for the third and the twentieth traffic light respectively. In the second column, Figs. 9(c) and 9(d), we show the associated fuel consumption. Fuel consumption is normalized by the maximum fuel consumption among all trajectories analyzed.

points to the difficulty in designing strategies to save fuel or time in city traffic, as optimizations in time traveled may conflict with fuel consumption considerations.

An interesting feature is shown in Fig. 10, for the green wave case, with $\alpha = 1.3$. For two trajectories, the difference in travel time after $n = 20$ traffic

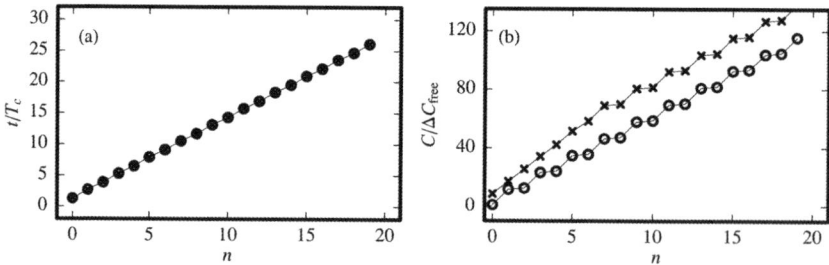

Fig. 10. The comparison of the (a) time traveled (normalized to T_c) and (b) fuel consumption (normalized to $\Delta C_{\text{free}} = F_r L$), for $\alpha = 1.3$, for two particular initial conditions, $v_0 = 18.02$ m/s and $v_0 = 4.55$ m/s, respectively. The rest of the parameters are those for Fig. 6.

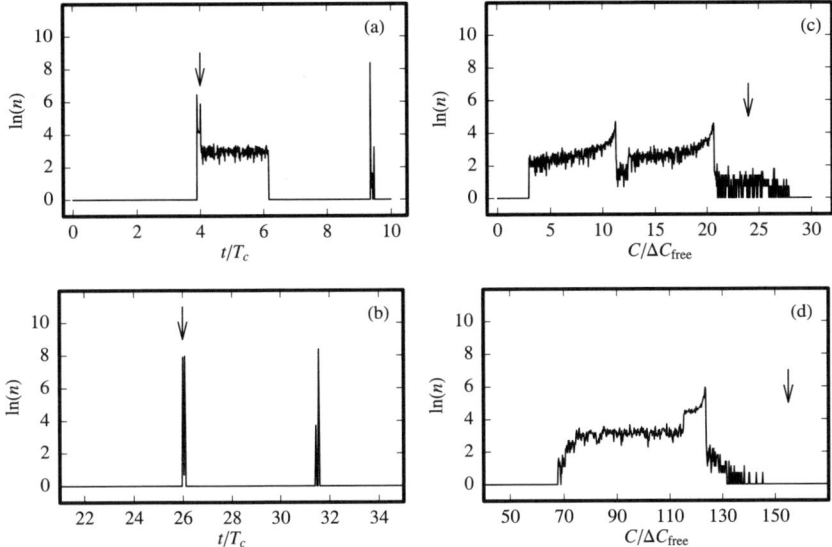

Fig. 11. Transient distributions as measured at different traffic lights for $\alpha = 1.3$, produced by different initial conditions distributed uniformly in the $v_0/v_{\max}$-$\omega t_0/2\pi$ plane. In Figs. 11(a) and 11(b), we show the distributions of time traveled for the third and the twentieth traffic light respectively. The time has been normalized by T_c. In the second column, Figs. 11(c) and 11(d), we show the associated distribution of fuel consumption. Fuel consumption has been normalized by $\Delta C_{\text{free}} = F_r L$. The vertical arrows are the predictions by the asymptotic formulation given by Eqs. (16) and (22). As expected from Fig. 7(b), the prediction for fuel consumption is not very good for $\alpha = 1.3$.

lights is negligible, whereas they vary by $\sim 20\Delta C_{\text{free}}$ in fuel consumption. These results show that fuel consumption can be a more sensible index to characterize the efficiency of the road system, as compared to travel time, and point out again the difficulty in devising general strategies for traffic control.

Another way to state this is to consider a set of initial conditions distributed uniformly in the v-t plane, and let the trajectories evolve. After $n = 3$ and $n = 20$ traffic lights, the distributions of time and fuel consumption are reconstructed and displayed in Fig. 11 with the same arrangement as in Fig. 9. We note that the distributions are highly asymmetrical and tend to be centered around a certain point that is related to the corresponding asymptotic expression for $\alpha = 1.3$, shown in Fig. 7(a). The width of the distribution for fuel consumption is larger than the width of the distribution for elapsed time, which is consistent with Fig. 10. This shows the high sensitivity of this variable and suggests its relevance in city traffic. On

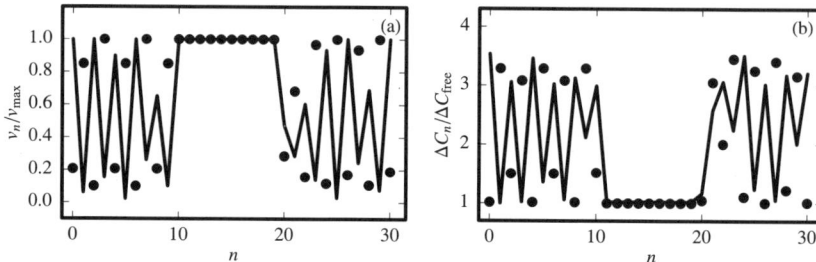

Fig. 12. Orbit collapsing due to the phase change $\phi : 0 \to \pi$ at the 10th traffic light. In both figures, the period-4 orbit is represented by dots, and the chaotic orbit by a line. (a) Period-4 to free motion and chaos to free motion collapsing, (b) Fuel consumption between lights, ΔC_n, normalized by its minimum value $\Delta C_{\text{free}} = F_r L$.

the other hand, let us remember that in this figure we are representing a statistical distribution, at a given time, of a big number of initial conditions randomly chosen over the whole phase space. The variations that we are seeing here characterize the nontrivial transient part of the trajectories. For the period-2 situation we are considering here, there exist a maximum asymptotic spread in time because of those cars that are caught by a red light during the transient part of the trajectory (remember that the average waiting time at the traffic light is $\sim 2T_c$). Therefore, we can see the convergence of the time distribution to two well defined peaks, whereas for the fuel distribution the two hills shown in Fig. 9(c) will merge into the one observed in Fig. 9(d).

If we are interested in short trips, we may devise strategies that can minimize certain variables by inducing certain transients. For instance let us take $\alpha = 1.19$ where we have a period-4 orbit, and $\alpha = 1.2$ where the orbit is chaotic. However, if at the 10th traffic light the phase is changed from 0 to π, a transition to free resonant motion is observed. This motion eventually collapses back to the period-4 or chaotic orbits respectively [see Fig. 12(a)], but only after going through a nice transient of p traffic lights, which is in close agreement with Eq. (4). As displayed in Fig. 12(b), the phase induced green corridor proposed above reduces fuel consumption because rolling friction is the only source of dissipation. This analysis may suggest another control strategy to improve traffic flow by adaptively changing traffic lights phases. It also gives further insight into the origin of complex solutions when the resonance condition is approached. As time progresses, a periodic or chaotic solution suddenly may spot a green corridor that changes completely its observed trajectory.

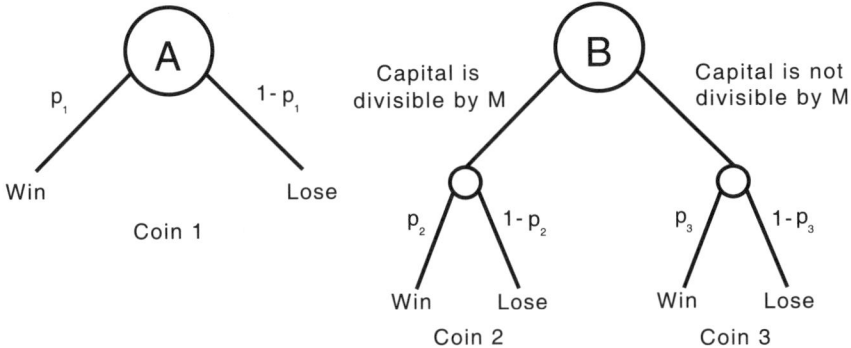

Fig. 13. Realization of a Parrondo game.

6. Parrondo-like Games for Controlling

The phase control of transients, as suggested above, brings us to the concept of treating traffic control as a game. In a game we have a number of agents bounded by a set of rules pursuing a definite goal. In our case, for simplicity, we consider the goal of maximizing the mean velocity, although other goals, such as minimizing fuel consumption, can be considered. Particular attention will be given to the two directional flow, through the same sequence of traffic cars. This problem is interesting, for if we apply the green-wave strategy in a given direction, we may be able to bring the traffic to resonate. But for the cars travelling in the opposite direction, the average speed will be reduced considerably, even compared with the $\phi_n = 0$ (or random) situation (see Fig. 15).

An interesting starting point could be found in the Parrondo paradox [8], in which two different games are defined so that the player always loses in both of them. But when combined, even in a random sequence, the player wins.

Let us consider the capital gained by a player. Figure 13 shows the decision tree of the standard Parrondo game consisting of only three biased coins, where p_1, p_2 and p_3 are the winning probabilities for the individual coins. We can define losing games as follow, let us take $\epsilon = 0.005$, then we have $p_1 = 1/2 - \epsilon$ such that the game A is losing in the long run. We play game A by generating a random number $0 \leq r \leq 1$. If $r < p_1$ then we increase our capital by one. Otherwise, we decrease our capital by one. If we play game A continuously, we obtain the curve showing in Fig. 14. For

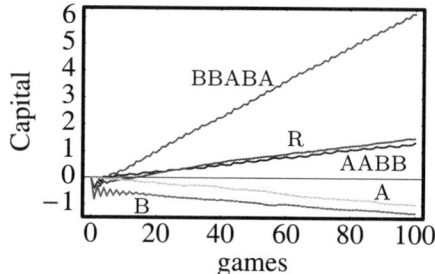

Fig. 14. Five simulations of a 100-game run, with 10,000 trials of each. The two losing games are just A and B. The two moderately winning games are AABB and R. The big winner is BBABA.

the game B, we set $p_2 = 1/10 - \epsilon$, $p_3 = 3/4 - \epsilon$ and $M = 3$. Game B is played by computing if our capital is divisible by M or not. If it is, then we play a game similar to A but with p_2. If our capital is not divisible by M, then we play a game similar to A but with p_3. Since the frequency of the coin 3 is higher than the coin 2, game B is also a losing game as shown in Fig. 14.

It is interesting to note that if we now choose a random sequence of game A and B, then we can obtain a winning game, even though A and B are losing games. Furthermore, we can show that certain particular deterministic sequences of games A and B can also produce winning games, as shown in Fig. 14. In particular, the sequence $BBABA$ is a very profitable game.

For the traffic problem, we can visualize the car going through the line of randomly distributed traffic lights, as a player flipping a coin to stop or to go through at a given traffic light. Although, is not clear how to perform the original Parrondo' game in this situation, it is possible to take the basic idea and combine it with a green-wave, which would be our winning game for the traffic going in a particular direction. For the traffic in the opposite direction, the "anti-green-wave" would be the winning strategy. As we now show in Fig. 15, our simulations show interesting results.

In Fig. 15, we assume a forward "green-wave" strategy which we will perturb in the following way. Let us take the sequence,

$\{0,0,0,0\}, 1, \{0,0,0\}, 1, \{0,0,0,0\}, 1, \{0,0,0,0,0\}, 1, \{0,0,0\}, 1, \{0,0,0,0\}, 1,$
$\{0,0,0,0\}, 1, \{0,0,0,0\}, 1, \{0,0,0,0\}, 1, \{0,0\}, 1, \{0,0,0,0,0,0\}, 1, \{0,0,0\}, 1,$
$\{0,0\}, 1, \{0,0,0,0\}, 1, \{0,0,0,0,0,0\}, 1, \{0,0,0\}, 1, \{0,0\}, 1, \{0,0,0,0\}, 1,$
$\{0,0\}, 1, \{0,0,0,0\}, 1, \{0,0\}, 1, \{0,0,0,0\},$

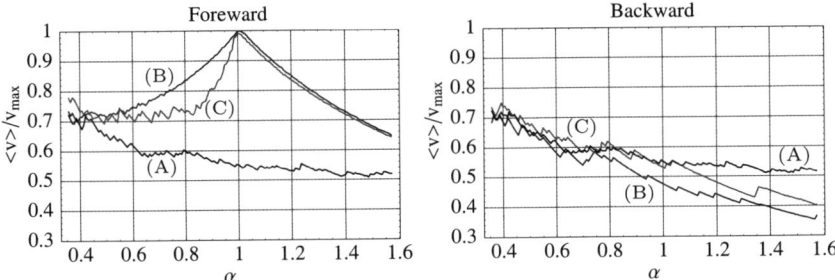

Fig. 15. Normalized average velocity for the cases $\phi_n = 0$ (A), $\phi_n = -\omega x_n/v_{wave}$ (B) and $\phi_n :=$ game (C). With the game we obtain approximately an 8% of improvement in the backward direction and almost no change in the forward direction (the reference is the green-wave strategy at $\alpha = 1$).

where a zero appears when the game is not played, namely we apply the regular forward "green-wave" strategy. If a one appears, then we apply the corresponding forward "green-wave" phase and we add $\Delta\phi = -\omega(\frac{2L}{v_{max}} - \frac{v_{max}}{a_-})$. This phase shift, is in favor of the backward direction, by increasing the range of initial conditions that can pass through the traffic light. This is an example of a successful game in this context, as shown in Fig. 15, since it improves on the average velocity for the backward direction, without a considerable reduction in the forward direction.

7. Conclusion

As a complex system, a traffic network has many interesting features. We have developed a minimal model, that consists of a single car going through a sequence of traffic lights, that displays some of the basic features present in city traffic. Even under these simple conditions, we observe a range of complicated behaviors that, for some critical parameters, display non-chaotic motion, chaos and intermittency (not considered here). Hence, certain perturbations may drive the system to a region where the average speed is reduced considerably. We have investigated control procedure through the manipulation of the phase in the traffic light system. We can perform this manipulation deterministically, which can be useful for few cars under very constrained conditions, or statistically, in the spirit of a game, if we maximize the probability of winning most of the time (maximizing the average velocity). This last approach seems to be fruitful given changing conditions in a city. Therefore, the best game (which maximizes the average velocity or minimizes the traveled time) will be dependent on the city conditions and hence spatial and time dependent. This is a work in progress.

Appendix: The $M(t,v)$ Map

It is convenient to construct an exact map that relates successive crossing of the traffic lights. Let L be the distance between origin O and next traffic light. After crossing the nth light, the car reaches $v_{\max}$ at

$$x_c = \frac{v_{\max}^2 - v_n^2}{2a_+} ,$$

$$t_c = t_n + \frac{v_{\max} - v_n}{a_+} ,$$

$$v_c = v_{\max} ,$$

and continues to move at constant velocity until the decision point

$$x_d = L_n - \frac{v_{\max}^2}{2a_-} ,$$

$$t_d = t_c + \frac{x_d - x_c}{v_{\max}} ,$$

$$v_d = v_{\max} .$$

At this point we have two choices depending on the sign of $\sin(\omega_n t_d + \phi_n)$. If $\sin(\omega_n t_d + \phi_n) > 0$, the car reaches the traffic light with a state

$$x_{n+1} = L_n ,$$

$$t_{n+1} = t_d + \frac{L_n - x_d}{v_{\max}} ,$$

$$v_{n+1} = v_{\max} .$$

If $\sin(\omega_n t_d + \phi_n) < 0$, the car must start slowing down with a_-, and it will take an extra time $\Delta t = v_{\max}/a_-$, to reach the $(n+1)^{\text{th}}$ traffic light and stop. This time must be compared with the next time the light turns green t_g, at which point the car can accelerate again. Defining the phase $\xi_d = \omega_n t_d + \phi_n$, we can compute

$$\xi_g = \omega_n t_g + \phi_n = 2\pi \left(\text{Int}\left[\frac{\xi_d}{2\pi}\right] + 1 \right) ,$$

where $\text{Int}[x]$ is the integer part of x. Therefore, if $t_d + \Delta t < t_g$, the car will cross the $(n+1)^{\text{th}}$ traffic light with

$$x_{n+1} = L_n$$

$$t_{n+1} = t_g ,$$

$$v_{n+1} = 0 .$$

In the other case, $t_d + \Delta t > t_g$, the car starts accelerating at the state

$$x_g = x_d + v_d(t_g - t_d) - a_-(t_g - t_d)^2/2 ,$$

$$t_g = t_g ,$$

$$v_g = v_d - a_-(t_g - t_d) ,$$

and again we have two cases before it reaches L. We need to determine if the car reaches $v_{\max}$ before the light. We compute the distance at which the car reaches $v_{\max}$, namely $x_m = x_g + (v_{\max}^2 - v_g^2)/2a_+$. Therefore, if $x_m > L$, then the car reaches the traffic light with

$$x_{n+1} = L_n,$$

$$t_{n+1} = t_g + \frac{v_{n+1} - v_g}{a_+},$$

$$v_{n+1} = \sqrt{v_g^2 + 2a_+(L_n - x_g)},$$

otherwise, it reaches $v_{\max}$ at

$$x_m = x_m ,$$

$$t_m = t_g + \frac{v_{\max} - v_g}{a_+} ,$$

$$v_m = v_{\max} ,$$

and the light at

$$x_{n+1} = L_n ,$$

$$t_{n+1} = t_m + \frac{L_n - x_m}{v_{\max}} ,$$

$$v_{n+1} = v_{\max} .$$

References

[1] A. Benyoussef, H. Chakib, and H. Ez-Zahraouy. *Phys. Rev. E*, 68:026129, 2003.
[2] G. Boffetta, M. Cencini, M. Falcioni, and A. Vulpiani. *Physics Reports*, 356:367, 2002.
[3] E. Brockfeld, R. Barlovic, A. Schadschneider, and M. Schreckenberg. *Phys. Rev. E*, 64:056132, 2001.

[4] R. H. Essenhigh. *Transportation Research*, 8:457, 1974.
[5] I. Evans. *British J. Appl. Phys.*, 5:187, 1954.
[6] N. V. Findler and J. Stapp. *Journal of Transportation Engineering*, 118:99110, 1992.
[7] B. D. Greenshields. *Highway Reserch Board, Proc.*, 14:458, 1935.
[8] G.P. Harmer and D. Abbott. *Nature (London)*, 402 (6764):864, 1999.
[9] K. Hasebe, A. Nakayama, and Y. Sugiyama. *Phys. Rev. E*, 68:026102, 2003.
[10] D. Huang. *Phys. Rev. E*, 68:046112, 2003.
[11] D. Huang and W. Huang. *Phys. Rev. E*, 67:056124, 2003.
[12] B. S. Kerner and S. L. Klenov. *Phys. Rev. E*, 68:036130, 2003.
[13] J. Kuhl, D. Evans, Y. Papelis, R. Romano, and G.S. Watson. *IEEE Computer*, 1995.
[14] N. Moussa. *Phys. Rev. E*, 68:036127, 2003.
[15] T. Nagatani. *Phys. Rev. E*, 60:1535, 1999.
[16] T. Nagatani. *Rep. Prog. Phys.*, 65:1331, 2002.
[17] T. Nagatani. *Phys. Rev. E*, 68:036107, 2003.
[18] K. Nagel and M. Paczuski. *Phys. Rev. E*, 51:2909, 1995.
[19] K. Nishinari, M. Treiber, and D. Helbing. *Phys. Rev. E*, 68:067101, 2003.
[20] J. B. Rundle, K. F. Tiampo, W. Klein, and J. S. Martins. *Proc Natl Acad Sci*, 67:2514, 2002.
[21] G. Stachowiak and A. W. Batchlor. *Engineering Tribology*. Butterworth-Heinemann, 2001.
[22] B. A. Toledo, E. Cerda, V. Muñoz, J. Rogan, R. Zarama, and J. A. Valdivia. *Phys. Rev. E*, 75:026108, 2007.
[23] B. A. Toledo, V. Muñoz, J. Rogan, C. Tenreiro, and J. A. Valdivia. *Phys. Rev. E*, 70:016107, 2004.
[24] M. Treiber and D. Helbing. *Phys. Rev. E*, 68:046119, 2003.

Chapter 10

CONTROLLING CHAOTIC BURSTING IN MAP-BASED NEURON NETWORK MODELS

R. L. Viana,[*] J. C. A. de Pontes and S. R. Lopes

*Departamento de Física, Universidade Federal do Paraná,
81531-990, Curitiba, Paraná, Brasil*
[*]*viana@fisica.ufpr.br*

C. A. S. Batista and A. M. Batista

*Departamento de Matemática e Estatística,
Universidade Estadual de Ponta Grossa,
84032-900, Ponta Grossa, Paraná, Brasil*

Neuron dynamics presents two timescales related to the spiking (fast) and bursting (slow), a feature already exhibited by simple, map-based models. Neuronal assemblies can be thus investigated using coupled map lattices, once a suitable coupling prescription is adopted to emulate aspects of the neuronal architecture. We have considered map-based neural networks using both scale-free and power-law couplings, in order to study strategies of controlling synchronized chaotic bursting. We have used both an external harmonic signal and a time-delayed feedback signal as control strategies, investigating in what extent bursting synchronization can be suppressed or maintained in such map-based network models.

1. Introduction

One of the landmark papers in the history of chaos control described the experimental study by Steven Schiff and coworkers of the control of chaos in brain using the Ott-Grebogi-Yorke method [1]. There was considered a spontaneously bursting neuronal network *in vitro*, for which a chaotic regime has been demonstrated by the presence of unstable fixed-point behavior. Control of chaos in this case was accomplished through application of small but judiciously chosen perturbations to an accessible parameter, according to the Ott-Grebogi-Yorke method [2], what has been shown to increase the periodicity content of the bursting activity. One technical problem reported by the authors of Ref. [1] was the difficulty on locating the stable manifold of the unstable fixed point detected in experimental return plots, and this was explained on account of the high-dimensionality of the dynamical system.

In fact, the brain, which consists of about a hundred specialized modules with different functions, each of them a complex network itself, is a paradigmatic example of high-dimensional dynamical system. The network unit, the neuron, receives excitatory inputs from a few thousands of other neurons and processes them according to some deterministic rules [3]. Models of biological neuronal networks must consider both the intrinsic dynamics at each neurons as well as their connection architecture.

A description of the bursting dynamics in a neuron requires the use of mathematical models possessing two timescales: (i) a fast time scale characterized by repetitive spiking; and (ii) a slow timescale with bursting activity, where neuron activity alternates between a quiescent state and spiking trains [4]. Many mathematical models emulate this spiking-bursting behavior, ranging from differential equations [5] to discrete-time maps [6, 7]. Map-based neural network models have some advantages over continuous-time models, chiefly due to the typically large number of network units necessary to perform numerical simulations. Accordingly, coupled map lattices have been considered as candidates for artificial neural networks for a long time [8].

Effective neural network modeling demands consideration of coupling prescriptions which describe desirable characteristics of biological neurons. While most theoretical works on coupled map lattices have been done on models with nearest-neighbor (local) interactions, the large connectivity of the biological neuronal networks requires non-local couplings [9]. The simplest model is a global coupling lattice, which considers the interaction of each site with the mean field of all the other sites, regardless of their position along the lattice [10]. A variant of such models takes into account a power-law decrease of the coupling strength with the lattice distance [11]. Moreover, there is a growing interest in scale-free networks, for which the number of connections per site obeys a power-law statistics [12].

The existence of a slow timescale in coupled bursting neurons enables us to define a bursting phase and frequency (its time rate) for each of them, even though on the spiking time scale they may behave asynchronously [10]. The adjustment of the bursting phases and frequencies of two or more neurons can be treated as an example of chaotic phase synchronization, which is defined as the occurrence of a certain relation between phases of interacting systems, bursting neurons in our case, while the amplitudes (related to the spiking time scales) can remain chaotic and uncorrelated [13].

The presence of synchronized rhythms has been experimentally observed in electroencephalograph recordings of electrical activity in the brain, in the form of an oscillatory behaviour generated by the correlated discharge of populations of neurons across cerebral cortex. The behavioral state alters the amplitudes and frequencies of these oscillations, such that high frequency and low amplitude rhythms tend to occur during arousal and attention; whereas low frequency and high amplitude activity occurs during slow-wave sleep [14].

Some types of synchronization of bursting neurons are thought to play a key role in Parkinson's disease, essential tremor, and epilepsies [15]. The synchronous firing of neurons located in the thalamus and basal ganglia appears to cause resting tremor in Parkinson's disease, in such a way that the firing frequency is in the same range ($3-6$ Hz) of the tremor itself [16]. The peripheral shaking results from the activation of cortical areas due to the existence of a cluster of synchronously firing neurons that acts as a pacemaker [17]. Hence a possible way to control pathological rhythms would be to suppress the synchronized behavior. This can be obtained through application of an external high-frequency (>100 Hz) signal, and it constitutes the main goal of the deep-brain stimulation technique [18].

Deep-brain stimulation consists of the application of depth electrodes implanted in target areas of the brain like the thalamic ventralis intermedius nucleus or the subthalamic nucleus [18]. The overall effects of deep-brain stimulations are similar to those produced by tissue lesioning and have proved to be effective in suppression of the activity of the pacemaker-like cluster of synchronously firing neurons, so achieving a suppression of the peripheral tremor [19].

While most progress in this field has come from empirical observations made during stereotaxic neurosurgery, methods of nonlinear dynamics are beginning to be applied to understand this suppression behavior. In this work we review some recent work using two types of control strategies which can be used to suppress synchronized chaotic bursting in assemblies of map-based neural network models. The first technique is to apply a time-periodic harmonic signal of fixed frequency and amplitude to one or more selected neurons. Another strategy, proposed by Rosenblum and Pikowsky [20], makes this external signal to depend on a time-delayed mean-field behavior of the lattice, what amounts to a feedback control procedure.

In the theoretical models to be considered here, the individual neuron dynamics will be described by a two-dimensional dissipative map proposed by Rulkov [7], which yields a spiking-bursting activity in the same way as more complex models do, with the advantage of less computer time, such that it is suitable for simulations using a large number of neuron units. More sophisticated models would include systems of differential equations, like the Hindmarsch-Rose equations, for example, but with the same qualitative behavior, as far as the bursting synchronization is concerned.

Previous works on the control of synchronized bursting have used globally coupled map lattices [10, 20]. In this work we will present results using two different neuronal connection architectures: (i) a power-law coupling, in which the coupling strength decreases with the lattice distance; and (ii) a scale-free coupling, for which the connections are randomly distributed along the lattice, with the number of connections per site satisfying a power-law probability distribution.

Recent experimental evidence suggests that some brain activities can be assigned to scale-free networks, as revealed by high-definition functional magnetic resonance imaging [21, 22]. On the other hand, Achard and co-workers [23] have found that the human functional network is dominated by a neocortical core of highly connected hub-like neurons which do not obey properly a scale-free but rather have an exponentially truncated power-law degree distribution. Humphries and co-workers [24] argue that the medial reticular formation of the brainstem is characterized by a neural network exhibiting small-world, but not scale-free properties.

The structure of this chapter is as follows: in Sec. 2 we present the properties of the map describing neuron dynamics, as well as the definition of a geometrical phase for the bursting dynamics. Section 3 outlines the kind of coupling considered between neurons and some of its properties. Section 4 deals with the dependence of chaotic phase synchronization on the network properties using suitable numerical diagnostics. In Sec. 5 we consider the synchronization between the bursting phases of neurons and the driving phase provided by a time-periodic external signal applied to one selected neuron. Our conclusions are left to the last Section.

2. Neuron Dynamics and Coupling

In the neuron models we consider here, the time evolution of the action-potential is supposed to exhibit two timescales. The fast timescale is related to the spiking neuron activity, whereas the slow timescale appears in the form of bursts characterized by the repetition of spikes [25]. Mathematical models of such bursting neurons may be built upon systems of three or more ordinary differential equations, like the Hindmarch-Rose equations [5] or discrete-time processes with at least two dimensions, like the map proposed by Rulkov [7]

$$x_{n+1} = \frac{\theta}{1+x_n^2} + y_n, \quad (1)$$

$$y_{n+1} = y_n - \sigma x_n - \beta, \quad (2)$$

where x_n is the fast and y_n is the slow dynamical variable.

The parameter θ affects directly the spiking timescale, its values being chosen so as to produce chaotic behavior for the evolution of the fast variable x_n, characterized by an irregular sequence of spikes. Since in neuronal assemblies the neurons are likely to exhibit some diversity, we choose randomly the values of θ within the interval $[4.1, 4.4]$ according to a uniform distribution. The parameters σ and β, on their hand, describe the slow timescale represented by the bursts, and take on small values so as to model the action of an external dc bias current and the synaptic inputs on a given isolated neuron [7].

We choose the parameter θ so as to yield chaotic behavior for the characteristic spiking of the fast variable x_n [Fig. 1(a)]. The bursting timescale, on the other hand, comes about the influence of the slow variable y_n. This can be understood by using a simple argument: since, from Eq. (1), y_n represents a small input on the fast variable dynamics its effect can be approximated by a constant value γ. The resulting one-dimensional map, $x_{n+1} = [\theta/(1+x_n^2)] + \gamma$, can have either one, two, or three fixed points $x_{1,2,3}^*$, depending on the value of the input γ. As the latter approaches a critical value γ_{SN} the fixed points $x_{1,2}^*$ (one stable and another unstable) undergo a saddle-node bifurcation, such that, for $\gamma \gtrsim \gamma_{SN}$, however, the fixed points $x_{1,2}^*$ disappear. For values of $\gamma > \gamma_{CR}$ there is also a chaotic attractor that, provided $\gamma_{CR} < \gamma < \gamma_{SN}$, coexists with the stable fixed point attractor. Actually, at $\gamma = \gamma_{CR}$ the chaotic attractor collides with the unstable fixed point x_1^* and is destroyed through a boundary crisis. The bursting regime then comes from a hysteresis between the stable fixed point (quiescent evolution) and the chaotic oscillations (fast sequence of spikes) [9].

In an assembly of bursting neurons, we do not expect synchronization in the spiking timescale, but we may look for a weaker form of synchronization in the bursting timescale. This is actually possible by conveniently defining

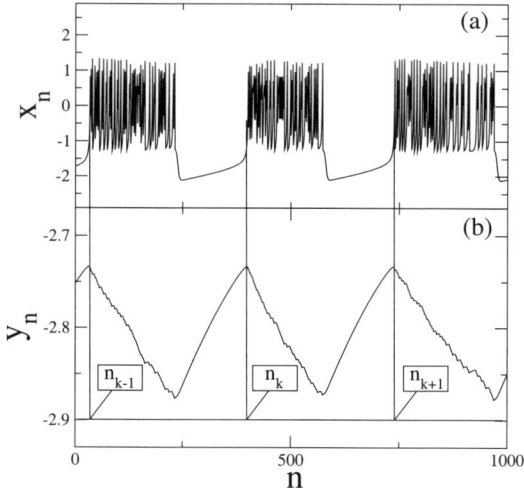

Fig. 1. Time evolution of the (a) fast and (b) slow variables in the Rulkov map (1) and (2) for $\theta = 4.1$, $\sigma = \beta = 0.001$.

a phase for the slow timescale: a burst begins when the slow variable y_n, which presents nearly regular saw-teeth oscillations, has a local maximum, in well-defined instants of time we call n_k [Fig. 1(b)]. A phase can be thus defined as

$$\varphi_n = 2\pi k + 2\pi \frac{n - n_k}{n_{k+1} - n_k}, \qquad (3)$$

and increases monotonically with time. However, due to the chaotic evolution in the fast (spiking) timescale, it turns out that the interval $n_{k+1} - n_k$ is different for each burst. Hence a bursting frequency,

$$\Omega = \lim_{n \to \infty} \frac{\varphi_n - \varphi_0}{n} \qquad (4)$$

gives the time rate of the phase evolution.

In the brain, neurons are connected by axons and dendrites, so that we can regard those neurons as embedded in a three-dimensional lattice. Such higher-dimensional lattices can be very difficult to work with in terms of computer simulations, specially if long-range interactions are present and the number of neurons is large enough. However, good insights are expected to come from simpler models, which can nevertheless retain some of the general characteristics of higher-dimensional lattices. It is from this point of view that we use one-dimensional lattices with N neurons, whose bursting dynamics is governed by the maps (1) and (2):

$$x_{n+1}^{(i)} = \frac{\theta^{(i)}}{1 + \left(x_n^{(i)}\right)^2} + y_n^{(i)} + \mathcal{C}_n^{(i)}(x_n^{(j \neq i)}), \tag{5}$$

$$y_{n+1}^{(i)} = y_n^{(i)} - \sigma x_n^{(i)} - \beta, \quad (i = 1, 2, \ldots N). \tag{6}$$

In neuron ensembles some diversity in the biophysical parameters describing each unit is always expected. Hence we consider the parameter θ to be different for each site and to take on values in the interval $[4.1, 4.4]$, which produces chaotic behavior in the fast (spiking) timescale. The parameteres σ and β, on the other hand, as describing the slow (bursting) timescale, are to take on small values only, and this value is the same for all neurons.

When choosing what kind of coupling model to use we focus on the mean connectivity, i.e. the average number of connections per neuron $\langle k \rangle$. When this number is large enough a global type of coupling has been frequently used [10]

$$\mathcal{C}_n^{(i)}(x_n^{(j)}) = \frac{\epsilon}{N} \sum_{j=1}^{N} x_n^{(j)}, \tag{7}$$

where each neuron is coupled to the "mean-field" generated by the entire lattice irrespective of the lattice distance between neurons. Other models, however, have been proposed to emulate specific features of neuronal assemblies from the point of view of a coupled map lattice. Two of them will be considered in this work, and are described in the following.

2.1. Scale-free coupling

In scale-free networks the connectivity k satisfies a power-law probability distribution $P(k) \sim k^{-\varpi}$, where $\varpi > 1$, in such a way that highly connected neurons are connected, on the average, with highly connected ones, a property also found in many social and computer networks [12, 26]. Functional magnetic resonance imaging experiments have suggested that some brain activities can be assigned to scale-free networks, with a scaling exponent ϖ between 2.0 and 2.2, with a mean connectivity $< k > \approx 4$ [21]. In fact, this scale-free property is consistent with the fact that the brain network increases its size by the addition of new neurons, and the latter attach preferentially to already well-connected neurons [3]. However, in some neural networks, it has been proposed a slightly different connection architecture, where the degree distribution obeys an exponentially truncated power-law scaling $P(k) \sim k^{\alpha-1} e^{k/k_C}$, where α and k_C are fitting parameters [23].

It is possible to build a scale-free network out of a coupled map lattice, in which we consider basically random interactions between neurons, or shortcuts, added to the lattice so as to yield a power-law probability distribution. Our using of a scale-free lattice does not necessarily imply that the shortcuts connect distant neurons, since the physical distance among neurons does not play any role in the scale-free model we have used. Hence our model can describe electrical synapses, where the coupling can only exist between neighboring neurons [27]. Chemical synapses, on the other hand, would need an on-off threshold in order to model neuron excitability [6, 28].

In this paper we use the Barabási-Albert coupling prescription to generate scale-free lattice of the form (5)-(6), where the coupling term is

$$C_n^{(i)}(x_n^{(i)}) = \frac{\epsilon}{k^{(i)}} \sum_{j \in I} x_n^{(j)}, \tag{8}$$

where $\epsilon > 0$ is the coupling strength and we assumed that each site i is coupled with a set I comprising $k^{(i)}$ other sites randomly chosen along the lattice.

We build the scale-free lattice by means of a sequence of steps $s = 0, 1, 2, \cdots s_{max}$, starting from an initial lattice with $N_0 = 11$ sites [Fig. 2(a)]. At each step s a new site is inserted in the lattice of size N_s, such that it is connected to $\ell \geq 2$ randomly chosen sites. According to the scale-free distribution, the connections occur preferentially with the more connected sites, what can be accomplished by using a different probability for each

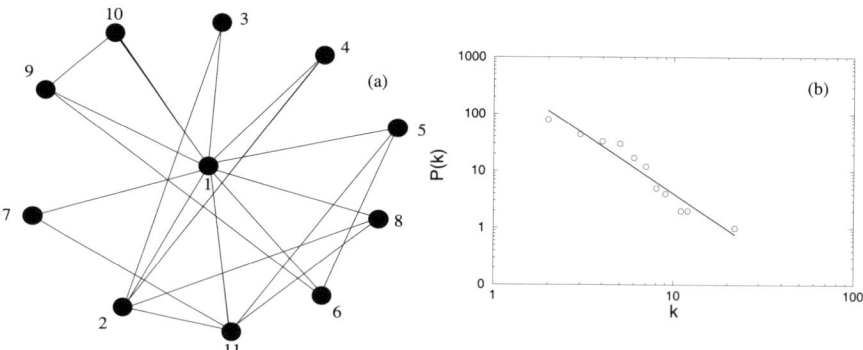

Fig. 2. (a) Scheme of the initial lattice with $N_0 = 11$ sites used to build a scale-free lattice, (b) Probability distribution for the connectivity of the final scale-free lattice with $N = 230$ sites. The solid line is a least-squares fit with slope -2.08.

site $P_s^{(i)} = k_s^{(i)}/N_s$, where $k_s^{(i)}$ is the number of connections *per* site at the step s. The process is repeated until we achieve a desired lattice size $N = 230$ [29]. After a number s_{max} of steps we have $k^{(i)}$ connections *per* site, corresponding to a probability $P^{(i)} = k^{(i)}/N$. Figure 2(b) shows a histogram for the number of sites with connectivity k, obtained through this procedure for $N = 230$ sites. The numerical approximation to the (non-normalized) probability is shown to display the scale-free signature of a power-law scaling $k^{-\varpi}$ with a slope $\varpi = 2.08$, which compares well with the experimental values reported in Ref. [21].

2.2. *Power-law coupling*

In contrast with the previous case, where the mean connectivity of the neuron network was supposed to be small, we now consider the case when the mean connectivity is large enough. In a first approximation we may use a global type of coupling, but a step further would be to include some dependence on the lattice distance (a non-local coupling form). Radhavachari and Glazier, based on the fractal character of the dendritic branching, have proposed to include the interactions among non-nearest neighbors in a way that the coupling strength decreases as a power law with the lattice distance [30]. The fractally coupled lattice is built according a given probability, as in the previous case, but now the probability is no longer constant throughout the lattice but decreases with the lattice distance as a power law:

$$p_{ij} = \frac{p_0}{|r_i - r_j|^\alpha}, \qquad (9)$$

where p_{ij} is the probability that the sites located at positions r_i and r_j be coupled, and p_0 is a constant. The parameter $\alpha > 0$ is sometimes quoted as the range exponent. When this type of probabilistic lattice is embedded in a d-dimensional Euclidean space, it turns out that the number of connections in a sphere of radius R scales as $R^{d-\alpha}$, where α is a semi-positive real number, what explains the fractal character of the coupling [31].

We have considered a simpler case, namely of a regular lattice, where the connections occur between all sites, but the coupling strength was made to decrease with the lattice distance as a power law with exponent $-\alpha$, as in Eq. (9), in the following form:

$$\mathcal{C}^{(i)}(x_n^{(j)}) = \frac{\epsilon}{\eta(\alpha)} \sum_{j=1, j \neq i}^{N'} \frac{1}{j^\alpha} \left[x_n^{(i+j)} + x_n^{(i-j)} \right], \qquad (10)$$

where

$$\eta(\alpha) = 2 \sum_{j=1}^{N'} \frac{1}{j^\alpha},\qquad(11)$$

is a normalization factor, with $N' = (N-1)/2$ for N odd.

The coupling term in Eq. (10) is a weighted average of discretized second spatial derivatives, the normalization factor being the sum of the corresponding weights. If $\alpha \to \infty$, only those terms with $j = 1$ will contribute to the summations present in the coupling term, which results in $\eta \to 2$, and in the Laplacian coupling, where a given site interacts only with its nearest-neighbors. This kind of lattice, however, is not suitable for describing neuronal networks. For $\alpha = 0$, on the other hand, we recover the mean-field coupling (7), such that we pass continuously from a local to a global coupling by varying the range parameter α.

3. Phase Synchronization of Bursting Neurons

The coupled system of Rulkov neurons, although not prone to exhibit complete synchronization in the fast (spiking) timescale, can present a non-trivial coherent behavior, since their bursting phases can synchronize through the interaction provided by the coupling term $\mathcal{C}^{(i)}$. If we had just two coupled neurons, chaotic phase synchronization would imply simply that their phases be approximately equal ($|\varphi^{(1)} - \varphi^{(2)}| \ll 1$). In the case of a large number N of systems, however, other diagnostics of phase synchronization need to be used.

3.1. Mean field

One such indicator is the mean field of the lattice,

$$X_n = \frac{1}{N} \sum_{j=1}^{N} x_n^{(j)}.\qquad(12)$$

If the neurons are weakly coupled, they burst at different times in a non-coherent fashion, and the mean field fluctuates irregularly with small amplitudes. Oppositely, if the neurons burst synchronously and a nonzero mean field is formed such that X_n presents regular oscillations of comparatively large amplitude. Only the slow time scale dynamics becomes coherent as the neurons burst synchronously, and the fast time scale spiking remains

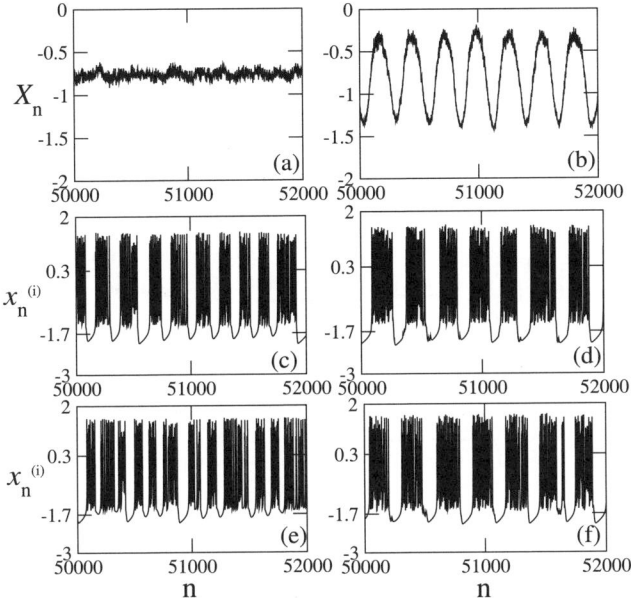

Fig. 3. Time evolution of the mean field for a scale-free lattice of Rulkov neurons with (a) $\epsilon = 0$, and (b) $\epsilon = 0.04$. Time evolution of the fast variable for a map with (c) $\theta^{(i)} = 4.1$ and (e) $\theta^{(j)} = 4.4$ for $\epsilon = 0.0$ where bursting is uncorrelated. (d) and (f) are the respective situations for $\epsilon = 0.04$, showing approximate synchronization of bursting.

incoherent and do not contribute to the mean field dynamics, which is kept close to a periodic regime [10].

As an example of this behavior in a scale-free network, we plotted in Fig. 3 the time evolution of the mean field and the fast variable of selected neurons for the coupled and uncoupled cases [29]. In the latter case ($\epsilon = 0$) the mean field indeed has small-amplitude noisy fluctuations [Fig. 3(a)] indicating that the neurons are not bursting in phase, as can be seen by comparing the uncorrelated bursting activity of two selected neurons [Figs. 3(c) and 3(e)]. For a sufficiently large coupling strength ϵ the mean field exhibits large-amplitude oscillations [Fig. 3(b)] since neurons burst at approximately the same time, in spite of their spiking evolution being poorly or not correlated at all [Figs. 3(d) and 3(f)]. While both maps present a monotonic increase of their phases with time, if they are uncoupled these evolutions are mutually independent since their phase difference grows with time and eventually becomes as large as the phases themselves. The phase difference is kept in a small value if the neurons are coupled in the network.

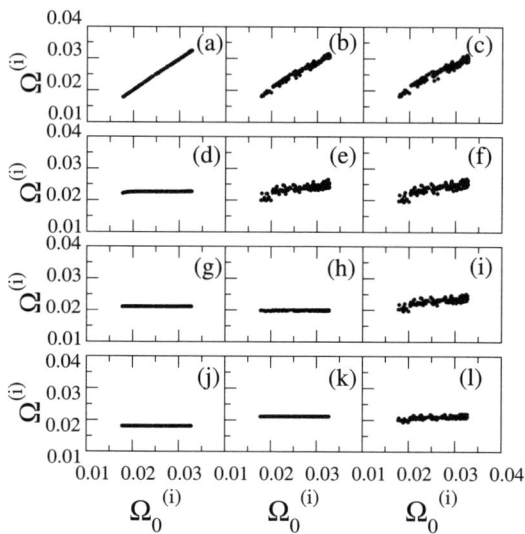

Fig. 4. Bursting frequency for the power-law lattice *versus* the zero-coupling frequencies for the following values of the coupling parameters (α, ϵ): (a) $(0.5, 0.015)$, (b) $(2.0, 0.015)$, (c) $(4.0, 0.015)$, (d) $(0.5, 0.035)$, (e) $(2.0, 0.035)$, (f) $(4.0, 0.035)$, (g) $(0.5, 0.045)$, (h) $(2.0, 0.045)$, (i) $(4.0, 0.045)$, (j) $(0.5, 0.07)$, (k) $(2.0, 0.07)$, (l) $(4.0, 0.07)$.

3.2. Frequency locking

When the neurons are uncoupled their unperturbed frequencies ($\Omega_0^{(i)}$, given by Eq. (4)), undergo random fluctuations. For the parameter values we adopted, these zero-coupling frequencies take on values within the interval $[0.0175, 0.0330]$. On the other hand, due to the coupling many neurons adjust their bursting phases so as to equalize their frequencies $\Omega^{(i)}$. We may think of these sites (not necessarily neighbors) as forming synchronization plateaus with constant values of the frequencies $\Omega^{(i)}$.

The behavior of the bursting frequencies $\Omega^{(i)}$ in a power-law coupling network is illustrated by Fig. 4, where we plotted them *versus* the zero-coupling frequencies $\Omega_0^{(i)}$ [34]. When the coupling strength is low enough there is no synchronization of bursting and the frequencies are distributed so as to have a linear trend $\Omega^{(i)} \approx \Omega_0^{(i)}$ [Fig. 4(a)]. Keeping the α parameter fixed at 0.5 and increasing the coupling strength indeed leads to phase and frequency synchronization of bursts, the frequencies having been locked at values around 0.02, the actual value decreasing slightly with the coupling strength used [Figs. 4(d), 4(g), and 4(j)]. This evolution is qualitatively the same for higher α, but frequency synchronization occurs earlier than in the previous case [Fig. 4(b), 4(e), and 4(h)]. Moreover, the common frequency achieved for strong coupling reaches a higher value (just above 0.02) than it does for lower coupling [Fig. 4(l)]. For even higher values of α, meaning local or nearest-neighbor coupling, it is clear that frequency synchronization does not occur [Figs. 4(c), 4(f), and 4(i)], unless a very strong coupling is used and, even so, frequency locking occurs with some dispersion coming from the imperfect character of the phase synchronized states [Fig. 4(l)].

3.3. *Order parameter*

Another diagnostic of phase synchronization is the complex phase order parameter [32]

$$z_n = R_n \exp(i\Phi_n) \equiv \frac{1}{N} \sum_{j=1}^{N} \exp(i\varphi_n^{(j)}), \qquad (13)$$

where R_n and Φ_n are the amplitude and angle, respectively, of a centroid phase vector for a one-dimensional lattice with periodic boundary conditions. If the neurons are uncoupled, for example, their bursting phases $\varphi_n^{(j)}$ would be so spatially uncorrelated such that their contribution to the result of the summation in Eq. (13) is typically small. In the thermodynamical limit of an infinite site ($N \to \infty$) we expect R_n to tend to zero. On the other hand, in a completely phase synchronized state the order parameter magnitude asymptotes to the unity, indicating a coherent superposition of the phase vectors for all sites with the same amplitudes R_n at each time. We usually analyze the time averaged order parameter magnitude $<R> = \lim_{T\to\infty} \frac{1}{T} \sum_{n=0}^{T} R_n$.

The usefulness of the order parameter to characterize the bursting synchronization in a scale-free network is illustrated in Fig. 5 [29]. The lattice we have considered was obtained by randomly adding sites with ℓ connections at each step, until reaching the final number of sites. We observed that a scale-free lattice with good synchronization properties is possible only for $\ell \geq 2$, since for $\ell = 1$ the average order parameter cannot achieve values larger than 0.75 even if strong coupling is used. On the other hand, for $\ell = 2$ and a coupling strong enough we have order parameter values fluctuating with time around *circa* 0.8 [like the case for which $\epsilon = 0.07$ in Fig. 5(a)]. As this coupling strength is decreased the phase synchronization of bursting becomes more imperfect, fluctuating around a smaller value of R and with

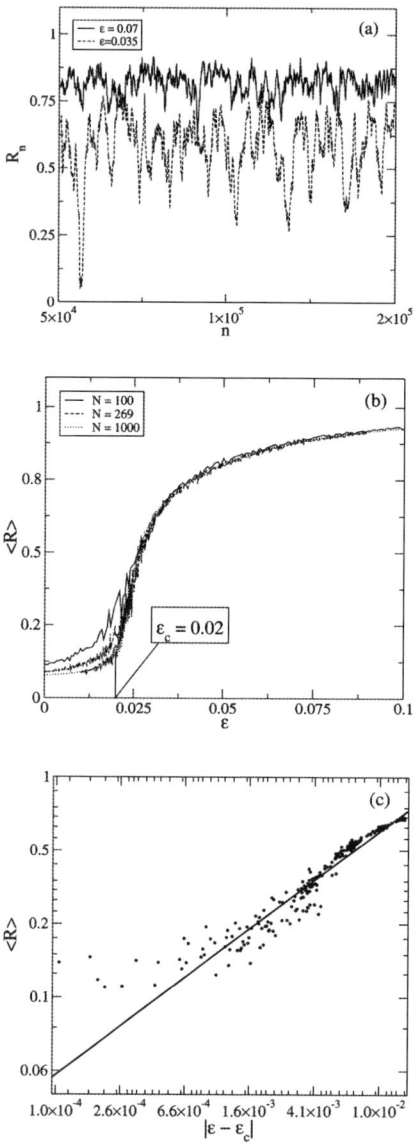

Fig. 5. (a) Time evolution of the order parameter magnitude for two values of the coupling strength in a scale-free lattice, (b) Average order parameter magnitude as a function of ϵ for different lattice sizes. The critical value of $\epsilon_c = 0.02$ results from an extrapolation to the thermodynamical limit, (c) Scaling behavior of the average order parameter magnitude near the critical point. The full line represents a least-squares fit giving a power-law with exponent 0.5.

more dispersion as well [as illustrated by the case of $\epsilon = 0.035$ in Fig. 5(a)]. The minimum value of the mean order parameter $<R>_{min}$ is obtained for zero coupling, and decreases as the lattice turns larger [Fig. 5(b)]. Supposing that $<R>_{min}$ goes to a small (yet probably nonzero) limiting value as we take the thermodynamical limit $(N \to \infty)$ we extrapolated the behavior illustrated by Fig. 5(b) for increasing lattice size and considered $\epsilon_c = 0.02$ as an approximate value for the critical coupling strength.

The order parameter is also an invaluable tool to verify general properties of the network. As an example, we can use the order parameter to show that the transition to chaotic phase synchronization of bursting is a second-order phase transition, by considering the behavior near criticality of the mean order parameter, as depicted by Fig. 5(c), where we obtain a power-law scaling which is analogous to the behavior observed in Kuramoto's model of mean field coupled phase oscillators [32]

$$<R> \sim |\epsilon - \epsilon_c|^\kappa, \qquad (14)$$

where a least squares fit leads to the exponent $\kappa = 0.502 \pm 0.007$, which agrees with the exponent $1/2$ characteristic of magnetic phase transition [32].

4. Control through External Phase Synchronization

Once the neurons in an ensemble present synchronization of bursting activity, it is a relevant question to investigate in what extent we can control this synchronized state through an external time-periodic signal with a given amplitude and frequency. Such perturbation, when applied to an ensemble of first-order phase oscillations, has been shown to produce global phase locking [33], and it has been observed in lattices of Rulkov neurons with global coupling [10] and scale-free coupling [29]. We have also implemented this procedure adding external time-periodic interventions to the power-law lattice (5) and (6).

An external harmonic signal is applied to one selected neuron $i = S$ (the remaining neurons remaining unchanged) in the following way [10]

$$x_{n+1}^{(S)} = \frac{\theta^{(S)}}{1 + \left(x_n^{(S)}\right)^2} + y_n^{(S)} + \mathcal{C}_n^{(S)}(x_n^{(j \neq S)}) + d\sin(\omega n), \qquad (15)$$

$$y_{n+1}^{(S)} = y_n^{(S)} - \sigma x_n^{(S)} - \beta, \quad (i = 1, 2, \ldots N), \qquad (16)$$

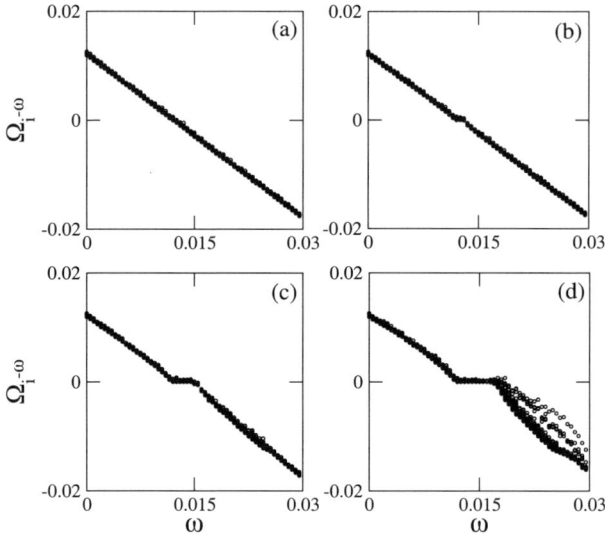

Fig. 6. Frequency mismatch of bursting neurons *versus* the external driving frequency for a scale-free lattice with $\epsilon = 0.2$ and an a driving signal with amplitude (a) $d = 0.05$, (b) $d = 0.09$, (c) $d = 0.15$, and (d) $d = 0.20$.

where d and ω are the external signal amplitude and frequency, respectively, and the site S in principle can be anyone chosen at random in the network, if it exhibits a global or power-law type of coupling. In scale-free networks this choice is not so arbitrary due to the strong dependence of the connectivity on the particular neuron. We have chosen the site with the greater number of connections. We may also apply the external signal to more than one neuron in a straightforward manner, with the same amplitude and frequency as in Eq. (15).

In Fig. 6 we present an example of this procedure for a scale-free network [29]. In order to investigate the effect of this external source we have used coupling strength values for which the unperturbed lattice ($d = 0$) exhibits bursting synchronization, their corresponding frequencies $\Omega^{(i)}$ locking approximately at a common value.

When the driving amplitude d is nonzero, we considered many situations for which the neurons synchronize at different common frequencies $\Omega^{(i)}$ within the range $[0, 0.03]$ and plotted in Fig. 6 the corresponding mismatches with the external signal frequency ω. If the signal amplitude is too low [Fig. 6(a)] the difference $\Omega^{(i)} - \omega$ vanishes for a particular value of ω, but for $d \gtrsim 0.09$ we obtain a narrow frequency locking interval around

$\Omega = 0.013$ [34]. The width of this locking interval, $\Delta\omega$, increases with the signal amplitude [Fig. 6(c) and 6(d)], in a situation akin to the Arnold tongue structure existing for periodically forced oscillators.

This synchronization between the bursting neurons and the external signal is possible due to the coupling effect on the triggering or termination of a burst in the individual neurons. A burst can be terminated (the neuron is driven to a quiescent state) if the external signal is positive. Conversely, a burst can be delayed if the signal is negative. The combination of these effects leads to the synchronization of the driven neuron with the signal. The effect of coupling, once it takes into account the mutual influences of all lattice sites, is to change the mean-field that each neuron feels. In the case of a power-law coupling, distant sites contribute less to coupling. Hence, the more local is the coupling (large α) the less a distant neuron feels the synchronization effect caused by the signal on the driven neuron. Hence we expect worse synchronizaion properties as α increases, and no synchronization at all for nearest-neighbor couplings ($\alpha \to \infty$).

The frequency locking interval is a cross-section of an Arnold-like tongue in the parameter plane amplitude *versus* frequency of the external driving signal. This tongue is clearly asymmetric for small amplitudes, for its left boundary is steeper than the right one, becoming more symmetric as the signal amplitudes are higher, as long as they do not exceed the saturation limit. In order to characterize quantitatively this asymmetry we define the width of this locking interval $\Delta\omega$, and the interval width $\delta\omega$ counted only from the unperturbed frequency (i.e. at $d = 0$) [see Fig. 7(a) for an example of asymmetric tongue in power-law lattices]. The former increases with the signal amplitude initially with a small rate, then more rapidly (provoking the abovementioned asymmetry), and finally at a slower rate [Fig. 7(b)] [34].

In power-law networks the dependence of the width with the amplitude can be fitted by a power-law $\Delta\omega \sim d^{\varpi}$ [Fig. 7(c)]. The asymmetry of the tongue manifests ifself in two similar regimes for the part of the tongue to the right of the zero-signal, with different constant for the fast and slow rate branches of the tongue [Fig. 7(d)]. In scale-free networks the tongue is qualitatively the same but exhibits a fuzzy boundary due to the characteristic randomness of its connection architecture [35].

The wider the frequency-locking interval is, the more robust is the external driving with respect to imperfect parameter determination and noise, which is a question of considerable experimental importance. However, it may well happen that, if only one pinning (i.e., a driving applied on only one neuron) is considered, the corresponding locking interval would be too small

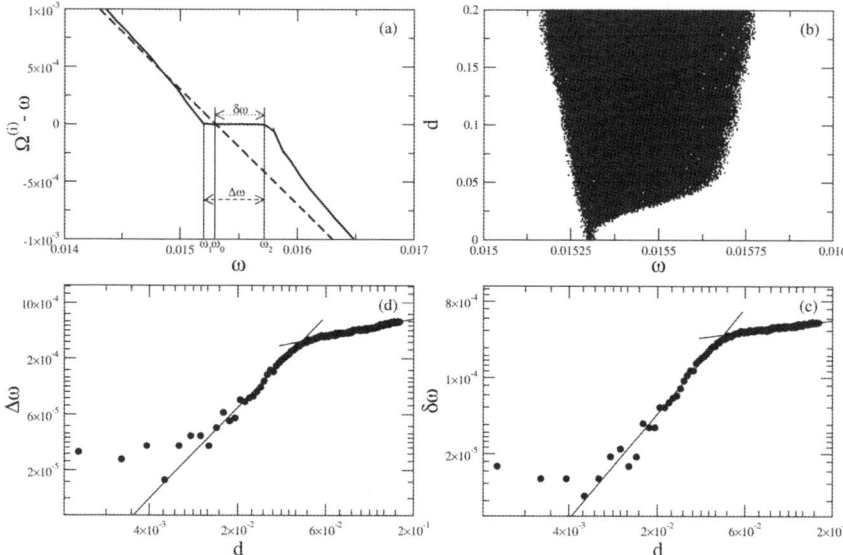

Fig. 7. Mode-locking for a power-law lattice with $N = 51$ neurons, $\alpha = 0.5$, $\epsilon = 0.07$, driving signal applied at a randomly chosen site $i = S$. (a) Frequency mismatch of bursting neurons *versus* the external driving frequency for amplitude $d = 0$ (dashed line) and $d = 0.15$ (full line), (b) Mode-locking tongue in the amplitude *versus* frequency of the driving signal, (c) Full width of the mode-locking tongue as a function of the driving amplitude, (d) The same as (c) for the partial width (to the right and with respect to the $d = 0$ point). The full lines in (c) and (d) are least-squares fits.

so as to ensure robustness against imperfect parameter determination. The latter problem is particularly severe when the lattice size increases, as illustrated by Fig. 8, where we plot the locking interval width as a function of the inverse lattice size. If only one pinning is used, the tongue width decreases linearly with $1/N$. One possible way to circumvent this problem would be to use more than one site to control at the same time (multiple pinnings) [10]. As depicted in Fig. 8, the tongues dilate as the number of pinnings (N_p) is augmented, and the dependence on the lattice size is more pronounced. In other words, using more pinnings results in larger tongues, as long as the lattice is small. For larger lattices, however, the use of more pinnings does not imply on a corresponding enlargement of the locking interval.

5. Control through Time-delayed Feedback

The use of an external input as a control device for neuron bursting involves a number of problems related with the choice of parameters, specially the

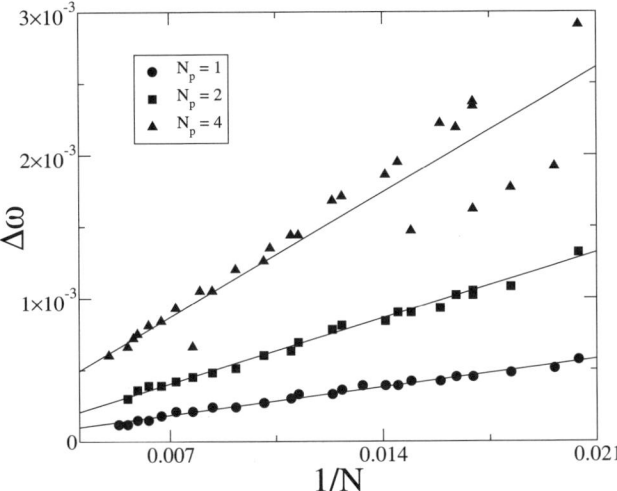

Fig. 8. Width of the frequency-locking interval *versus* inverse lattice size for different number of driving inputs, all of them with the same amplitude $d = 0.15$, in a power-law lattice with $\alpha = 0.5$ and $\epsilon = 0.1$.

amplitude and frequency. If the amplitude of the external signal is too large, for example, we could have neuron damage, and if the frequency falls outside a given mode-locking tongue we would have practically no effect in terms of control. Another procedure to accomplish bursting control consists on using a time-delayed feedback signal. This has the advantage of always working with signals of appropriate intensity, and it has been shown to be capable to suppress chaotic bursting synchronization in mean-field lattices [20].

Let us consider an assembly of N neurons with mean field given by Eq. (12). The original proposal of Pikowsky and Rosenblum, who have considered a mean-field coupling type (equivalent to take the $\alpha \to 0$ limit in the power-law coupling) includes both the current mean field X_n and its value τ iterations before, $X_{n-\tau}$, with different intensities ϵ and ϵ_f, respectively:

$$x_{n+1}^{(i)} = \frac{\theta^{(i)}}{1 + \left(x_n^{(i)}\right)^2} + y_n^{(i)} + \epsilon X_n + \epsilon_f X_{n-\tau} - \epsilon'_f X_n, \qquad (17)$$

$$y_{n+1}^{(i)} = y_n^{(i)} - \sigma x_n^{(i)} - \beta, \quad (i = 1, 2, \ldots N). \qquad (18)$$

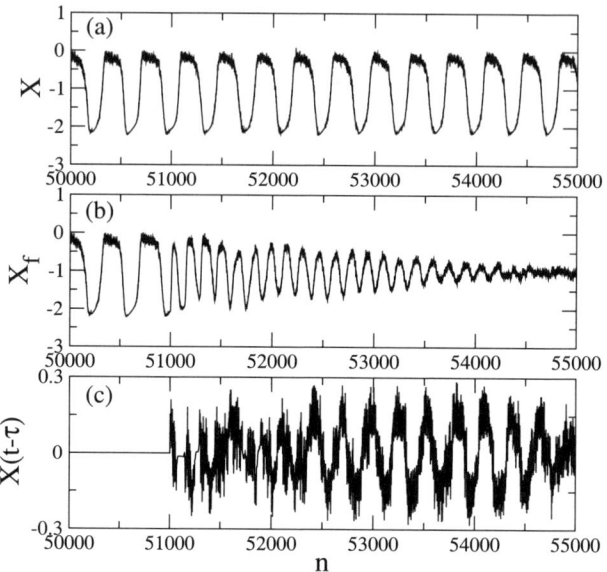

Fig. 9. (a) Time evolution of the mean field for a scale-free lattice of Rulkov neurons with $\epsilon = 0.08$, (a) $\epsilon_f = 0$, (b) $\epsilon_f = 0.08$, (c) Time evolution of the time-delayed feedback signal $X_{n-\tau}$ with $\tau = 140$.

We applied the Pikowsky-Rosemblum control procedure to the scale-free network, after redefining the mean-field as the coupling term given by Eq. (8):

$$X_n = \frac{\epsilon}{k^{(i)}} \sum_{j \in I} x_n^{(j)}, \qquad (19)$$

Moreover, we have set $\epsilon'_f = \epsilon_f$ in such a way that the control is given by

$$x_{n+1}^{(i)} = \frac{\theta^{(i)}}{1 + \left(x_n^{(i)}\right)^2} + y_n^{(i)} + \epsilon X_n + \epsilon_f \left[X_{n-\tau} - X_n\right]. \qquad (20)$$

As a representative example of the usefulness of time-delayed feedback to control bursting oscillations, we depict in Fig. 9(a) the evolution of the mean field of a scale-free lattice with $\beta = \sigma = 10^{-3}$ and $\epsilon = 0.08$, showing the characteristic bursting activity already expected from such a coupled system. In Fig. 9(b) we show the same mean field but now with the addition of a time-delayed signal with $\epsilon_f = 0.08$ and $\tau = 140$ [the latter shown in Fig. 9(c)] applied at the instant $n = 51000$. After three thousand iterations the mean field amplitude was considerably reduced and its oscillations became irregular, characterizing the loss of bursting synchronization, which is the aim of the control procedure. Moreover the amplitude of the feedback signal has been kept as less than one-third of the mean field amplitude itself, which is a desirable feature.

The effectiveness of the control procedure on reducing or suppressing synchronization can be measured by the *suppression coefficient* [20]

$$S = \sqrt{\frac{\text{Var}(X)}{\text{Var}(X_f)}}, \qquad (21)$$

where X and X_f are the values of the mean field in the absence and presence of the control, respectively. The feedback scheme is ideally efficient when the variance of the controlled mean field vanishes, irrespectively of its value without control, corresponding thus to an infinite value of S. As a general rule, the larger is the value of S, the more efficient will be the feedback on suppressing synchronization. It is convenient to consider regions with large values of S, or domains of control, in the control parameter plane where the control strength ϵ_f (normalized by the coupling strength ϵ) is plotted against the time delay τ (normalized by the mean bursting period T).

Figure 10(a) displays such domains (the values of S are indicated using a grayscale) for a differential control scheme. For small values of the time delay the control achieves a good suppression of synchronized bursting for decreasing values of ϵ_f until, for $\tau \gtrsim 1.0$ the boundary of the control domain stabilizes at $\epsilon_f/\epsilon \gtrsim 1.0$. These findings are in marked contrast with the result obtained for a global (mean-field) coupling given by Eq. (7) with the same number of neurons as in the scale-free network [see Fig. 10(b)]. In the latter the suppression coefficient takes on small values only (the grayscale is the same for both panels for ease of visualization), and we can barely distinguish a control domain centered at $\tau \sim T$. The results, for direct feedback, do not differ much from the differential case [36].

6. Conclusions

While controlling chaos in the brain is experimentally feasible, much less is known about the mathematical modeling of this process. Models of brain functions are nowadays limited to mimetize some limited sort of phenomena, whose choice indicates the kind of neuronal architecture one tries to imitate with spatially extended dynamical systems. A large number of works along this line of research use a mean-field type of coupling, but more real-

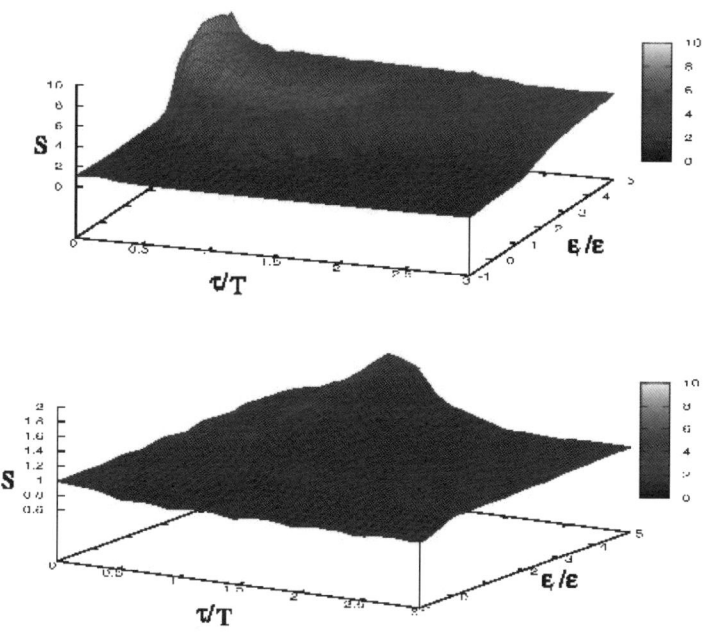

Fig. 10. Suppression coefficient as a function of the normalized control parameters (strength and time delay) for differential feedback control of a network of $N = 230$ Rulkov neurons with (a) scale-free and (b) global mean-field coupling architectures.

istic neuronal architectures would need non-local and possibly fractal types of coupling. In this work we presented results for two types of couplings, namely scale-free (random) and power-law (regular) coupling schemes. For the neuron dynamics itself, we concentrated on simple map-based models which nevertheless are able to exhibit two timescales for the spiking (fast) and bursting (slow) oscillations.

The particular phenomenon we have focused on is the chaotic bursting synchronization, which is possible due to the mutual coupling, and is thought to be related with abnormal rhythms and even neurological problems. Hence it would be helpful to develop methods to control this bursting synchronization in the sense of suppressing it whenever necessary through some external intervention. We have analyzed two possible ways of accomplishing this task. One of them is the application of one or more localized harmonic signals with suitable frequency and amplitude, whereas the second one is the use of a time-delayed feedback. The application of an external signal makes the synchronized bursting to adjust the common frequency so as to match the driving frequency. However, if the latter is higher than the mode-locking frequencies and the driving amplitude is large enough, we may suppress bursting synchronization in the lattice.

For both controlling procedures it has been proved feasible to suppress undesirable chaotic bursting synchronization and, at least when the external intervention is a harmonic signal, the overall characteristic of the phenomenon is the familiar mode-locking tongue in the parameter space. We hope that, in the near future, other coupled map lattice models using more sophisticated neuronal architectures could be investigated as well, in order to shed some light on the dynamical mechanisms underlying the control of chaotic neural oscillations achieved in laboratory.

Acknowledgments

This work was made possible with help of CNPq, CAPES, and Fundação Araucária (Brazilian Government Agencies).

References

[1] S. J. Schiff, K. Jerger, D. H. Duong, T. Chang, M. L. Spano, W. L. Ditto, Nature **370**, 715 (1994).
[2] E. Ott, C. Grebogi, and J. A. Yorke, Phys. Rev. Lett. **64**, 1196 (1990).
[3] M. F. Bear, B. W. Connors, and M. A. Paradiso, *Neuroscience: Exploring the Brain*, 2nd ed. (Lippincott Williams & Wilkins, Philadelphia, 2002).
[4] I. Belykh, E. de Lange, and M. Hasler, Phys. Rev. Lett. **94**, 188101 (2005).
[5] J. L. Hindmarsch, and R. M. Rose, Philos. Trans. R. Soc. London, Ser. B **221**, 87 (1984).
[6] D. Chialvo, Chaos, Solit. & Fract. **5**, 461 (1995).
[7] N. F. Rulkov, Phys. Rev. Lett. **86**, 183 (2001); Phys. Rev. E **65**, 041922 (2002); N. F. Rulkov, I. Timofeev, and M. Bazhenov, J. Comp. Neurosci. **17**, 203 (2004).
[8] H. Nozawa, Chaos **2**, 377 (1992); S. Ishii, M. Sato, Physica D **121**, 344 (1998).
[9] G. Tanaka, B. Ibarz, M. A. F. Sanjuan, and K. Aihara, Chaos **16**, 013113 (2006).
[10] M. V. Ivanchenko, G. V. Osipov, V. D. Shalfeev, and J. Kurths, Phys. Rev. Lett. **93**, 134101 (2004).
[11] S. E. de S. Pinto, R. L. Viana, Phys. Rev. E **61**, 5154 (2000); S. E. de S. Pinto, S. R. Lopes, and R. L. Viana, Physica A **303**, 339 (2002).
[12] A.-L. Barabási and R. Albert, Science **286**, 509 (1999).

[13] M. G. Rosenblum, A. S. Pikovsky, and J. Kurths, Phys. Rev. Lett. **76**, 1804 (1996).
[14] A. M. Thomson, Current Biology **10** (2000) R110.
[15] J. Milton and P. Jung (Eds.), *Epilepsy as a Dynamic Disease* (Springer Verlag, Berlin, 2003).
[16] Y. L. Maistrenko, O. V. Popovych, and P. A. Tass in *Dynamics of coupled map lattices and of related spatially extended systems*, J.-R. Chazottes and B. Fernandez (Eds.), Lecture Notes in Physics Vol. 671 (Springer Verlag, Heidelberg-Berlin, 2005).
[17] A. Nini, A. Feingold, H. Slovin, and H. Bergman, J. Neurophysiology **74**, 1800 (1995).
[18] A. L. Benabid, P. Pollak, C. Gervason, D. Hoffmann, D. M. Gao, M. Hommel, J. E. Perret, and J. de Rougemont, Lancet **337**, 403 (1991).
[19] S. Blond, D. Caparros-Lefevre, F. Parker, R. Assaker, H. Petit, J.-D. Guieu, and J.-L. Christiaens, J. Neurosurgery **77**, 62 (1992).
[20] M. Rosenblum and A. Pikowsky, Phys. Rev. E **70**, 041904 (2004).
[21] D. R. Chialvo, Physica A **340**, 756 (2004); O. Sporns, D. R. Chialvo, M. Kaiser, and C. C. Hilgetag, Trends Cogn. Sci. **8**, 418 (2004); V. M. Eguiluz, D. R. Chialvo, G. A. Cecchi, M. Baliki, and A. V. Apkarian, Phys. Rev. Lett. **94**, 018102 (2005).
[22] M. P. van der Heuvel, C. J. Stam, M. Boersma, and H. E. Hulshoff Pol, NeuroImage **43** 528 (2008).
[23] S. Achard, R. Salvador, B. Whitcher, J. Suckling, and E. Bullmore, J. Neuroscience **4** 63 (2006).
[24] M. D. Humphries, K. Gurney, and T. J. Prescott, T. J., Proc. Royal Soc. (London) B **273** 503 (2006).
[25] M. Dhamala, V. K. Jirsa, and M. Ding, Phys. Rev. Lett. **92**, 028101 (2004).
[26] R. Albert and A.-L. Barabási, Rev. Mod. Phys. **74**, 47 (2002); S. N. Dorogovtsev and J. F. F. Mendes, Adv. Phys. **51**, 1079 (2002).
[27] J. R. Gibson, M. Beierfein, and B. W. Connors, Nature **402**, 75 (1999).
[28] E. M. Izhikevich, Int. J. Bifurcat. Chaos **10**, 1171 (2000).
[29] C. A. S. Batista, A. M. Batista, J. A. C. de Pontes, R. L. Viana, and S. R. Lopes, Phys. Rev. E **76**, 016218 (2007).
[30] S. Raghavachari, J. A. Glazier, Phys. Rev. Lett. **74**, 3297 (1995).
[31] C. Wagner and R. Stoop, Int. J. Bifurcat. Chaos **17**, 3409 (2007).
[32] Y. Kuramoto, *Chemical Oscillations, Waves, and Turbulence* (Springer Verlag, Berlin, 1984).
[33] H. Sakaguchi, Prog. Theor. Phys. **79**, 39 (1988).
[34] J. C. A. de Pontes, R. L. Viana, S. R. Lopes, C. A. S. Batista, and A. M. Batista, Physica A **387**, 4417 (2008).
[35] C. A. S. Batista, A. M. Batista, J. C. A. de Pontes, S. R. Lopes, and R. L. Viana, Chaos, Solit. & Fract. **41**, 2220 (2009).
[36] C. A. S. Batista, A. M. Batista, R. L. Viana, and S. R. Lopes, Neural Networks, to be published.

Chapter 11

PARTIAL CONTROL OF CHAOTIC SYSTEMS

Samuel Zambrano* and Miguel A. F. Sanjuán
*Department of Physics, Universidad Rey Juan Carlos,
Tulipán, s/n, 28933, Móstoles, Madrid, Spain*
samuel.zambrano@urjc.es

1. Introduction

Some dynamical systems do not present a chaotic attractor but, instead, they present a zero-measure nonattractive set in phase space where the dynamics is chaotic: a chaotic saddle. In fact, it is common to find systems with a chaotic attractor for which, by varying one of the system's parameters, a boundary crisis [5] might occur. When this phenomenon takes place, the attractor is replaced by a nonattractive chaotic set, a *chaotic saddle*, from which nearly all trajectories escape. Before they do so, trajectories typically expend a finite amount of time in the vicinity of the saddle, whose duration typically depends on their initial condition, during which their behaviour is chaotic. This type of behaviour is known as *transient chaos*.

This type of situation is ubiquitous in nonlinear dynamical systems. Considering this, it is not strange to find that there are many different contexts where we might be interested in confining the trajectory of a dynamical system in the region of the phase space where the chaotic saddle lies, from which, in absence of control, nearly all the trajectories (except a zero measure set) escape after a finite amount of time. Considering this, following the tradition on control of dynamical systems initiated by the seminal paper by Ott, Grebogi and Yorke [12], different control techniques have been proposed in order to control this particular type of dynamics. This type of control is known as control of transient chaos [1, 4, 8, 16, 20], but also as chaos maintenance [7] or chaos preservation [21]. And, as we said, it can be useful in many different situations. For example, it has potential applications to prevent electric collapses in certain electric power systems [4], in thermal pulse combustors, preventing flameouts and enhancing fuel/air mixture [7] and even in ecology [17].

This type of control is achieved by applying certain corrections (that we refer generically here as *control*) on the trajectories. In most experimental situations, the system might be affected by noise, which adds an extra degree of difficulty to this control task, which is by itself nontrivial due to the nonattractive nature of the chaotic saddle. It is relatively easy to convince oneself that if the control that we apply on the system is big enough, for example if it is *bigger* than the effect of environmental noise in the system, then it will be easy to keep trajectories close to the chaotic saddle. If the control that we apply to the system is *equal* to the effect of noise, it can be shown that some of the techniques explored in the references given above, allow to achieve this goal.

What happens if the control is smaller than the effect of noise? In that situation, apparently this control task becomes impossible. However, in this Chapter we are going to show that the existence of a chaotic saddle provides indirectly a solution to this problem. We are going to show here that chaotic saddles are sometimes related with the existence of a "stretching and folding" geometrical action on regions of the phase space, which indirectly implies the existence of certain *safe sets* that allow to achieve this goal. First, we are going to show this for a variety of one-dimensional maps with a chaotic saddle. After this, we are going to show that those safe sets can also be found for an important class of maps: the horseshoe maps [18]. The advantages of this approach from a controlling point of view are evident, as long as our control technique allows to keep the system's trajectories close to the saddle with a control smaller than the noise. However, we are going to see that our control strategy does not determine exactly where trajectories will go. Thus, we call it *partial control* of a chaotic system.

The structure of this Chapter is as follows. In Sec. 2 we use a simple one-dimensional map, the slope-three tent map, to describe the type of dynamical system that we want to control. With this simple example in mind, in Sec. 3 our control problem is stated precisely. Using again the slope-three tent map, in Sec. 4 we show that some of the strategies previously proposed to partially control this type of dynamical systems require a control at least equal to the noise. After this, in Sec. 5 we show how the safe sets needed to achieve this goal with a control smaller than the noise are generated for this type of map. The safe sets are used in next section, Sec. 6, to describe our partial control strategy. After this, in Sec. 7 we describe a numerical example of application of our control technique to the well-known logistic map.

In Sec. 8 the link between these one-dimensional maps and the important class of horseshoe maps is described. In Sec. 9 we show how the safe sets are built for this type of dynamical systems, and our partial control strategy for horseshoe-like maps is outlined. In Sec. 10 a numerical example of our technique is given with the aid of the paradigmatic bouncing ball map [6]. Section 11 is devoted to the main conclusions of this Chapter.

2. One-dimensional Maps with a Chaotic Saddle

The first half of this Chapter deals with the control of a variety of one-dimensional maps with a chaotic saddle. The control technique that we use for this map is explored in Ref. [1], where the main objective is to control the dynamics of the slope-three tent map $x_{n+1} = 3(1-|x_n|)-1 = T(x_n)$, which is a paradigmatic example of this kind of maps. In this Section, we use this example to explain the phenomenon of transient chaos and the concept of chaotic saddle in one-dimensional maps, which can be then easily extended to higher-dimensional dynamical systems.

Let's focus now on the description of the map $x_{n+1} = T(x_n)$. The graph of this map is sketched in Fig. 1. The dynamics of this system can be easily understood as follows. First, we have to note for any point x' out of I we will have that $T^n(x') \to -\infty$ as $n \to \infty$. Note also that points in the middle-third interval of $I = [-1, 1]$ escape from I after one iteration. If we call I_1 to the set of points such that $T(I_1) = I$, we find that it consists on two segments of length $2/3$, and obviously all the points starting in this set need at least one iterate to escape from I. If we define the set I_2 as the set of points such that $T(I_2) = I_1$ we obtain the four $2/9$ length intervals shown in Fig. 1, and points belonging to this set need at least two iterates to escape from I. Following this procedure, we can define inductively the set I_k as the set of points such that $T(I_k) = I_{k-1}$. It consists on 2^k intervals of length $2/3^k$, and points on this set need at least k iterates to escape from I. The set resulting in the limit as $k \to \infty$, I_∞ is a *Cantor* set [2], a very particular set that has played an important role in set theory. And obviously the points on this set never escape from I under iterations of T.

The set I_∞ is a zero-measure set as long as the total length of I_k, that is $2^k \cdot 2/3^k$, goes to zero as $k \to \infty$. It is self-similar, as long as I_k contains 2^k copies of itself scaled down by a factor 3^k. The Hausdorff dimension of I_∞ is $d = \ln 2/\ln 3$ which is obviously noninteger, so this object is a fractal. On the other hand, it can be proved that I_∞ is an invariant set under T, and that the dynamics of any trajectory inside it is topologically equivalent

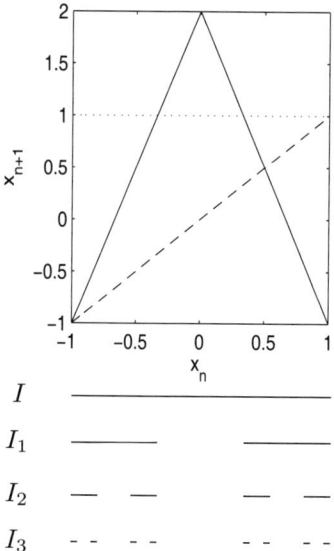

Fig. 1. Inductive construction of the chaotic saddle for the slope-three tent map. The set $I_\infty = \lim_{k \to \infty} I_k$ is a Cantor set, that is, a fractal.

to a full shift on two symbols. From this, it can be proved that inside I_∞ there are unstable periodic orbits of all the periods, and that trajectories present sensitive dependence on the initial conditions. Thus, the dynamics in I_∞ under T is chaotic. The set I_∞ is a nonattractive chaotic set, that is, it is a chaotic saddle.

Probably for the reader it is clear now that maps similar to the slope-three tent map will also present a chaotic saddle. In fact it can be proved [13] that any one-dimensional map $x_{n+1} = f(x_n)$ that fulfils the following conditions will present a nonattractive chaotic Cantor-like set Λ where the dynamics is chaotic:

(a) There is an interval $I = [a, b]$ such that $I \subset f(I)$. The interval I can be divided in three subintervals A_1, and A_2 such that $f(A_1) = f(A_2) = I$ and $f(A_0) \notin I$.
(b) The map f is continuous and differentiable in $A_1 \cup A_2$ and for all $x_0 \in A_1 \cup A_2$, $|f'(x_0)| > 1$.
(c) For all $x_0 \notin I$, $|f^n(x_0)| \to \infty$ as $n \to \infty$.

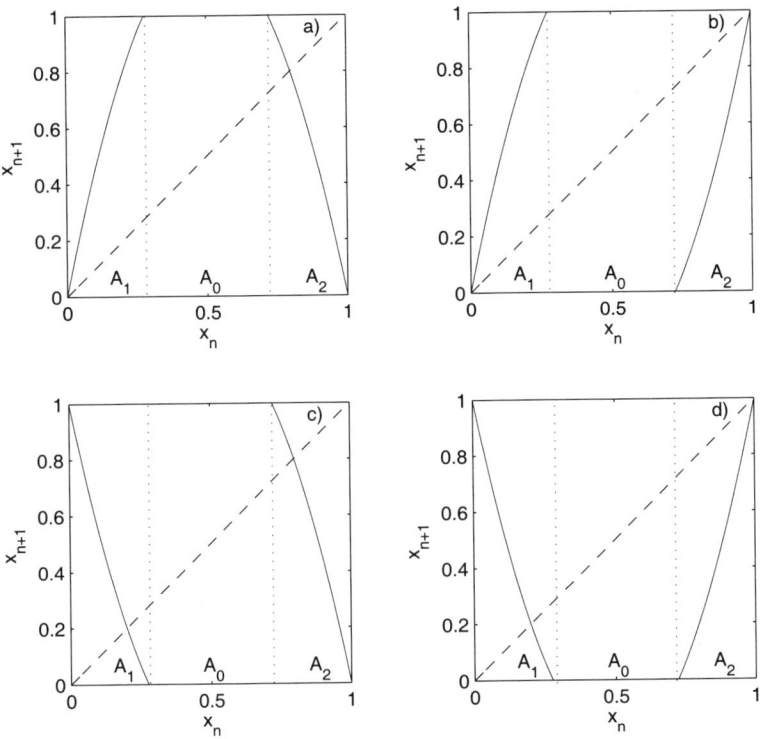

Fig. 2. Four possible configurations of a map $x_{n+1} = f(x_n)$ satisfying conditions (a)–(c). Note that each point in I has just one preimage in A_1 and other in A_2.

The slope-three tent map obviously satisfies these conditions. We must point out that condition (b) is not a necessary condition for the existence of a chaotic saddle.

Once we have defined the type of dynamical system that we will deal with, we can define in a more precise way the type of situation that we want to control. We have seen that nearly all the trajectories in I, except those starting in the zero-measure set I_∞, escape from it under iterations of f. Trajectories starting close to I_∞ will behave chaotically before diverging from it, and escaping from I. It is easy to see that if we apply a random perturbation each iteration to the system, playing the role of noise, all the trajectories escape from I. This situation occurs also in a wide variety of higher-dimensional dynamical systems, and our aim is to avoid it by using small and accurately chosen corrections on the system. In next section we state this problem in a mathematically precise way.

3. Problem Statement

Here we are going to state our problem in a general way using the ideas given in the former Section. We consider that the dynamics of the system considered are given by a map $\mathbf{p}_{n+1} = \mathbf{f}(\mathbf{p}_n)$, that can also be a Poincaré map.

We assume that this map has a chaotic saddle that is inside a region Q, from which nearly all trajectories will escape under iterations. Thus, the dynamics of the map $\mathbf{f}$ is somehow similar to the dynamics observed in the example explained in the former Section, for which the considered region Q would be the $I = [-1, 1]$ interval.

If we add a random perturbation playing the role of environmental noise to the system, it is easy to see that all trajectories escape from Q. We can model the dynamics of this kind of system by the equation $\mathbf{p}_{n+1} = \mathbf{f}(\mathbf{p}_n) + \mathbf{u}_n$, where $\mathbf{u}_n$ is a bounded random perturbation, $||\mathbf{u}_n|| \leq u_0$, that plays the role of *noise*.

As we said before, the goal of partial control is to keep the trajectories in the region Q where the chaotic saddle lies, although we might not need to determine exactly where the trajectory will go in Q. With this purpose, we can design a control strategy based on applying an accurately chosen *control* $\mathbf{r}_n$ each iteration, that we assume also bounded by a positive constant $||\mathbf{r}_n|| \leq r_0$, in such a way that the global dynamics is given by

$$\begin{cases} \mathbf{q}_{n+1} = \mathbf{f}(\mathbf{p}_n) + \mathbf{u}_n \\ \mathbf{p}_{n+1} = \mathbf{q}_{n+1} + \mathbf{r}_n, \end{cases} \tag{1}$$

so the control $\mathbf{r}_n$ depends on $\mathbf{p}_n$ and $\mathbf{u}_n$, as in other paradigmatic control methods [12].

Our aim is to achieve this goal with $r_0 < u_0$. In next Section we are going to state this problem in the particular case in which the the one-dimensional tent map T that was described in Sec. 2 plays the role of $\mathbf{f}$. We will explore a classical approach to keep trajectories close to the saddle, and it will become evident that it needs control at least equal to the noise, that is $r_0 \geq u_0$, to keep trajectories bounded. However, the use of safe sets will allow us to achieve this goal with $r_0 < u_0$.

4. A Control Strategy for One-dimensional Maps with $r_0 = u_0$

In this section we are going to consider how trajectories can be kept bounded for the slope-three tent map $x_{n+1} = T(x_n)$ described in Sec. 2. In this particular case, the control problem described in Eq. (1) is mathematically formulated as:

$$\begin{cases} q_{n+1} = T(p_n) + u_n \\ p_{n+1} = q_{n+1} + r_n. \end{cases} \qquad (2)$$

The goal here would be to keep the trajectories, that is the points p_n, inside the interval $I = [-1, 1]$.

A paradigmatic approach to this problem, inspired in the method given in Ref. [20] would be to stabilize the system in one of the unstable periodic orbits embedded in I_∞. Let's explore it here in some detail. To do this, assume first that we can place a point of the trajectory p on the fixed point $p^* = T(p^*) = -1$. In that case, we will have that $q' = T(p) + u = T(p^*) + u = p^* + u$. Assume that $u = u_0$. Then, we have to apply the control r to correct that deviation. There are two cases that we can consider here:

- If we apply a control such that $r = -u_0$ the next point of the trajectory is $p' = p^* + u_0 - u_0 = p^*$ and this can be repeated forever. Thus, using these ideas it is easy to see that trajectories can be kept bounded forever if the amplitude of the corrections satisfies the equality $r_0 = u_0$, that is, if the control is equal to the noise.
- If we use a control smaller than the noise, we will have that the next point of the trajectory, p' can be written as $p' = p^* + \delta$. Then, assuming again that $u' = u_0$, $q'' = T(p') + u_0 = p^* + C\delta + u_0$, where C is a constant whose modulus is bigger than one due to the fact that p^* is an unstable periodic orbit. Thus, it is easy to see that in the worst possible scenario, this new point q'' will be *further* away from p^* that q', so if we apply again a control equal to the noise again, p'' will be further from p than p'. In other words, with a control smaller than the noise, using this approach, it is not possible to keep trajectories close to p^* forever.

The alternatives given by most of the strategies cited in the Introduction would lead to an analogous result. Thus, if we want to keep trajectories of this system bounded with control smaller than noise, that is with $r_0 < u_0$, we need to tackle this problem using an alternative approach. Our approach here, that is the same that is described in Ref. [1], will be to steer each iteration the trajectories to an adequate set of points, those forming a *safe set*. In particular, in the next Section we are going to describe a family of safe sets $\{S^k\}$ of different order k, that will be those needed for different values of u_0.

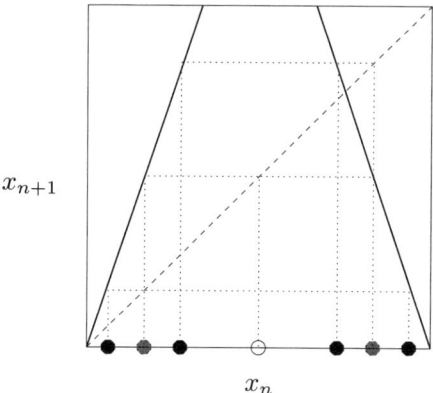

Fig. 3. The safe sets for the slope three tent map S^0 (white dot), S^1 (grey dots) and S^2 (black dots).

5. Safe Sets for One-dimensional Maps

Let's address how safe sets are built for one-dimensional maps, making use again of the slope-three tent map. The safe sets of order k here are obtained inductively. The set S^0 here is taken to be $x = 0$. From this, any safe set is defined inductively by the formula:

$$S^k = T^{-1}(S^{k-1}). \tag{3}$$

The structure of the sets S^0, S^1 and S^2 is shown in Fig. 3, where we can see graphically how they are generated. From this picture, and proceeding analogously, we can see that the set S^k is formed by 2^k points, which are of the form:

$$\pm \frac{2}{3^1} \pm \frac{2}{3^2} \pm \ldots \pm \frac{2}{3^k}. \tag{4}$$

So it is easy to see that the safe sets S^k verify the two following properties:

(i) Any point of S^k has two adjacent points of S^{k+1}, one to its left and other to its right, that are closer to it that any other point of S^k.
(ii) The maximum distance Δ^k and the minimum distance δ^k between any point of S^{k-1} and its two adjacent points of S^k goes to zero as k goes to infinity.

It is not difficult to see that for any map $x_{n+1} = f(x_n)$ fulfilling the conditions (a)–(c) given in Sec. 2, and taking as only point of S^0 the middle point of A_0, the set generated inductively as

$$S^k = f^{-1}(S^{k-1}) \tag{5}$$

has a structure analogous to the one described by properties (i) and (ii). The proof of this result is given in Ref. [22]. Note that $\Delta^k \neq \delta^k$ in general, but for the slope-three tent map these two quantities are equal. With these properties in mind, we can now see how these sets are used in our partial control strategy.

6. Partial Control Strategy for One-dimensional Maps

Once that we have given the key properties of the sets S^k, we can now explain why for each value of u_0 there is a S^k that is a safe set, which allows to keep trajectories bounded inside I even if $r_0 < u_0$.

The strategy is the following: Given u_0, we just have to chose k in such a way that $u_0 > \Delta^k$ which, by property (ii) is always possible if k is sufficiently big. Then, we put the initial condition p on a point on S^k. After this, the point is taken by the joint action and of the noise to $q' = T(p) + u$, a point that is at most u_0 away from certain point of S^{k-1} (the point $T(p)$). If that point lies between those adjacent points, then a correction equal to Δ^k allows to put the trajectory back on S^k. If not, it is easy to see that a correction smaller or equal than $u_0 - \delta^k$ will allow to put the point $p_1 = q_0 + r_0$ again back on S^k, and this can be repeated forever.

Thus, by applying an adequate correction r_n to the trajectory each iteration, bounded by $r_0 = \max\{u_0 - \delta^k, \Delta^k\} < u_0$, trajectories starting on S^k can always be kept in S^k. Furthermore, using the above ideas, it can be found that the optimal ratios of control and noise are bounded by:

- $\dfrac{r_0}{u_0} \leq \dfrac{\Delta^k}{u_0}$ if $u_0 \in (\Delta^k + \delta^{k+1}, \Delta^k + \delta^k]$.

- $\dfrac{r_0}{u_0} \leq \dfrac{u_0 - \delta^k}{u_0}$ if $u_0 \in (\Delta^k + \delta^k, \Delta^{k-1} + \delta^k]$.

Again, it should not be strange for the reader to know that the same strategy [22] could be applied if instead of the map T we would have a map f like those described in Sec. 2. In fact, in the next Section we are going to give a numerical example of application of our technique to one of such maps: the well-known logistic map.

7. An Example of Application: Safe Sets and Control of the Logistic Map

We show now a numerical example of application of our technique in a simple situation, using the well known logistic map $x_{n+1} = \mu x_n(1 - x_n)$. Although it is well known that for $\mu > 4$ this map presents a chaotic saddle [5], which is formed after a boundary crisis, we consider here the $\mu = 5$ case. For this value of μ it is easy to see that this map satisfies conditions (a)–(c). For this map, the points bounding A_1, A_2 and A_0 are
$$x_- = \frac{1}{2} - \frac{\sqrt{\mu^2 - 4\mu}}{2\mu} \text{ and } x_+ = \frac{1}{2} + \frac{\sqrt{\mu^2 - 4\mu}}{2\mu}, \text{ so } x_0^1 = \frac{1}{2}.$$

We are going to use an example to show how the system is controlled when it is perturbed with noise such that $u_0 = 0.25$. The first thing that we need to do is to find a k such that $u_0 \geq \Delta^k$. We observe numerically that with $k = 2$, this condition is fulfilled. The sets S^2 and S^1 are shown in Fig. 4(a), and we can appreciate how they clearly present the expected structure: each point of S^1 has two adjacent points of the set S^2.

In Fig. 4(b) we can observe a controlled trajectory. As we said, the idea is to put the initial condition in S^2 and to adjust the applied control r_n each iteration in such a way that the resulting p_{n+1} lies always on a point of S^2. We can see in that figure that the trajectory is kept bounded after 75 iterations (and it could be bounded forever). Note that, in absence of perturbations (even of noise), considering that the initial condition lies on a safe point of order two, after three iterations the trajectory would lie out of $[0, 1]$, and then go to infinity, In Fig. 4(c) we also show the value of the correction applied each iteration, clearly smaller than $u_0 = 0.25$. In fact, we observe that $\max_n (|r_n|)/u_0 \approx 0.15/0.25 = 0.6$, so with a control that is approximately 60% of the noise amplitude the trajectories are kept bounded.

Finally, in Fig. 5 we show the ratios r_0/u_0 that allow to keep the trajectories in I using our technique, obtained both numerically using trajectories of 1000 steps and with the expressions obtained in previous Section (for which a numerical estimation of the Δ^k and δ^k is also needed). There is a good agreement between them. Note that these ratios are always smaller than one, but their value depends strongly on the value of u_0.

Now that we have explored in depth our partial control strategy for one-dimensional maps, we can describe how to extend this strategy to an important class of two dimensional-maps: the horseshoe maps. To do this, first we will describe the analogies existing between these maps and the one-dimensional maps considered so far.

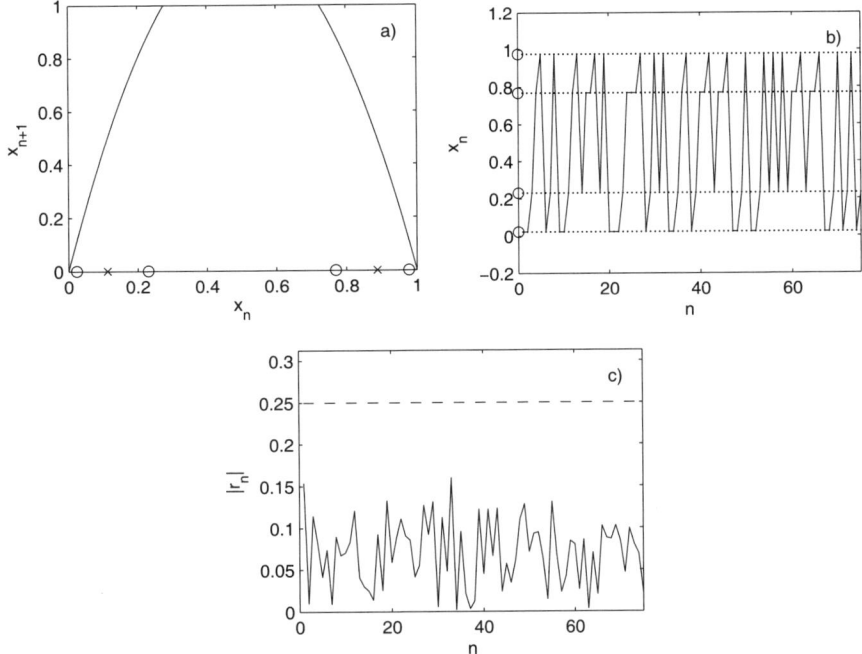

Fig. 4. (a) The points of S^2 ('o') and of S^1 ('×') plotted in the x_n axis together with the graph of the logistic map $x_{n+1} = 5x_n(1-x_n)$, (b) A trajectory controlled with $u_0 = 0.25$, which is always kept in the safe set S^2 (marked by 'o' in the x_n axis), (c) The correction applied each iteration, which is always smaller than the maximum perturbation applied $u_0 = 0.25$, marked with a dashed line.

8. Stretching and Folding: From One-dimensional Maps to the Paradigmatic Horseshoe Map

Up to now we have considered one-dimensional maps with a chaotic saddle. We can analyze the dynamics of this kind of maps from a geometrical point of view. Take for example the paradigmatic slope-three tent map on the $[-1, 1]$ interval. It can be interpreted as Fig. 6 suggests: it stretches the interval by a factor 3 and then folds it back into itself, in such a way that part of it is mapped inside it and part outside it. For the maps discussed in Sec. 2 an analogous interpretation can be given, just by changing "folding" by "cutting" or by adding an inversion.

This kind of action clearly reminds of the action of an important class of maps that has played an important role in Nonlinear Dynamics: horseshoe

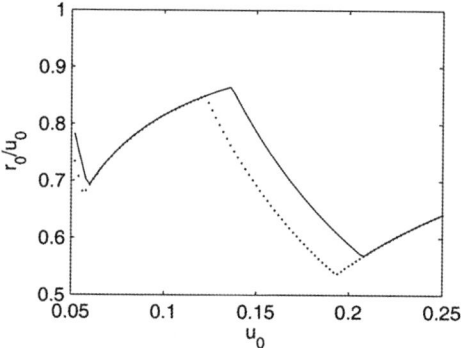

Fig. 5. The ratio r_0/u_0 obtained for different values of u_0 numerically ('··') and from the analytical expressions ('—') obtained. Note that this ratio is always smaller than 1.

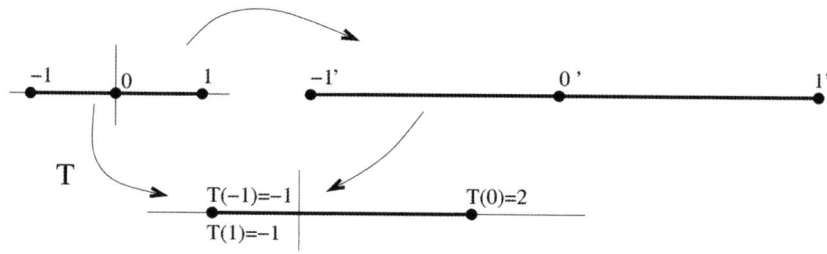

Fig. 6. The action of the slope-three tent map $x_{n+1} = T(x_n)$ on the $I = [-1, 1]$ interval. It can be interpreted as if it stretches it by a factor 3 and then folds it across itself, in a way that reminds to the action of horseshoe maps on areas.

maps [18]. In fact, in the remaining of this Chapter we are going to show that the "stretching and folding action" of horseshoe-like maps implies that they are both an adequate model of map **f** where our partial control technique can be applied, and that safe sets and an advantageous partial control strategy can also be found for them, as it is shown in [23].

Before we do that, it is worth to say something about horseshoe maps. The typical action of this kind of maps is shown in Fig. 7. We can see there how a square Q is stretched and then folded back into itself under the action of **f**. Thus, its geometrical action somehow reminds of the action of the one-dimensional maps described so far. Moreover, it can be proved [6] that this geometrical type of geometrical action (which is well captured by the Conley-Moser conditions [10]) implies the existence of a zero-measure chaotic saddle inside the square Q. Thus, nearly all the trajectories except

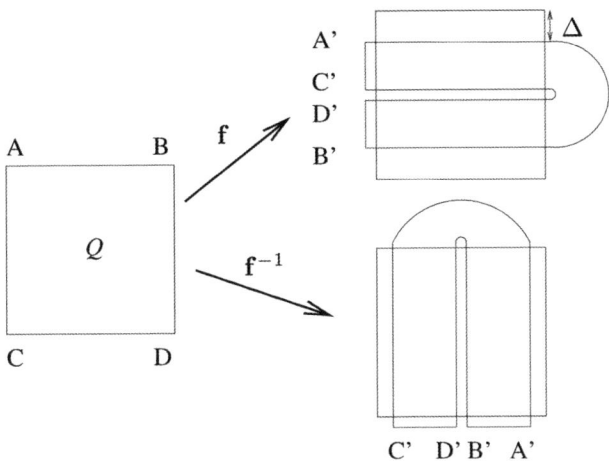

Fig. 7. The action of a horseshoe map. We denote by Δ the minimum distance between the top and bottom sides of Q and $\mathbf{f}(Q)$.

a zero measure set escape from Q under iterations of $\mathbf{f}$. Considering this, this map is a good example of the type of maps that we want to control through a partial control scheme.

The importance of generalizing our previous result on one-dimensional maps to this type of horseshoe maps is due to the fact that they are widespread in Nonlinear Dynamics. Poincaré, in his study of the three-body problem, realized that the existence of transverse homoclinic points (points where the stable and the unstable manifold of saddle-type periodic orbit of a map intersect transversally) implied the existence of complex dynamics. Smale [18] was able to show that this is due to the fact that close to a transverse homoclinic point of a map $\mathbf{g}$ the dynamics is topologically equivalent to the dynamics of a horseshoe map. This is shown in Fig. 8, and it can be briefly explained as follows. If we choose an adequate square R enclosing the saddle point $\mathbf{p}^*$, it is easy to see that certain number of forward iterates k and backward iterates l on the map will deform the square R in the way that is shown in Fig. 8. Thus, if we chose the square Q to be $\mathbf{g}^{-l}(R)$ and we define the map $\mathbf{f} = \mathbf{g}^{k+l}$, then the map $\mathbf{f}$ clearly acts like a horseshoe map on Q.

Transverse homoclinic points are widespread in nonlinear dynamical systems, so either by detecting them or by direct verification of the geometrical action of a dynamical system on certain square Q, horseshoes can be easily found to be present in a wide variety of dynamical systems. For example,

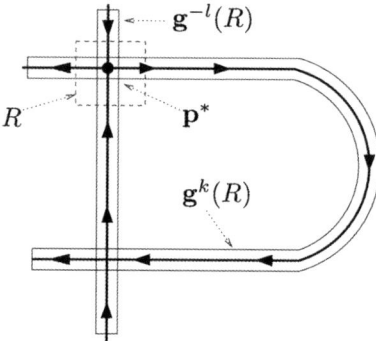

Fig. 8. Sketch of the geometrical action of a map **g** close to a saddle point **p*** with a homoclinic intersection.

they can be found for the following dynamical systems of interest in physics: a particle bouncing on an oscillating table [6], a driven laser [15], a wide class of nonlinear oscillators [9], of an atom-field interaction [11], a bistable optical system [19], a Josephson junction [3] and in fluid advection [14], just to cite some examples.

As we said, for any dynamical system with a horseshoe we can find a region Q with a chaotic saddle from which nearly all trajectories escape, and the ubiquity of this type of geometrical action suggests that we might be interested in avoiding those escapes following the control scheme described by Eq. (1) with $r_0 < u_0$. Although at first sight it might also seem impossible, we are going to show that this can be done using a partial control strategy that is analogous to the one shown for one-dimensional maps.

But first, we can use the picture of the simplest horseshoe map shown in Fig. 7 to describe briefly why classical control strategies can keep trajectories inside Q only if $r_0 = u_0$. For example, an option would be to use $\mathbf{r}_n$ to steer the trajectories to points with long-lived chaotic transients (here, a Cantor set of vertical segments), analogously to what is done in Ref. [4]. But the presence of noise implies that trajectories can fall u_0 away from these points (i.e. if $\mathbf{p}_n$ falls in the leftmost segment), so we need $r_0 = u_0$ to make it work. Another possibility would be also to try to stabilize the trajectory in one of the saddle-type periodic orbits embedded in the chaotic saddle [20]. This can be done by using $\mathbf{r}_n$ to place the trajectory each iteration on the stable manifold (that can be locally approximated by a segment) of the saddle periodic orbit selected. But again here, the presence of noise makes this only possible if $r_0 = u_0$. Thus, these strategies would fail if $r_0 < u_0$.

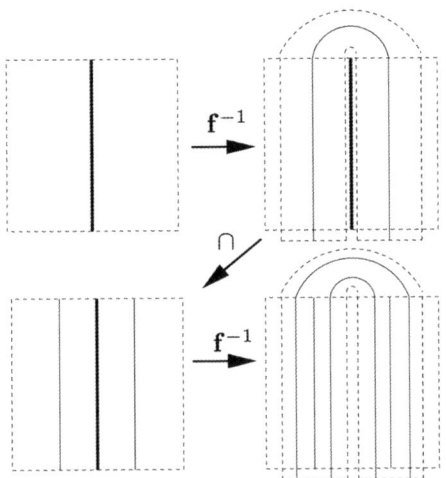

Fig. 9. The set S^0 (thick line) consists of a vertical segment in Q. Its preimage in Q consists of two vertical segments, that form the safe set S^1 (black line). If we take the preimage in Q of S^1, we obtain the safe set S^2 (gray). The arrow with the label $\mathbf{f}^{-1}$ indicates that we take the inverse map, and the arrow with the label "∩" indicates that we take the intersection with the square Q. Both Q and $\mathbf{f}^{-1}(Q)$ are also shown (- -).

With these ideas in mind we can now explain how to implement our partial control strategy to these horseshoe maps.

9. Safe Sets and Partial Control Strategy for Horseshoe Maps

Thus, our aim is to extend our partial control strategy for a system of the type given by Eq. (1) to the particular case in which $\mathbf{f}$ acts like a horseshoe map on a square Q, from which nearly all trajectories escape, and we want to prevent those escapes even in presence of noise. To do this, we need to build the adequate safe sets $\{S^k\}$ for this map. Safe sets for this system can be described by considering the simple example of horseshoe map shown in Fig. 7. It is easy to see that the vertical segment S^0 dividing the square Q into two equal squares is mapped out of the square under one iteration of the map. From this set, though, we can define inductively the sets $\{S^k\}$ as follows:

$$S^k = \mathbf{f}^{-1}(S^{k-1}) \cap Q. \tag{6}$$

As we shall see later, the sets obtained following this rule are the adequate safe sets. The sets S^0, S^1 and S^2 can be observed in Fig. 9, and it is easy

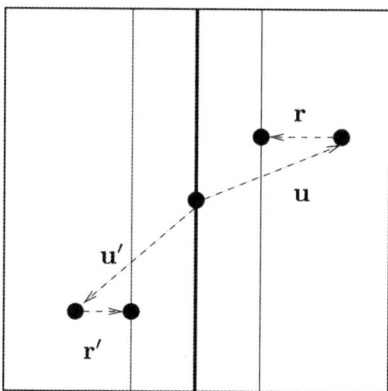

Fig. 10. Illustration of the partial control strategy for a horseshoe map. The trajectory that lies in S^k (thin black line) is mapped into a point on S^{k-1} (thick black line). However, noise (here **u** or **u'**) deviates it. But the fact that any curve of S^{k-1} is surrounded by two curves of S^k allows to achieve this goal even if the correction applied (here **r** or **r'**) is smaller than the noise amplitude.

to see that any set S^k obtained this way is formed by 2^k curves going from the left side to the right side of Q, with these properties:

(i) Each curve of S^k is surrounded by two curves of S^{k+1}. In other words, each curve of S^k has two curves of S^{k+1} that are closer to it than any other curve of S^k.
(ii) We call the maximum and the minimum distance between any curve of S^{k-1} and the two curves of S^k that surround it Δ_k and δ_k respectively. Then,

$$\lim_{k\to\infty} \Delta^k = \lim_{k\to\infty} \delta^k = 0. \tag{7}$$

From this, we can see that the strategy that allows to keep the trajectory inside the square Q with $r_0 < u_0$ will be analogous to the strategy found for one-dimensional maps. We can describe it briefly as follows:

For simplicity we assume here that u_0 is smaller than the minimum distance between $\mathbf{f}(Q)$ and the left and right sides of Q. Given u_0, we have to find the set S^k such that $\Delta^k < u_0$, which is always possible by Eq. (7). Then, put the initial condition **p** in any point on S^k. The action of the map will take the trajectory to $\mathbf{f}(\mathbf{p})$, that by definition will lie in one of the 2^{k-1} curves of S^{k-1}. After this, the noise acts. But, as we can see in Fig. 10, the fact that any curve of S^{k-1} is surrounded by two adjacent curves of S^k allows to use a correction $||\mathbf{r}|| < u_0$ to put the resulting point $\mathbf{f}(\mathbf{p}) + \mathbf{u} + \mathbf{r}$

back on a point of S^k, and this can be repeated forever. Note that this implies that we can find a value of the control r_0 such that trajectories can be kept inside Q even if $r_0 < u_0$.

As for one-dimensional maps, the power of our technique can be evaluated by estimating the value of the ratio r_0/u_0 needed to control the system. For horseshoe maps, an upper bound of the value of those ratios is given by the same expressions as those obtained for one-dimensional maps on Sec. 6, with the only difference of taking as Δ^k and δ^k those quantities as defined for safe sets of horseshoe maps.

It is easy to see that for more general horseshoe maps the sets S^k obtained using Eq. (6) will have an analogous structure. It can be proved that for maps satisfying the Conley-Moser conditions, safe sets with properties analogous to properties (i) and (ii) can be found. In fact, in order to make our exposition clearer and to stress the generality of this method, we are going to describe next Section a map presenting a horseshoe map, the bouncing ball map. After this, we will calculate its safe sets and use them to apply our partial control strategy.

10. Partial Control Strategy for a Bouncing Ball Map

The map that we deal with is the bouncing ball map [6], which is defined as:

$$(\phi_{n+1}, v_{n+1}) = (\phi_n + v_n, \alpha v_n + \beta \cos(\phi_n + v_n)). \tag{8}$$

This map will play the role of the map $\mathbf{f}$ in Eq. (1). This map displays very different types of dynamics depending on the values of parameters α and β. However, we are interested here in the values $\alpha = 1$ and $\beta = 12.6$. For these parameters values, it can be proved rigorously that for certain parallelogram Q of the phase space, the map $\mathbf{f}$ acts like a horseshoe map. This particular geometrical action is shown in Fig. 11.

As a consequence of this stretching and folding, nearly all the trajectories escape from Q under iterations of the map. If we add noise to the system, then all trajectories eventually escape from Q. As we said, our aim here is to avoid those escapes, and to show that this type of partial control is possible even if we use control that is smaller than the noise amplitude. To do that, we just need to generate the safe sets using Eq. (6), that are shown in Fig. 11(b). They obviously fulfil the properties (i)–(ii) given in the former Section.

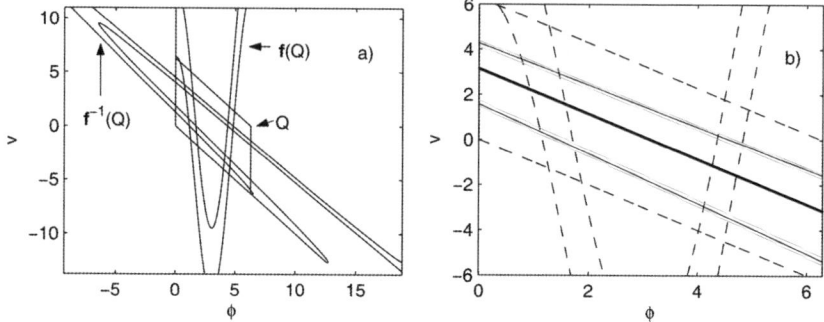

Fig. 11. (a) Action of the bouncing ball map on the parallelogram Q and (b) the resulting safe sets S^0 (thick black), S^1 (black) and S^2 (grey), whose geometrical structure will allow to partially control the system inside Q even if $r_0 < u_0$.

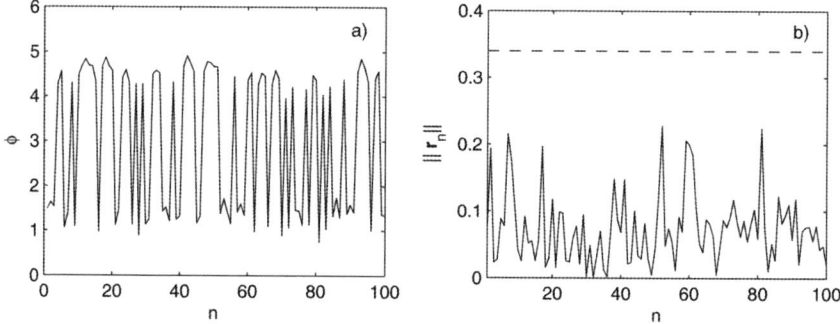

Fig. 12. (a) A controlled series of the bouncing ball map for $u_0 = 0.32$ and (b) the control applied each iteration, that is obviously smaller than u_0 (marked with a dashed line).

We can show how our control technique works for a given value of u_0, for example $u_0 = 0.35$. For this value of the noise amplitude the adequate safe set turns out to be S^2, that is shown in Fig. 11. An example of a controlled time series of the variable ϕ is shown in Fig. 12(a), and it is easy to see that trajectories are kept inside the parallelogram. On the other hand, in Fig. 12(b) we can see that, in order to steer the trajectories into S^2, a control such that $||\mathbf{r}_n|| \leq 0.2$ is enough. Thus, with a control that is approximately 57% of the noise amplitude, trajectories can be kept bounded forever.

We have also made some simulations in order to see the validity of the approximations of the ratios r_0/u_0 needed to keep trajectories bounded

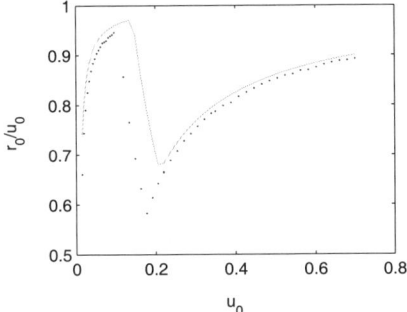

Fig. 13. Ratio r_0/u_0 needed to partially control the system for different values of u_0, computed numerically (black dots) using 100 different time series of 10000 time steps and its analytical estimate (grey line). Note that the estimates are quite similar to the real values and that the computed values of r_0/u_0 are always smaller than one, as claimed.

done in Sec. 6. We have picked 100 different time series of the system for different values of the noise amplitude u_0 and we have estimated the minimum value of r_0 needed to keep trajectories bounded. The results of these calculations can be seen in Fig. 13. We can notice that there is a good agreement between numerical estimates and the values obtained from those upper bounds.

11. Practical Issues and Conclusions

Some remarks are necessary. In order to apply our technique in a dynamical system, we need to locate the safe sets. This implies a previous knowledge of the region of the phase space where the horseshoe acts like a horseshoe map. We expect this to be feasible by doing a time series analysis of the system, although we do not know clear references on how this should be exactly done. Location of homoclinic points and homoclinic intersections can be a good starting point. However, it is important to notice that just an approximate knowledge of the position of the safe sets is needed to adequately partially control the considered system. This can be easily proved by using some of the ideas sketched in this Chapter. From our point of view this problem is equivalent to an inaccuracy of the applied control. However, even in presence of moderate inaccuracies the ratio r_0/u_0 needed to partially control the system can be kept smaller than one.

In summary, we have described an advantageous partial control strategy that can be used for a wide variety of one-dimensional maps and for an important class of two dimensional dynamical systems, those related with horseshoe maps. This control strategy allows to keep trajectories in a region of phase space from which they typically escape even with a control that is smaller than the noise. Paradoxically, this is due to the same general geometrical conditions that make nearly all the trajectories diverge from that region, that also implies the existence of certain sets, the safe sets, with a very interesting structure. We have also shown numerical examples of application of our method, for the one-dimensional logistic map and for the bouncing ball map. In our opinion, the wide presence of this kind of maps in dynamical systems, and more precisely of horseshoe maps, allow to apply this technique in many different contexts. A more general lesson that can be extracted from this Chapter is that the particular geometrical action that is responsible for the appearance of certain types of complex dynamics may be useful in order to control it.

This research has been financed by the Spanish Ministry of Education and Science under Project Number FIS2006-08525 and by Universidad Rey Juan Carlos and Comunidad de Madrid under Project Number URJC-CM-2006-CET-0643.

References

[1] J. Aguirre, F. d'Ovidio, and M. A. F. Sanjuán. Controlling chaotic transients: Yorke's game of survival. *Phys. Rev. E*, 69:016203, 2004.
[2] Kathleen T. Alligood, Tim D. Sauer, and James A. Yorke. *Chaos, an introduction to dynamical systems*. Springer-Verlag, New York, 1996.
[3] M. Bartuccelli, P. L. Christiansen, N. F. Pedersen, and M. P. Sorensen. Prediction of chaos in a josephson junction by the melnikov-function technique. *Phys. Rev. B*, 33:4686–4691, 1986.
[4] M. Dhamala and Y. C. Lai. Controlling transient chaos in deterministic flows with applications to electrical power systems and ecology. *Phys. Rev. E*, 59:1646, 1999.
[5] C. Grebogi, E. Ott, and J. A. Yorke. Chaotic attractors in crisis. *Phys. Rev. Lett.*, 48:1507–1510, 1982.
[6] J. Guckenheimer and P. Holmes. *Nonlinear Oscillations, Dynamical Systems, and Bifurcations of Vector Fields*. Springer-Verlag, New York, 1983.
[7] V. In, M. L. Spano, and M. Ding. Maintaining chaos in high dimensions. *Phys. Rev. Lett.*, 80:700–703, 1998.
[8] T. Kapitaniak and J. Brindley. Preserving transient chaos. *Phys. Lett. A*, 241:41–45, 1998.
[9] F. G. Moon and G. X. Li. Fractal basin boundaries and homoclinic orbits for periodic motion in a two-well potential. *Phys. Rev. Lett.*, 55:1439 – 1442, 1985.

[10] J. Moser. *Stable and Random Motions in Dynamical Systems*. Princeton University Press, Princeton, 1973.
[11] A. Nath and D. S. Ray. Horseshoe-shaped maps in chaotic dynamics of atom-field interaction. *Phys. Rev. A*, 36:431 – 434, 1987.
[12] E. Ott, C. Grebogi, and J. A. Yorke. Controlling chaos. *Phys. Rev. Lett.*, 64:1196, 1990.
[13] R. C. Robinson. *Dynamical systems. Stability, Symbolic Dynamics and Chaos*. CRC Press, Boca Raton, 1999.
[14] M. A. F. Sanjuán, J. Kennedy, C. Grebogi, and J. A. Yorke. Indecomposable continua in dynamical systems with noise: Fluid flow past an array of cylinders. *Chaos*, 7:125–138, 1997.
[15] I. B. Schwartz. Sequential horseshoe formation in the birth and death of chaotic attractors. *Phys. Rev. Lett.*, 60:1359–1362, 1988.
[16] I. B. Schwartz and I. Triandaf. Sustaining chaos by using basin boundary saddles. *Phys. Rev. Lett.*, 77:4740, 1996.
[17] L. Shulenburger, Y.-C. Lai, T. Yalcinkayac, and R. D. Holt. Controlling transient chaos to prevent species extinction. *Phys. Lett. A*, 260:156–161, 1999.
[18] S. Smale. Differentiable dynamical systems. *Bull. Amer. Math. Soc.*, 73:747–817, 1967.
[19] M. Taki. Horseshoe chaos in a bistable optical system under a modulated incident field. *Phys. Rev. E*, 56:6036–6041, 1997.
[20] T. Tél. Controlling transient chaos. *J. Phys. A: Math. Gen.*, 24:L1359–L1368, 1991.
[21] W. Yang, M. Ding, A. Mandell, and E. Ott. Preserving chaos: Control strategies to preserve complex dynamics with potential relevance to biological disorders. *Phys. Rev. E*, 51:102–110, 1995.
[22] S. Zambrano and M. A. F. Sanjuán. *Differential Equations, Chaos and Variational Problems, chapter Control of transient chaos using safe sets in simple dynamical systems*. Birkhauser Publishing, New York, 2007.
[23] S. Zambrano, M. A. F. Sanjuán, and J. A. Yorke. Partial control of chaotic systems. *Phys. Rev. E*, 77:055201(R), 2008.

Chapter 12

CONTINUOUS AND PULSIVE FEEDBACK CONTROL OF CHAOS

Grzegorz Litak
Department of Applied Mechanics,
Technical University of Lublin,
Nadbystrzycka 36, PL-20-618 Lublin, Poland
g.litak@pollub.pl

L. M. Saha
Mathematical Sciences Foundation,
N-91, Greater Kailash I,
New Delhi 110048, India
lmsaha.msf@gmail.com

Mohamad Ali
Department of Mathematics,
Faculty of Mathematical Science,
University of Delhi,
Delhi 110007, India
mali_homs@yahoo.com

Chaotic evolution is the property of nonlinear systems when the parameters incorporated within it attain a certain set of critical values. The system becomes sensitive to initial conditions and shows unpredictability and disorder. It is beneficial if one can find an effective method to control such chaotic evolution. Numerous recent researches in nonlinear dynamics are directed towards controlling or stabilizing or ordering chaos and have achieved commendable success. As nonlinear systems hardly obey a single rule during motion, a number of chaos control methods (or techniques), emerge through these investigations. Due to this chaos control researches are receiving increasing attention at present. The control techniques, though quite specific for particular systems, may provide significant rule to understand evolutionary phenomena of chaos. The present chapter is prepared with an idea to explain some effective methods of chaos control. Some chaos control techniques have been explained mathematically with proper analysis. Also, we have reviewed numerous chaos control techniques emerged through recent studies in nonlinear dynamics. We examine a strange chaotic attractor and its unstable periodic orbits in cases of two-dimensional discrete maps and two different single-degree of freedom nonlinear oscillator. We discuss various approaches to chaos control and propose an efficient method of stabilization unstable periodic orbits by a pulsating feedback technique. Discrete set of pulses enable us to transfer the system from one periodic state to another.

1. Introduction

A system is said to be chaotic if it shows sensitivity to initial condition, i.e, a very small change (perturbation) in initial condition results in divergence in behavior. In such a case it seems to make the system difficult to forecast, and hopeless to control. A chaotic motion appearing in specific physical systems may show various positive and negative effects[1] depending on applications. Opportunity of its control triggered a new field of nonlinear research[2]. The chaos controlling procedure may be explained in a simple mathematical theory shown in the book by Chen and Dong[3]. But recently, the very sensitivity of chaotic systems itself is used to change the chaotic behavior of a dynamical system and to control chaos[3]. This is possible because of the following two properties of chaotic attractors:

(a) Any small neighborhood of a given point on a chaotic attractor contains points on orbits that will visit arbitrarily closely to all other points on the attractor[4].
(b) Infinite number of unstable periodic orbits are imbedded into any chaotic attractor of a dynamical[5].

For controlling chaotic behavior in a variety of dynamical systems a number of theoretical and experimental methods have been successfully applied recently[6-11,2]. Several methods of chaos control, including time delayed feedback method, in nonlinear systems have been discussed widely by Andrievskii and Fradkov[12,13]. The chaos control method suggested by Saha et al.[14] applied to two dimensional systems, though very effective, but have some limitations. Starting with a strange chaotic attractor to system control is a comfort situation because of infinite number of unstable periodic orbits included in it. This novel idea of using unstable periodic orbits to chaos control has been invented by Ott, Grebogi, and Yorke[15] (OGY method). But, ever since the publication of the OGY technique, a number of research articles have been published by scientists and mathematicians, namely, Fradkov and Evans[2], proposing some modifications and improvements on the original technique. The resonant parametric method applied to control chaos has been explained in detail by Zhou et al.[16] They discussed different aspect of this technique. Nonlinear approximation approach to control chaos by Hill[17] increases the effectiveness of the OGY technique. It has been shown that this technique is a way of further increasing effectiveness over OGY technique for many systems.

The aim of this chapter is to revise and compare the most recent techniques used in controlling chaos in dynamical systems. We have demonstrated different techniques of chaos control with appropriate examples in the following sections of this article.

In order to compare the outcome of these techniques and for the sake of clarity, we apply these controlling techniques to the following map

$$x_{n+1} = \rho - x_n(x_n + \alpha + \tfrac{2}{5}) + \beta\, y_n,$$
$$y_{n+1} = x_n, \tag{1}$$

where $\alpha = 6/5$, and $\beta = 3/10$. For $\rho = 0$, this system is chaotic and has unstable fixed point $(x^*, y^*) = (0,0)$ embedded within its chaotic attractor which is shown in Fig. 1.

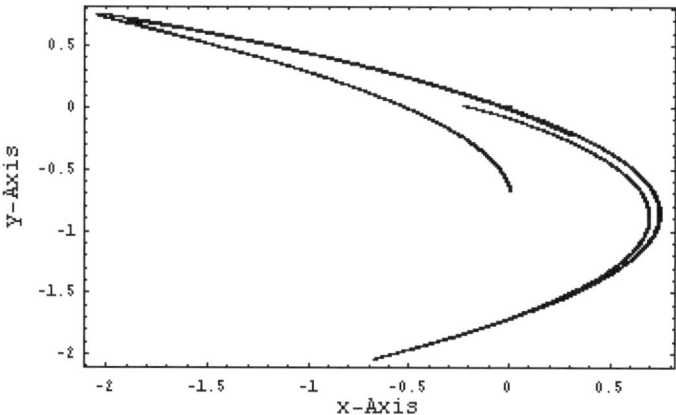

Fig. 1. The chaotic Attractor of the map corresponding to Eq. (1).

1.1. *Local control of chaos (stabilization)*

Local control of chaos means controlling the chaotic orbit of a given dynamical system once it enters a predetermined neighborhood of a desired non-chaotic orbit of the same system (generally an unstable fixed point or unstable periodic orbit). This would allow the linearization of the given dynamical system and the use of the well-known linear theory which makes the system easier to predict within that neighborhood. It is the process of stabilizing an unstable periodic orbit of the dynamical system, the reason why it is alternately called a stabilization technique. For local control of chaos, we discuss the following technique in the next section.

1.2. *Linear feedback control (pole placement technique)*

This is the main tool borrowed from the traditional control theory. Let us consider a following map

$$\mathbf{x}_{n+1} = \mathbf{T}_p(\mathbf{x}_n), \quad \mathbf{x} \in \Re^d, \tag{2}$$

where $p \in \Re$ is an adjustable system's parameter and the control should be designed such that only small variations ($\delta p_{\max} \ll 1$)

of the system's parameter around the nominal value p^* are allowed. Assume the map (Eq. (2)) to have an unstable fixed point $\mathbf{x}^*$ at the nominal value p^* of the parameter. The aim is to stabilize $\mathbf{x}^*$ using only the allowed range of parameter's perturbation. A general case for stabilizing an unstable periodic orbit can be derived similarly.

Linearizing Eq. (2) around $(\mathbf{x}^*, p_0) \in \mathfrak{R}^{d+1}$, we get

$$\mathbf{x}_{n+1} - \mathbf{x}^* \approx \mathbf{A}(\mathbf{x}_n - \mathbf{x}^*) + \delta p \mathbf{B}, \qquad (3)$$

where $\mathbf{A}$ is the $d \times d$-dimensional Jacobian matrix and $\mathbf{B}$ is a $d \times 1$ dimensional matrix:

$$\mathbf{A} = \left.\frac{\partial \mathbf{T}_p}{\partial \mathbf{x}}\right|_{\mathbf{x}=\mathbf{x}^*, p=p^*}, \quad \mathbf{B} = \left.\frac{\partial \mathbf{T}_p}{\partial p}\right|_{\mathbf{x}=\mathbf{x}^*, p=p^*}. \qquad (4)$$

Let us assume

$$\delta p = \mathbf{D}^t (\mathbf{x}_n - \mathbf{x}^*), \qquad (5)$$

where $\mathbf{D}^t$ is a $1 \times d$-dimensional matrix which is called "*feedback gain matrix*". On substituting into Eq. (3), we get

$$\mathbf{x}_{n+1} - \mathbf{x}^* = (\mathbf{A} + \mathbf{B} \cdot \mathbf{D}^t)(\mathbf{x}_n - \mathbf{x}^*). \qquad (6)$$

The point $\mathbf{x}^*$ is stable if and only if the map (Eq. (6)) is a contracting map (i.e. the $d \times d$-dimensional matrix $\mathbf{A} + \mathbf{B} \cdot \mathbf{D}^t$ has eigenvalues of modulus less than one). The existence of the regulator matrix $\mathbf{D}$ was resolved by the well known "*Pole Placement Technique*"[18] and the use of Ackermann's formula. Whereas, the use of Eq. (5) will allow us determine the control window around $\mathbf{x}^*$ in terms of the maximum allowed parameter's perturbation δp_{max}.

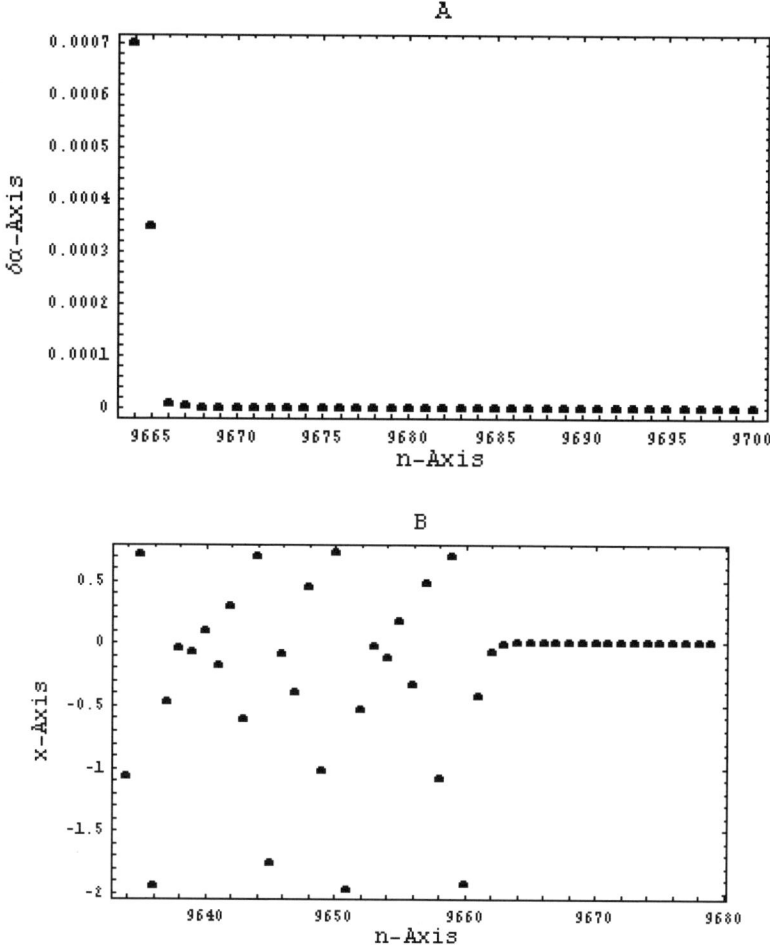

Fig. 2. Stabilization using Linear Feedback Control. (a) The required parameter perturbation for stabilization, (b) The stabilized orbit of the system (Eq. (1)) after applying the parameter perturbation.

For the system given by Eq. (1) with $\delta\alpha_{max} = 0.001$, we chose the eigenvalues of the matrix $\mathbf{A} + \mathbf{B} \cdot \mathbf{D}^t$ to be $\mu_1 = -\mu_2 = 0.1$, and we find $\mathbf{D} = (-8/5, 29/100)$.

Starting with an initial point $(x_0, y_0) = (0.01, 0.015)$, the point (x_{9664}, y_{9664}) fell into the control window which is described by the equation:

$$\left| -\frac{8}{5}x + \frac{29}{100}y \right| = |\delta\alpha| \leq \frac{1}{1000}. \tag{7}$$

Applying continuously the perturbation $\delta\alpha$ given in Eq. (5), the unstable fixed point was stabilized and the orbit marched exponentially towards the fixed point. The parameter's perturbations required at each iterate of the map (Eq. (1)) along with the stabilize orbit are shown in Figs. 2(a) and 2(b), respectively.

1.3. *Stabilizing a hyperbolic object (OGY technique)*

A saddle fixed point $\mathbf{x}^*$ is said to be *hyperbolic* if the stability matrix $\mathbf{A}$ at $\mathbf{x}^*$ does not have eigenvalues of modulus one, i.e. the tangent manifold of the map $\mathbf{T}$ (given by the matrix $\mathbf{A}$) at $\mathbf{x}^*$ has a decomposition into stable subspace E^s (generated by eigenvectors corresponding to eigenvalues of the matrix $\mathbf{A}$ of modulus less than one) and unstable subspace E^u (generated by eigenvectors corresponding to eigenvalues of the matrix $\mathbf{A}$ of modulus bigger than one). Similarly, we can define a *hyperbolic* periodic orbit.

Accordingly, the right choice of the regulator poles is to keep the stable eigenvalues unchanged and to make the unstable ones vanish. This was the original OGY technique[15]. A similar technique was proposed by Blollt and Meiss[19]. We will explain this technique through a simple case in which E^s and E^u are one-dimensional. A generalization of this technique to higher dimensions will be described briefly at the end.

Without loss of generality and using a change in the variables, we may assume the unstable fixed point $\mathbf{x}_{p^*}^*$ to be at the origin of the coordinates and the nominal value of the parameter p^* to be zero. On linearizing Eq. (2) near $\mathbf{x}_{\delta p}^* \in \Re^d$, we get

$$(\mathbf{x}_{n+1} - \mathbf{x}_{\delta p}^*)^T \approx \mathbf{A}(\mathbf{x}_n - \mathbf{x}_{\delta p}^*)^T. \tag{8}$$

But, from the matrix algebra, we have

$$A = \mu_u \, e_u^T \cdot f_u + \mu_s \, e_s^T \cdot f_s, \qquad (9)$$

where e_u & e_s are the unstable and stable eigenvectors corresponding to the unstable and stable eigenvalues $\mu_u > 1$ and $\mu_s < 1$ of the matrix A, respectively, and f_u & f_s are the contravariant basis vectors defined by

$$\begin{aligned} f_u e_u = f_s e_s = 1, \\ f_u e_s = f_s e_u = 0. \end{aligned} \qquad (10)$$

For stabilization in this case, x_{n+1} must lie on the stable direction of x_0^*, i.e.

$$f_u x_{n+1} = 0. \qquad (11)$$

However, the following approximation is always valid

$$x_{\delta p}^* \approx B \delta p, \text{ where } B = \left. \frac{\partial x^*}{\partial p} \right|_{p=0}. \qquad (12)$$

After substituting Eqs. (9) and (12) into Eq. (8) and doting both sides with f_u and using Eq. (11), we get the following relation which specifies the amount of parameter's perturbation needed for the stabilization

$$\delta p = \frac{\mu_u}{\mu_u - 1} \frac{x_n f_u}{B f_u}. \qquad (13)$$

Once the iteration $x_{n+1} = T_{\delta p}(x_n)$ lies on the stable manifold of x_0^*, we retain the nominal value $p^* = 0$ of the parameter by applying T_0 which will cause the following iterates to march exponentially towards x_0^*. By using Eq. (13), we can determine the control window around x_0^* which will be given by

$$\|x_n\| \leq |x_n e_u| = (1 - \frac{1}{\mu_u})|Be_u|\delta p \leq (1 - \frac{1}{\mu_u})\delta p_{\max}|Be_u|. \qquad (14)$$

Figure 3 shows how this technique drags the point from the real orbit to lie on the stable manifold (the dashed line) of the unstable fixed point $(x^*, y^*) = (0, 0)$ of the system (Eq. (1)); where the initial condition was taken to be $(x_0, y_0) = (0.0001, 0.0003)$ and the control window was given by the equation $\|\mathbf{x}\| \leq 0.0029$ for $\delta \rho_{max} = 0.001$.

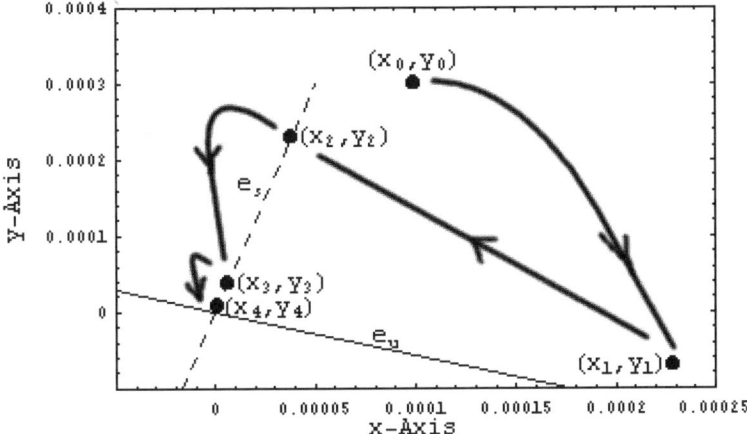

Fig. 3. Targeting the stable manifold using the OGY Technique. The dashed and the solid lines are the stable and unstable manifolds respectively.

Accordingly, we show the amount of the parameter's perturbation (Fig. 4(a)) required for this stabilization technique and the stabilized orbit (Fig. 4(b)). When the unstable subspace E^u is m-dimensional, we can still stabilize $\mathbf{x}_0^*$ by performing this technique m times over m iterates by stabilizing one unstable direction at a time. The stabilization techniques described above can also be applied to periodic orbits by applying it to the composed map $\mathbf{T}^q$ where q is the period of the orbit. But, this solution will not be stable for periodic orbits with long periods. It is advised then to control it by driving the iterates of $\mathbf{x}_n$ to land on the corresponding iterates of $\mathbf{x}_{p^*}^*$.

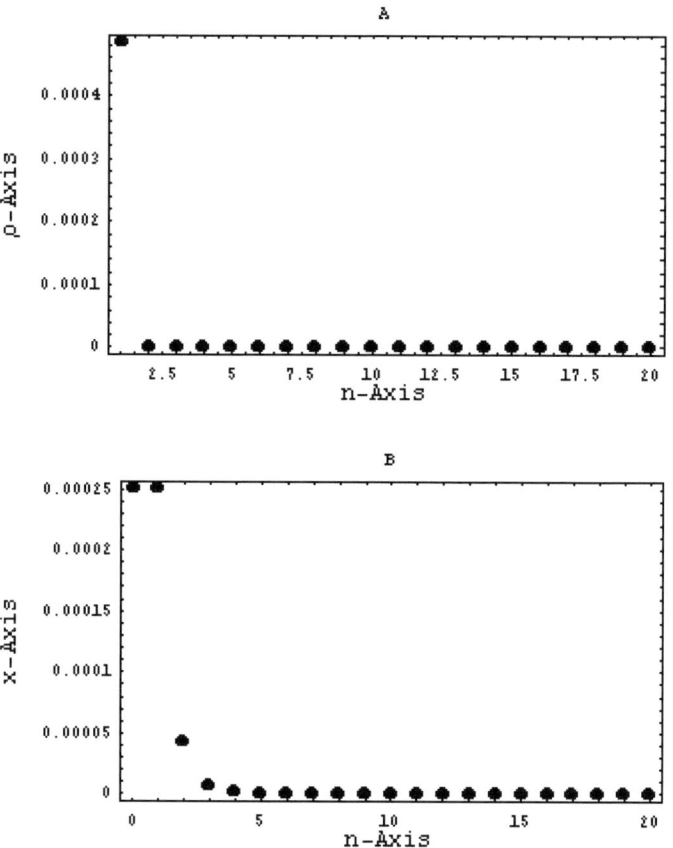

Fig. 4. Stabilization using the OGY Technique. (a) The required parameter perturbation for stabilization, (b) The stabilized orbit of the map (Eq. (1)).

1.4. *Global control of chaos (targeting)*

Considering an initial point a in the phase space, the real orbit often takes a long time to enter the neighborhood of the target point b where the stabilization technique can be activated. Practically, this can be very costly. Therefore, the issue of control should go farther beyond the neighborhood's boundary of the unstable periodic orbit which is required to be stabilized and to investigate the whole area of the phase space in order to gain powerful and less time-consuming control techniques. In other words, we seek to target the neighborhood of the target point from any place in the phase space.

All targeting techniques have common characteristics such as understanding the system in order to know "where does an initial condition go?" and finding an initial condition $\mathbf{x}_a$ near the starting point a and a control procedure $\{p_i\}_{i=1}^n$ which will cause $\mathbf{x}_a$ to iterate into the neighborhood of the target point b as quickly as possible by using only small perturbations. The technique of targeting a time-optimal path technique which was discovered and applied by Shinbrot et al..[20,21], and the cell map technique which is explained in the book of Hsu[22] and applied further by Bradley[23] are good examples of such targeting techniques.

1.5. *Targeting the meshes on a net of paths*

These meshes are nothing but observed points in the phase space associated with paths leading to the target point b. Once the orbit starting from the point a enters a trap (i.e. the control window surrounding any mesh), the control technique is designed to make the orbit follow the corresponding path leading to the target b. E. Kostelich et al.[24] realized that for a higher-dimensional system, there is no hope for finding enough orbits leading to the target in order to select the best one among them. Instead they designed this technique knowing that, because of the chaotic nature of the orbit starting from a, this orbit will eventually wander into the neighborhoods (traps) on these paths. Kostelich's net is organized in a tree hierarchy, where the root of that tree is at b.

We fix an integer m to be the length of every trunk (level), then we determine the first trunk of the tree by waiting for an arbitrary initial condition to eventually iterate close enough to b (inside $\mathbf{B}_\varepsilon(b)$) and storing the last m iterates of that orbit. The next level of the tree can be determined similarly by applying the same to the meshes of the first trunk, and so on for building higher levels in the tree. The branches of the tree are eventually sets of epsilon chains leading to b. The procedure of control here is quite similar to the OGY technique, and the only difference here is the existence of many control windows covering significant portion of the phase space in comparison to only one control window in the OGY technique.

In order to target the fixed point $(x^*, y^*) = (0, 0)$ of the map (Eq. (1)), we built a tree of three levels of 10^3 points on 10^2 paths of length 30 leading to the target point. Once the orbit is inside any trap, it is guaranteed that the target is within 30 iterates. The control was activated once the point on the test orbit is within a distance 0.007 from any mesh on the tree. The distribution of the traps in the phase-space is shown in Fig. 5. Considering an initial point $(x_0, y_0) = (-1.5, -1.5)$, the orbit enters the trap around the point $(x_s, y_s) = (0.718147, -0.677173)$ on the tree at the eleventh iterate at which the path to the target point was set accordingly. The control then was activated by targeting the stable direction of the next point on that path at each iterate, leading finally to the trap around the target point.

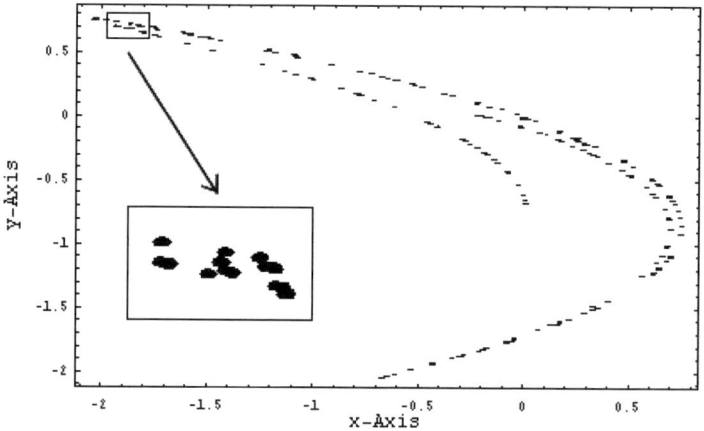

Fig. 5. Distribution of the traps in the phase space of the system (Eq. (1)).

We have plotted the values of the parameter's perturbation required at each step to bring the real orbit into the neighborhood of the unstable fixed point (Fig. 6(a)). Accordingly, in Fig. 6(b), we have shown the real orbit directed towards the neighborhood of the unstable fixed point. The orbit was brought to the neighborhood of the target state after 25 iterates whereas, without this targeting, it requires 4422 iterates till the real orbit falls into the same neighborhood of the unstable fixed point.

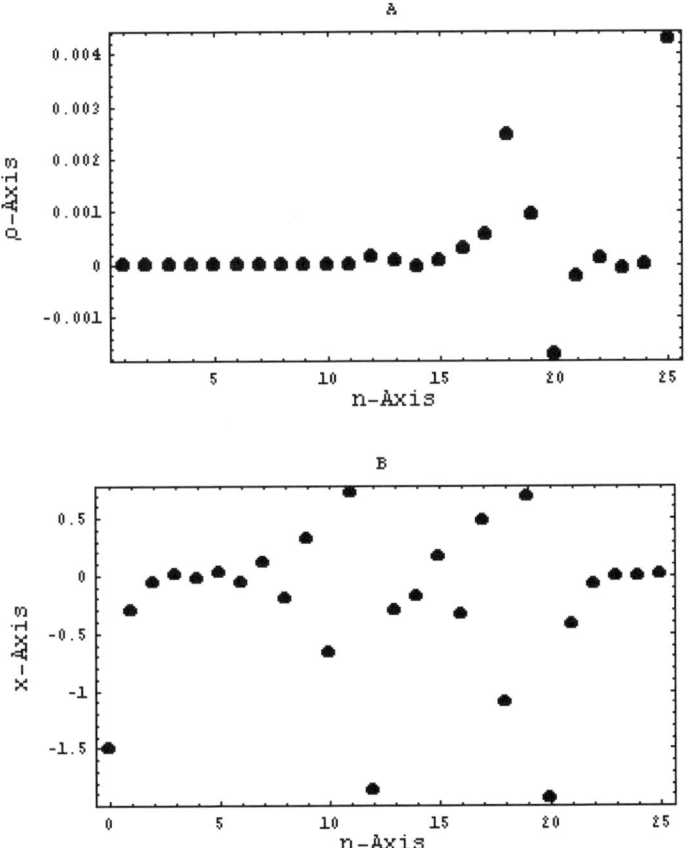

Fig. 6. Targeting the Meshes on a Net of Paths. (a) The parameters perturbation required at each step, (b) The real orbit targeted towards the unstable fixed point of system (Eq. (1)).

1.6. Optimal chaos control

Optimal control is the main issue concerning scientists and mathematicians during the designing of a targeting technique in order to find a control procedure $\{p_i\}_{i=1}^{n}$ which directs a real orbit to the target dynamics. Here, a control technique is considered optimal if some property of the control procedure $\{p_i\}_{i=1}^{n}$ is minimized. Optimality is usually addressed either by minimizing the time required to reach the target or by minimizing the total energy which is defined by

$$\sum_{i=1}^{n} p_i^{2} w_i \; , \qquad (15)$$

where w_i is a weighting function[25].

Variations of the OGY technique were suggested by scientists and mathematicians in order to achieve a kind of optimality in it. Reyl et al.[26] designed a control rule which enables the perturbation at step m to minimize the deviation of the system state at the step $m+1$ from the target orbit. This was a type of energy-optimal control, in the sense that every perturbation aimed, roughly, to minimize the energy required in the next perturbation. Epureanu and Dowel[27,28] proposed a time-optimal control for continuous chaotic systems using a time-varying parameter perturbation to steer a system state directly to a target orbit once the real orbit is within a specific controllable region. We will demonstrate, in the followings, the optimality issue and some of the methods used to achieve this optimality during the design of a control procedure.

1.7. Optimal stabilization: central manifold targeting technique

Starrett[29] proposed a modification of the OGY technique. He aimed to target the unstable hyperbolic fixed point directly instead of only targeting the stable manifold. Considering a two-dimensional map, Eq. (13) represents the parameter perturbation required to put the next iterate of the real orbit on the stable manifold of the target point. He derived a similar relation for the parameter perturbation that puts the next iterate on the unstable manifold and found

$$\delta p = \frac{\mu_s}{\mu_s - 1} \frac{\mathbf{x}_n \mathbf{f}_s}{\mathbf{B} \mathbf{f}_s}. \tag{16}$$

There are certain points in the control window that can be mapped, after one iterate, to the central manifold of the target point. These points are those points for which the parameter's perturbations given in Eqs. (13) and (16) are equal. Therefore, these points are the points satisfying the following condition

$$C \, \mathbf{x} \mathbf{f}_s = \mathbf{x} \mathbf{f}_u, \tag{17}$$

where

$$C = \frac{\mu_s (\mu_u - 1)}{\mu_u (\mu_s - 1)} \frac{\mathbf{B} \mathbf{f}_u}{\mathbf{B} \mathbf{f}_s}. \tag{18}$$

Equation (17) represents a segment of a straight line $L^{CMT}(\mathbf{x}^*)$ which lies in the control window around the target point. Provided that $\tilde{\mathbf{x}}$ is a point on $L^{CMT}(\mathbf{x}^*)$ and $\tilde{\tilde{\mathbf{x}}}$ is a point in the phase space such that $\tilde{\tilde{\mathbf{x}}} \tilde{\mathbf{x}} = 0$, this segment can be targeted easily by letting $\tilde{\tilde{\mathbf{x}}} \mathbf{x}_{n+1} = 0$ and the required parameter's perturbation will be

$$\delta p = \frac{\mu_s \tilde{\tilde{\mathbf{x}}}((\mathbf{e}_s{}^T \mathbf{f}_s) \mathbf{x}_n{}^T)^T + \mu_u \tilde{\tilde{\mathbf{x}}}((\mathbf{e}_u{}^T \mathbf{f}_u) \mathbf{x}_n{}^T)^T}{\mu_s \tilde{\tilde{\mathbf{x}}}((\mathbf{e}_s{}^T \mathbf{f}_s) \mathbf{B}^T)^T + \mu_u \tilde{\tilde{\mathbf{x}}}((\mathbf{e}_u{}^T \mathbf{f}_u) \mathbf{B}^T)^T - \tilde{\tilde{\mathbf{x}}} \mathbf{B}}. \tag{19}$$

The control strategy in the CMT technique is to send the system state from $\mathbf{x}_n$ into $\tilde{\mathbf{x}}$ using the above control rule (Eq. (19)) at the iterate $n+1$, then to the central manifold of the target point $\mathbf{x}^*$ at the iterate $n+2$ using the other rule (Eqs. (13) or (16)).

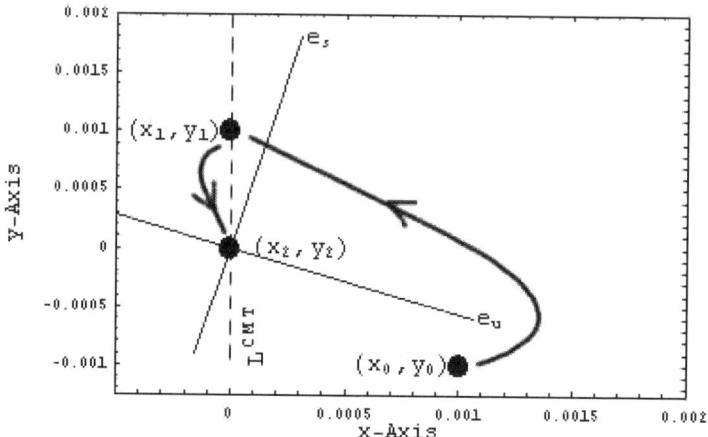

Fig. 7. The CMT control.

Figure 7 demonstrates the way in which the CMT control lands the real orbit on the unstable fixed point of the system (Eq. (1)) after two iterates of the map where $(x_0, y_0) = (0.001, -0.001)$. We observed that $L^{CMT}(\mathbf{x}^*)$ is along the x-axis. Therefore, $\tilde{\tilde{\mathbf{x}}}$ was chosen to be $(0, 1)$. We have shown, in Fig. 8, the parameter's perturbations required for the CMT control technique in this case along with the stabilized orbit.

Comparing the control procedures in Figs. 3 and 7, we notice the time optimality of the CMT technique over the OGY technique as it is faster to bring the real orbit to the target point. Therefore, for a two-dimensional map, any initial condition in the control widow can be sent directly to the target orbit over two iterates of the map by applying the parameter's perturbations (Eqs. (19) and (13)).

Generally, if the dynamical system is m-dimensional, the same can be done using m steps variation of the system's parameter over m iterates of the map. To see this, we note that the mth preimage of the preimages of $\mathbf{x}^*$ will be an m-dimensional slab and given by

$$\mathbf{x}(p_m, p_{m-1}, ..., p_1) = \sum_{i=1}^{m} p_{m-i+1}((\mathbf{A}^{-1})^{m-i} - (\mathbf{A}^{-1})^{m-i+1})\mathbf{B}. \quad (20)$$

This slab is parameterized by p_i; $i = 1, 2, ..., m$ which are linearly dependent on the original adjustable parameter $p = p_1$. Hence, we can direct any orbit in the m-slab into the $(m-1)$-slab in one iteration of the map via a perturbation of the control parameter $p = p_1$. Similarly, we direct the result into the $(m-2)$-slab, and so on until the last step in which we direct the result into the 0-slab, which is nothing but the target orbit.

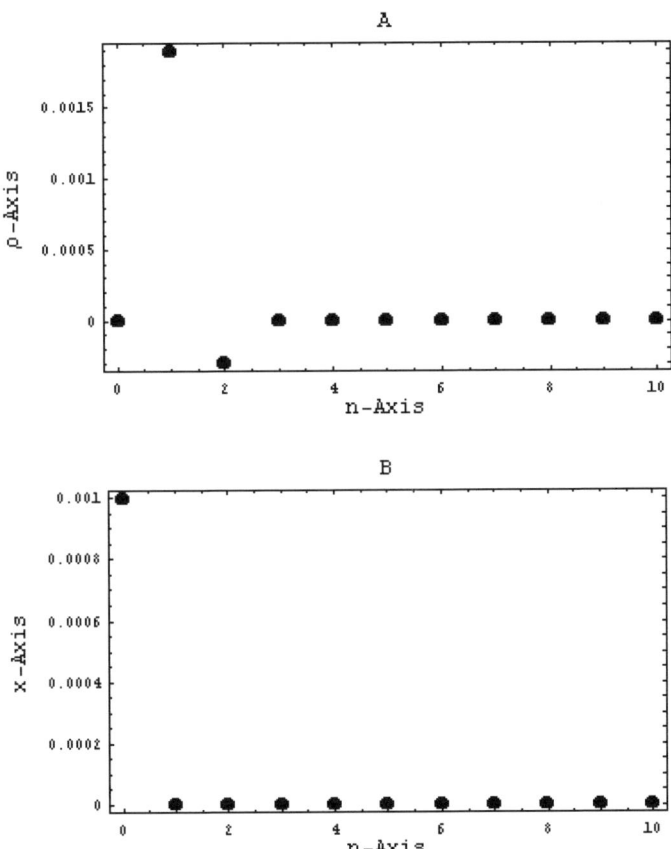

Fig. 8. The CMT control for system (1-1). (a) The parameter's perturbation required for CMT control, (b) Stabilized fixed point after applying the parameter's perturbation.

1.8. *Cutting recurrent loops on a path*

This technique aims to skip all recurrent loops encountered on a path between a starting point a and a target point b. On the other hand, the layered resonance and barrier structures found in Hamiltonian maps of the plane[30,31] make the fraction of points which get substantially far from a in a short time so small so that it may virtually be invisible to a computer search. By contrast, the algorithm introduced by Bollt and Meiss[32] lets a shorter path reveal itself as the "shadow" of an easily found slow orbit which nonetheless makes the desired transport.

Suppose that a point $\mathbf{x}_i$ on that path recurs, after s iterates of the map, with $\mathbf{x}_{i+s}$, i.e. $\|\mathbf{x}_{i+s} - \mathbf{x}_i\| < \delta_1$. Exploiting the instability, we can find a patch that skips the, often very long, recurrent loop. We, eventually, search for a point $\mathbf{x}'_i$, in the neighborhood of the points $\mathbf{x}_i$ and $\mathbf{x}_{i+s}$, which converges to the preorbit of $\mathbf{x}_i$ and to the orbit of $\mathbf{x}_{i+s}$. Practically, any point on an intersection between the unstable manifold of $\mathbf{x}_i$ and the stable manifold of $\mathbf{x}_{i+s}$, as shown in Fig. 9, will have the desired convergence property, and so may be used as $\mathbf{x}'_i$.

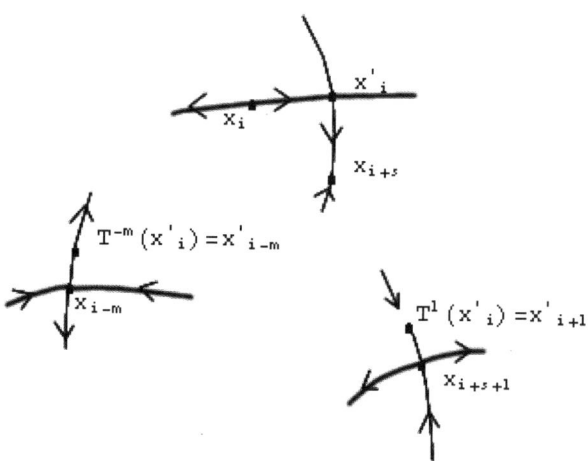

Fig. 9. Constructing a patch at the points of recurrence $\mathbf{x}_i$ and $\mathbf{x}_{i+s}$ after cutting the recurrence loop.

This technique was used successfully[33] to target the Moon from a point near to the Earth in order to let a satellite, launched from the earth, reach an orbit around the moon. The goal was to beat the energy requirements of the standard Hohmann transfer from a parking orbit around the Earth to a parking orbit around the Moon.

Since it is essential to find the unstable direction of the test orbit at $\mathbf{x}_i$ while constructing the patch for the backward iterates, a high level accuracy is required in finding the unstable direction of this orbit (which is usually not periodic) at the point of recurrence $\mathbf{x}_i$, especially when the patch to be constructed is too long ($i \gg 1$). A small deviation of the point $\mathbf{x}'_i$ from the unstable manifold $W^U(\mathbf{x}_i)$ at $\mathbf{x}_i$ would cause $\mathbf{T}^{-i}(\mathbf{x}'_i)$ to escape the initial point a (and may even leave the whole portion of the phase space which contains the chaotic attractor). Alternatively, we propose here a modified method for constructing the required patch as shown in Fig. 10.

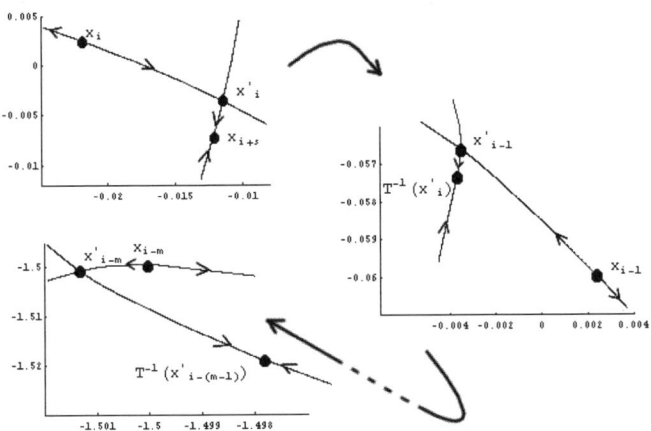

Fig. 10. Modified method for constructing a patch after cutting the recurrence loops.

After finding the point $\mathbf{x}'_i$, which is the principal intersection of the local unstable manifold $W^U(\mathbf{x}_i)$ at $\mathbf{x}_i$ and the local stable manifold $W^S(\mathbf{x}_{i+s})$ at $\mathbf{x}_{i+s}$, we proceed to find $\mathbf{x}'_{i-1}$ which is the principal intersection of $W^U(\mathbf{x}_{i-1})$ and $W^S(\mathbf{T}^{-1}(\mathbf{x}'_i))$, and so on till we find $\mathbf{x}'_0$ which is the principal intersection of $W^U(\mathbf{x}_0)$ and $W^S(\mathbf{T}^{-1}(x'_1))$. The orbit of $\mathbf{x}'_0$ is guaranteed to construct the required patch from near the initial point a to near the target point b.

Figure 10 shows the construction of the patch from the neighborhood of the point $a = (-1.5, -1.5)$ to the neighborhood of the unstable fixed point $b = (0, 0)$ of the system (Eq. (1)). In this illustrative example we found a 25 iterate test orbit starts at a and ends near to b. Choosing the maximum recurrence distance to be $\delta_1 = 0.02$, we achieved a 6 iterate pseudo-orbit having only one line segment at its end point; where the point (x_5, y_5) recurred with the point (x_6, y_6). Finally, by mapping $\mathbf{x}'_0$ five times, we were able to get the shadow orbit of length 5 which starts from near the point a and lands near the unstable fixed point b.

1.9. *Conclusion*

It is too exciting and meaningful to use the term "*Useful Chaos*". We have seen how chaos can be used and made helpful for transportation between different periodic orbits of a dynamical system using the targeting techniques illustrated herein-above. Everything and everyone have its own good and can be made helpful and productive depending on the way we deal with it and manipulate it.

All the recent and most used control techniques listed above are based on the OGY technique or it is a modification of that technique. It can be applied to stabilize any hyperbolic periodic orbit which is often found in dissipative chaotic dynamical systems. Therefore, the problem remains open and more challenging for conservative or what are better known as Hamiltonian systems.

2. Pulsive Feedback Technique to Control Chaos

An important step towards chaos control is the self controlling feedback method introduced by Pyragas[34], where the small perturbations were continuous in time. Impulsive methods for dynamical systems' control and synchronization are some known approaches in the field of chaos[35-39]. It was used successfully for controlling Roessler system[35] and the Duffing oscillator[36] to periodic motions. More recent paper about impulsive control was more successful in establishing more conservative and sufficient conditions for the stabilization and synchronization of Lorenz systems via impulsive control. In their recent work, Sun and Zhang[38,39,40], presented some new theorems on the stability of impulsive control systems, which was applied successfully to the Chua's oscillator. Based on stability theory of impulsive differential equation and new comparison theory, the authors of Refs. [38-40] studied the chaos impulsive synchronization of two coupled chaotic systems using the unidirectional linear error feedback scheme. Moreover, in the most recent work[41], this approach was used with non-linear partial differential equations. The authors determined a criterion for the solutions of these partial differential equations to be equi-attractive in the large and estimated the basin of attraction in terms of the impulse durations and the magnitude of the impulses. Luo et al.[42] have used a random proportional pulse feedback of system variable to Roessler and Lorenz-Harken hyperchaotic systems. They observed that one perturbed system variable is enough to obtain a stabilized periodic orbit. An adaptive feedback control method has been proposed by Hong and Qin[43] for synchronization of chaos in nonlinear systems. In our paper, we apply the same impulsive method with a linear feedback strategy based on the knowledge of unstable periodic orbit embedded within the chaotic attractor of the original system.

The efficient method of chaos control that stabilizes these orbits using a pulsive feedback technique has been proposed recently through two articles[44,45]. In these articles, a discrete set of pulses is able to transfer the system from one periodic state to another. They examined the strange chaotic attractor and its unstable periodic orbits for a one degree of freedom nonlinear oscillator, in each case, with a non-symmetric potential. In one case nonlinear oscillator models a quarter car forced by the road profile. Also, Marin and Sole[46,47], in their two research articles, introduced a search algorithm for exploring the parameter on periodically perturbed discrete maps to find desired orbits through chaos control. They confined to two types of control involving proportional pulses in the system variable and constant feedback.

2.1. *Feedback and pulsive feedback techniques for stabilizing unstable periodic orbits*

Chaos does exist in many engineering systems and some kind of feedback mechanism has been proposed for such problems to control chaos. In fact, feedback is prevalent in modern chaos control processes. In some problems, feedback technique being used to control chaos by utilizing various features of chaos. In other problems its utilization is to synchronize chaotic motion. Some noteworthy works in feedback control of chaos are to be referred here with references[48-55].

Let us consider the flow described by the following second-order differential equation

$$\ddot{x} + \alpha \dot{x} + \delta x + \gamma x^2 = \mu \cos(\omega t), \tag{21}$$

where x denotes a displacement, t time, while α, δ, γ, μ, ω define the system parameters.

The above equation has been extensively studied by Thompson[56], Thompson and Hunt[57], who found chaotic behavior in the system and examined transitions to chaos through a global homoclinic bifurcation and a cascade of period doubling bifurcations just before escape from the potential well. Such systems (Eq. (21)) have been also a subject of studies for many other researchers, inspired by possible applications in description of mostly mechanical systems[58-64] and the catastrophe theory[65]. They were also linked to possible meta-stable states of atoms and they appeared in problems within the elastic theory[38-39].

This equation is equivalent to the following autonomous system of three first-order differential equations:

$$\dot{x} = y$$
$$\dot{y} = -\alpha y - \delta x - \gamma x^2 + \mu \cos(\omega z), \quad (22)$$
$$\dot{z} = 1$$

where $z = t$. Therefore, whenever we attempt to integrate the system (Eq. 22), we must pay attention to the fact that the initial conditions must be such that $z_0 = t_0$.

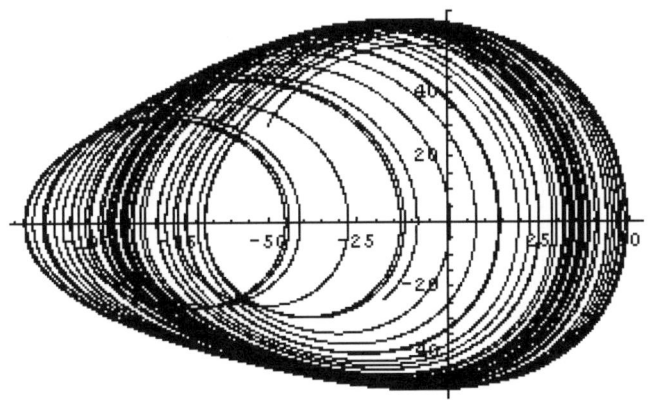

Fig. 11. The chaotic attractor of the system (Eq. (22)) potted in the phase plane x-y for the parameters values: $\alpha = 1.0$, $\delta = 1.0$, $\omega = 0.85$, $\gamma = 0.01089$, and $\mu = 60.8$.

From the papers by Litak et al.[63,64,66] it is clear that this system (Eq. (21)) exhibits a chaotic attractor for the parameters' values $\alpha = 1.0$, $\delta = 1.0$, $\omega = 0.85$, $\gamma = 1.089$ and $\mu = 0.608$ shown in Fig. 11. However, rescaling the variables by acquiring the following variables' changes $x \to \frac{x}{100}$ and $y \to \frac{y}{100}$, the system (Eq. (22)) will remain the same while the parameters' values of γ and μ will become 0.01089 and 60.8, respectively.

The system (Eq. (22)) exhibits an unstable periodic orbit $\mathbf{X}^*(t)$,

$$\mathbf{X}^*(t) = \left[x^*(t), y^*(t)\right], \quad (23)$$

embedded within its chaotic attractor of period $2\pi/\omega$. This orbit was obtained numerically by a method of recurrence and is shown in Fig. 12.

The basic idea of recurrence is to wait two successive iterations of the designed Poincare map of sections to fall in a sufficiently small neighborhood. In our case, for the sake of more accuracy, we have used the same concept but with a little modification. Given the dimension of the phase space and the range of the variables, we are able to determine a rectangle in the phase space where points of the unstable periodic orbit are suspected to be within it.

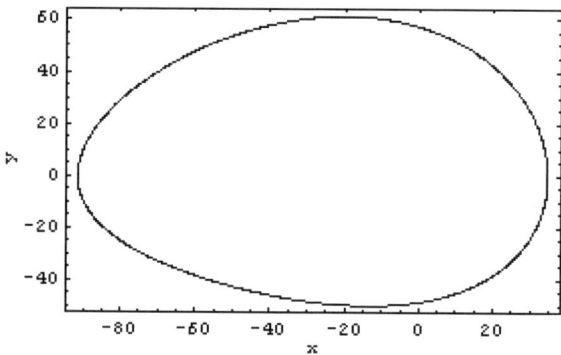

Fig. 12. Unstable period-1 orbit of the system (Eq. (22)).

Using Mathematica, we were able to develop a code that can detect a smaller rectangle within the previous one in which points of the unstable periodic orbit are lying within it. This was done by dividing the previous rectangle over a net of 10000 smaller and identical rectangles then integrating the given flow starting with each mesh on the net and finding the mesh at which the smallest recurrence occurs. Repeating the same procedure successively finitely many times, we were able to determine a point (mesh) at which an arbitrarily small recurrence occurs. Integrating and plotting the orbit initialized at this mesh over the same period of the Poincare map, would give us the best approximation of the required unstable periodic orbit of the given period.

Using the feedback technique, we were able to stabilize the aforementioned unstable period-1 orbit of this system which is embedded within its chaotic attractor. This was done by adding a perturbation $\varepsilon\,(\mathbf{X}(t) - \mathbf{X}^*(t))$ to the analyzed system (Eq. (22)).

The stabilized period-1 orbit of the given system is shown in Fig. 13 where $\varepsilon = -0.5$.

An alternative technique which can be thought of is to apply the same feedback technique on the system but on a discrete scale as pulses (pulsive feedback technique). An interesting and very important outcomes were noted while applying this technique; where the time span between two successive pulses was taken to be $\pi/(3\,\omega)$.

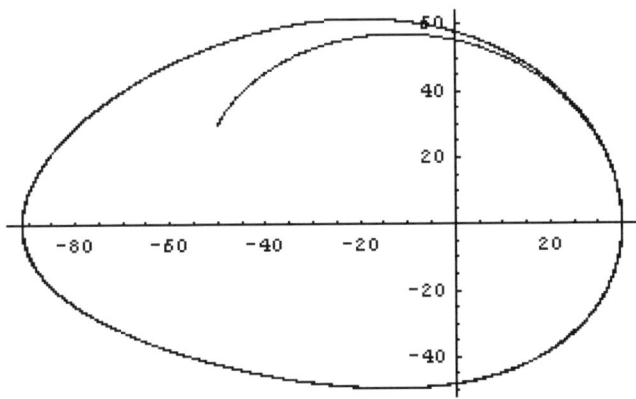

Fig. 13. Stabilized period-1 orbit of the system (Eq. (22)) potted in the phase plane x-y using the continuous feedback technique with $\varepsilon = -0.5$.

Beside the fact that we were able to stabilize the same unstable period-1 orbit of this system, we were able to transfer the system from one periodic state to another by varying the pulses strength ε. Figures 14, 15, and 16 show stabilized period-1, period-2 and period-4 orbits of the dynamical system (Eq. (22)) using the pulsive feedback technique with $\varepsilon = -0.05$, $\varepsilon = -0.03$, and $\varepsilon = -0.009$, respectively.

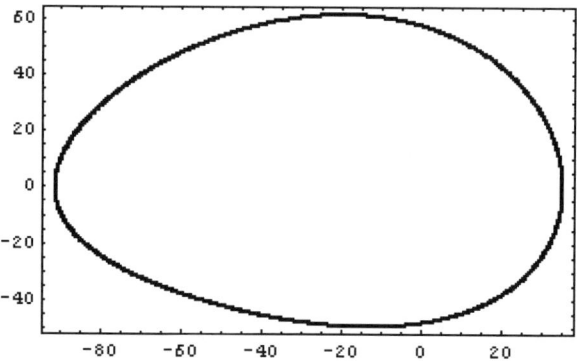

Fig. 14. Stabilized period-1 orbit of the system (Eq. (22)) potted in the phase plane x-y using pulsive feedback technique with $\varepsilon = -0.05$.

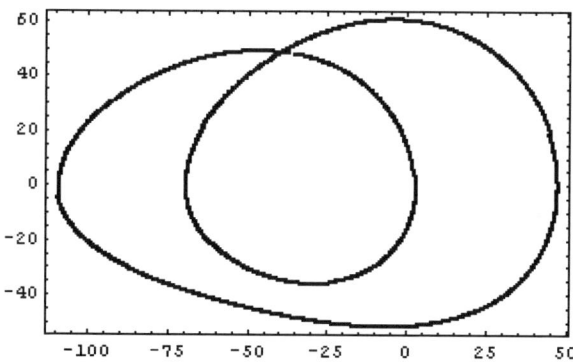

Fig. 15. Stabilized period-2 orbit of system (Eq. (22)) potted in the phase plane x-y using the pulsive feedback technique with $\varepsilon = -0.03$.

In fact, we have found that the system (Eq. (22)) with the pulsive feedback undergoes a period doubling bifurcation as the pulses strength ε increases in the interval $[-0.05, 0]$. This may be very useful in practice and real life experiments as it enables us switching between one state and another of the same system by simply controlling the pulses' strength provided for the system.

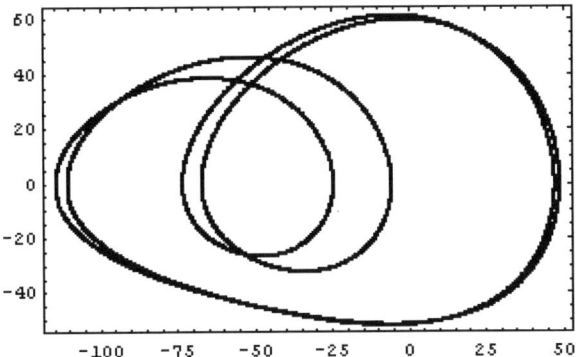

Fig. 16. Stabilized period-4 orbit of the system (Eq. (22)) plotted in the phase plane x-y using the pulsive feedback technique with $\varepsilon = -0.009$.

This may be also considered as an advantage of the pulsive feedback technique over the well known feedback technique which does not enable this switching from one state a dynamical system to another without re-engineering the whole technique in order to suit a particular state.

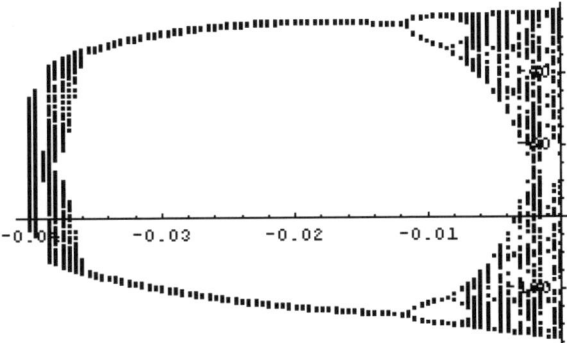

Fig. 17. Bifurcation diagram of the system (Eq. (22)) against the pulsive feedback coupling $\varepsilon \in [-0.04, 0]$.

A rough bifurcation diagram of the system (Eq. (20)) with the pulsive feedback control is shown in Fig. 17 with $\varepsilon \in [-0.04, 0]$. Note that starting from a single period unstable orbit one can obtain many different unstable periodic orbits included in the strange chaotic attractor.

2.2. *Pulsive control applied to a quarter car model*

Considering the system representing a quarter car forced by a road profile we assume that it can be described by the following single-degree-of-freedom second order differential equation[45,66]

$$\ddot{x} + \alpha \dot{x} + \beta x^3 + kx^3 = -g' + A\Omega^2 \sin(\Omega t), \qquad (24)$$

where x is a displacement and other system parameters: $k = -1.875$, $\alpha = -0.04034$, $\beta = 2.68957$, $g' = 0.014715$, $A = 0.41$, and $\Omega = 0.8$, respectively.

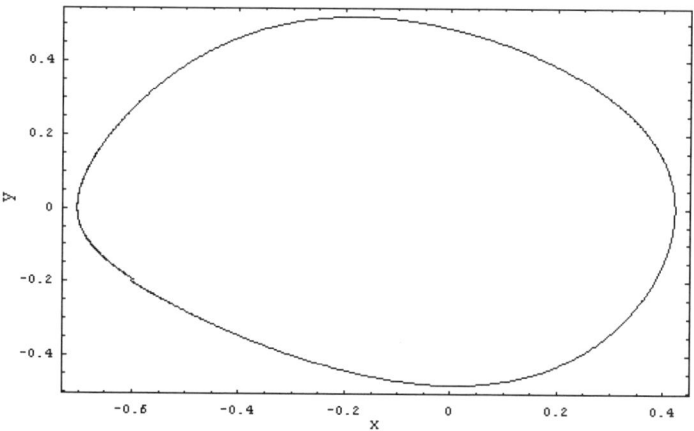

Fig. 18. Unstable period-1 orbit of the system (Eq. (24)).

By using the above equation we have obtained numerically the unstable period-1 orbit (Fig. 18) embedded within its chaotic attractor (Fig. 19).

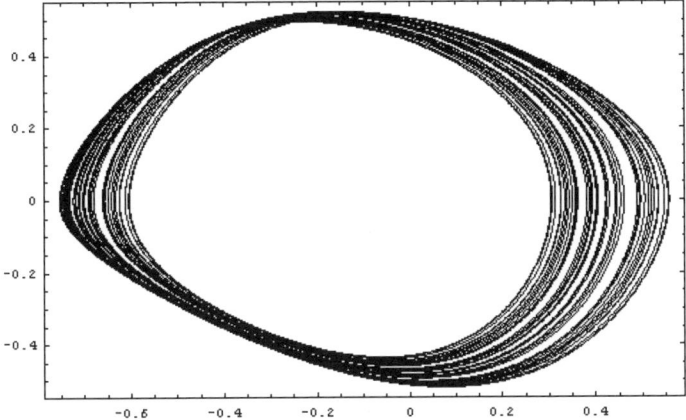

Fig. 19. Chaotic attractor of the system in the phase plane $x - y$ for fixed parameters values (Eq. (24)).

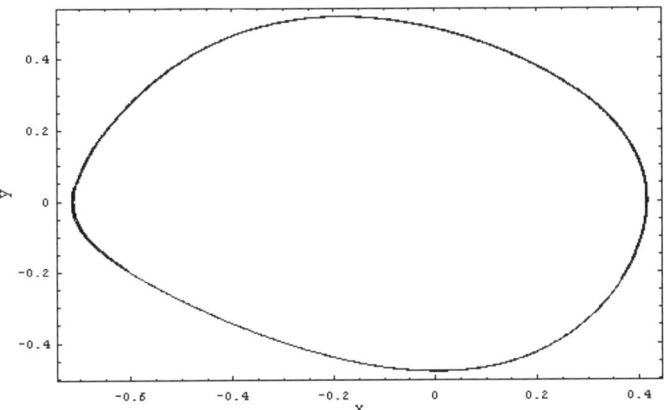

Fig. 20. Stabilized period-1 orbit of the given system (Eq. (24)) in the phase plane $x - y$ using pulsive feedback technique with $\varepsilon = -0.3$.

For integration, we used the Runge-Kutte method of second order with a step size equal to $2\pi/1000\Omega$ i.e. one cycle of period $\tau = 2\pi/\Omega$ was divided into 1000 equal time intervals for the integration purpose. With pulse strength $\varepsilon = -0.3$, we were able to stabilize a period-1 orbit

(Fig. 20). While for $\varepsilon = -0.05$, we were able to stabilize a period-2 orbit (Fig. 21).

Furthermore, the bifurcation diagram (Fig. 22) shows a variety of periodic orbits that can be stabilized using the pulsive feedback technique through changing the strength ε of pulses.

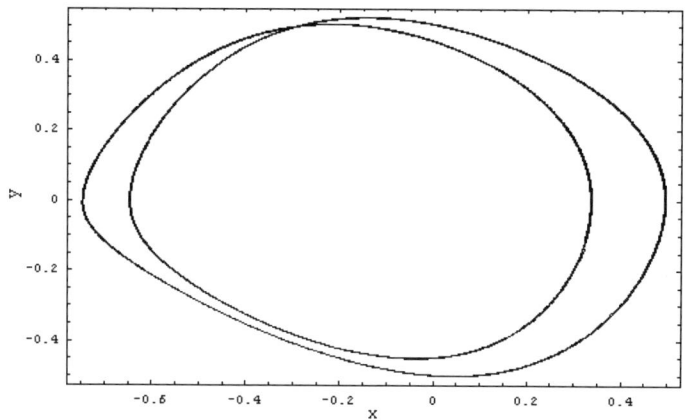

Fig. 21. Stabilized period-2 orbit of the given system (Eq. (24)) in the phase plane $x - y$ $(y = \dot{x})$ using the pulsive feedback technique with $\varepsilon = -0.05$.

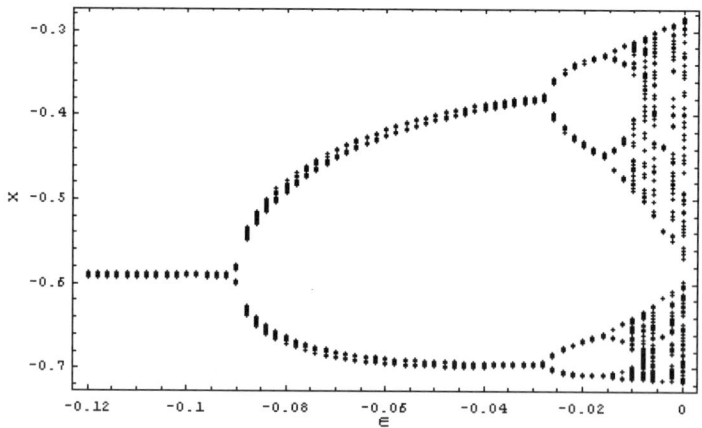

Fig. 22. Bifurcation diagram of the controlled system x versus the small parameter ε using the pulsive feed back technique.

3. Summary and Conclusions

In summary, we have successfully demonstrated the application of the pulsive control method to two different models including a single-degree-of-freedom nonlinear oscillator with a non-symmetric potential and a quarter car model. Our proposed method efficiently stabilizes unstable periodic orbits included in the strange attractors by a pulsating feedback technique.

Note also, our approach to system control differs from so-called impulsive control methods[35-37] where the main issue was suppressing chaotic motion. In their cases the system is stabilized by an adequately strong impulsive signal which leads to periodic motion with various properties. These periodic orbits, in contrast to the present consideration, were not related to unstable periodic orbits of a strange attractor. In our case, the pulsating control method is based on unstable periodic orbit and it makes use of a feedback control in a discrete way and the stabilized orbits are strongly related to other unstable orbits of a strange attractor with multiple period.

Interestingly, one can estimate ε_i for corresponding period doubling bifurcations[67] we obtained Feingenbaum universal constant converging to 4.669201 $\cdots$ (known with higher precision)[68] as for maps having a quadratic maximum. This value has been also found in many different experiments[68] Interestingly implementing the proper Feingenbaum number to our algorithm would provide possibility to predict the corresponding n-period orbit for given ε.

By changing a single small parameter, we were able to choose one of the periodic unstable orbit embedded in the chaotic strange attractor and stabilize it. However the practical problems involved in the implementtation of the method on a real system are left to future investigations.

References

1. J.M.T. Thompson, *Proceedings of the Royal Society of London* A 421, 195 (1989).
2. A.L. Fradkov and R.L. Evans, *Annual Reviews in Control* 29, 33 (2005).

3. G. Chen and X. Dong, *From Order to Chaos, Methodologies, Perspectives and Applications*, World Scientific Series on Nonlinear Sciences, Series A, Vol. 24, (World Scientific, Singapore 1998).
4. T. Shinbrot, E. Ott, C. Grebogi, and J.A. Yorke, *Phys. Rev. Lett.*, 65, 3215 (1990).
5. J. Farmer and J. Sidorowich, in *Evolution, Learning, and Cognition*, Ed. Y.C. Lee (World Scientific Press, Singapore 1988) p. 277.
6. F. Romeiras, C. Grebogi, E. Ott, and W. Dayawansa, *Physica* D 58, 165 (1992).
7. W.L. Ditto, S.N. Rauseo, and M.L. Spano, *Phys. Rev. Lett.*, 65, 3211 (1990).
8. A. Azevedo and S.M. Rezende, *Phys. Rev. Lett.*, 66, 1342 (1991).
9. P. Parmanada, P. Sherard, and R.W. Rollins, *Phys. Rev.* E 47, R3003 (1993).
10. W.X. Ding, H.Q. She, W. Huang, and C.X. Yu, *Phys. Rev. Lett.* 72, 96 (1994).
11. I Z Kiss, V. Gaspar, L. Nyikos, and P. Parmananda, *J. Phys. Chem.* A 101, 8668 (1997).
12. B.R. Andrievskii and A.L. Fradkov, *Automation and Remote Control* 64, 505 (2004).
13. B. R. Andrievskii and A.L. Fradkov, *Automation and Remote Control* 65, 673 (2004).
14. L. M. Saha, G.H. Erjaee, and M. Budhraja, *Iranian J. Sci. & Technology*, 28, 219 (2004).
15. E. Ott, C. Grebogi, and J.A. Yorke, In *CHAOS, Soviet-American Perspective on Nonlinear Science*, Ed. D. Campbell, (American Institute of Physics, N.Y. 1990), p. 153.
16. Y. Zhou, C.K. Tse, S.-S. Qui, and F.C.M. Lau, *Int. J. Bifurcation and Chaos* 13, 3459, (2003).
17. D. L. Hill, *Int. J. Bifurcation and Chaos* 11, 207 (2001).
18. K. Ogata, *Control Engineering*, 2nd edition.(Prentice-Hall, Englewood, NJ 1990).
19. E.M. Bollt, and J.D. Meiss, *Physica* D 66, 282 (1993).
20. T. Shinbort, C. Grebogi, E. Ott, and J.A. Yorke, *Phys. Rev.* A 45, 4165 (1992).
21. T. Shinbort, W. Ditto, C. Grebogi, E. Ott, M. Spano, and J.A. Yorke, *Phys. Rev. Lett*, 68, 2863 (1992).
22. C. Hsu, *Cell-to-Cell Mapping*. (Springer-Verlag, New York 1987).
23. E. Bradley, *Taming chaotic circuits*. M.I.T. Ph.D. thesis. M.I.T. A-TR 1338 (1992).
24. E. Kostelich, C. Grebogi, E. Ott, and J. A. Yorke, J. A.,*Phys. Rev.* E 47, 305 (1993).
25. G. Chen, *Int. J. of Bifurcation and Chaos* 4, 461 (1992).
26. C. Reyl, L. Flepp, R. Badii, and E. Brun, *Phys. Rev.* E, 47, 267 (1992).
27. B. Epureanu and E. Dowel, *Physica* D 116, 1 (1998).
28. B. Epureanu and E. Dowel, *Physica* D, 139, 87 (2000).
29. J. Starrett, *Phys. Rev.* E, 66, 6206 (2002).
30. Y. Lai, M. Ding, and C. Grebogi, *Rev.* E 47, 86 (1992).
31. J.D. Meiss, *Rev. Mod. Phys.* 64, 795 (1992).
32. E.M. Bollt and J.D. Meiss, *Physica* D, 81 (3) 280 (1995).
33. E.M. Bollt, and J.D. Meiss, *Phys. Lett.* A 204, 373 (1995).
34. K. Pyragas, *Phys. Lett.* A 170, 421 (1992).

35. T. Yang, C-M. Yang, C-M., and L-B. Yang, *Phys. Lett.* A 232, 356 (1997).
36. G.V. Osipov, L. Glatz, and H. Troger, *Chaos Solitons & Fractals* 9, 307 (1998).
37. G.V. Osipov, A.K. Kozlov, and V.D. Shalfeev, *Phys. Lett.* A 247, 119 (1998).
38. J.T. Sun, Y.P. Zhang, and Q.D. Wu, *Phys. Lett.* A 298, 153 (2002).
39. J.T. Sun and Y.P. Zhang, *Math. Comp. Sim.*, 66 499 (2004).
40. J. T. Sun, Y.P. Zhang, F. Qiao, and Q.D. Wu, *Chaos, Solitons & Fractals* 19, 1049 (2004).
41. A. Khadra, X. Liu, and X. Shen, *Chaos, Solitons & Fractals*, 26, 615 (2005).
42. X.-S. Luo, B-J Wang, F. Jiang, and Y. Gao, *Chinese Phys.*, 10, 17 (2001).
43. Y. Hong and H. Qin, *Int. J. Bifurcation and Chaos* 11, 1149 (2001).
44. G. Litak, M. Ali, and L.M. Saha, *Int. J. Bifurcation and Chaos* 17, 2797 (2007).
45. G. Litak, M. Borowiec, M. Ali, L. M. Saha, and M. I. Friswell, Chaos, Solitons & Fractals, 33, 1672, (2007).
46. J. Marin and R.V. Sole, *Phys. Rev.* E51, 6239, (1995).
47. J. Marin and R.V. Sole, Phys. Rev. Lett.,72, 1455, (1994).
48. K. Yagasaki, *Nonlin. Dynam.* 6, 125 (1994).
49. K. Yagasaki, *Nonlin. Dynam.* 9, 391 (1996).
50. J. Z. Yang, Z. L. Qu and G. Hu, *Phys. Rev.* E 53, 4402 (1996).
51. T. Ushio and K. Hirai, *Int. J. of Control* 38, 1023 (1983)
52. C. T. Sparrow, *J. of Theo. Biol*, 83, 93 (1980).
53. C. T. Sparrow, *J. of Math. Anal. Appl.* 83, 275 (1981).
54. W. J. Graantham and A. M. Athalye, *Contr. Dynam. Sys.*,34, 205 (1990).
55. R. Genesio and A. Tesi, *Int. J. Bifur. Chaos,* 2, 61 (1992)
56. J.M.T. Thompson, *Proc. Roy. Soc. London* A 421, 195 (1989).
57. J.M.T. Thompson and G.W. Hunt, *Elastic Instability Phenomena* (Wiley, Chichester 1984).
58. K. Szabelski and W. Samodulski, *Mechanika Teoretyczna i Stosowana* 23, 223 (1989).
59. W. Szemplinska-Stupnicka and J. Rudowski, *Physica* D 66, 368 (1993).
60. W. Szemplinska-Stupnicka, *Nonlinear Dynamics* 7 129 (1995).
61. G. Rega, A. Salvatori, and F. Benedettini, *Nonlinear Mechanics* 7, 249 (1995).
62. R.H. Rand, *Lecture Notes on Nonlinear Vibrations* (Ithaca:The Internet-First University Press, Ithaca 2005), http://www.tam.cornell.edu/randdocs/.
63. G. Litak, A Syta, and M. Borowiec, *Chaos, Solitons & Fractals,* 32, 694 (2007).
64. G. Litak, A. Syta, M. Borowiec, and K. Szabelski, *Chaos, Solitons & Fractals* 40, 2414 (2009).
65. T. Poston and J. Stewart, *Catastrophe Theory and its Applications* (Pitman, London 1978).
66. G. Litak, M. Borowiec, M.I. Friswell, and K. Szabelski, *Communication in Nonlinear Science and Numerical Simulations* 13, 1373 (2008).
67. M.J. Feingenbaum, M. J. *Los Alamos Sci.* **1**, 4 (1980).
68. J.C. Sprott, *Chaos and Time Series Analysis* (Oxford University Press, Oxford 2003).

Chapter 13

CHAOS CONTROL

Luiz Felipe Ramos Turci
*Instituto Tecnológico de Aeronáutica — ITA,
12228-900 — São José dos Campos — SP, Brazil*

Elbert E. N. Macau
*Instituto Nacional de Pesquisas Espaciais — INPE,
12227-010 — São Jóse dos Campos — SP, Brazil*
elbert@lac.inpe.br

1. Introduction

Henri Poincaré by the end of the 19th century, formulated and analyzed the now famous circular-restrict three-body-problem, aiming to understand the stability of the Solar System [Poincaré (1899)]. The analysis of this problem gave to him the first glimpse of the chaotic behavior by the observance that complex trajectories in the neighborhood of periodic orbits were apparently unpredictable, while trapped to a region of the state space. Despite great effort made by many mathematicians that have given their important contribution based on Poincaré's work, only in present days, mainly due to computational and graphical improvements, the knowledge about Chaotic Dynamic was consolidated, open the way to a wide number of important applications. Consequently it is known nowadays that the chaotic behavior is predominant in Nature and its understanding is of fundamental importance in various physical phenomena [Arecchi *et al.* (1982)], in biology [Schiff *et al.* (1994)], in chemistry [Hudson and Mankin (1981)], and so many others.

In recent times, the ambition to controlling natural phenomena governed by chaotic dynamic has motivated enthusiastic discussions in the scientific community regarding the possibility of controlling a chaotic dynamic. In 1990, a revolutionary work entitled *Controlling Chaos* [Ott *et al.*

(1990)], shows that chaotic evolution can be controlled, and moreover, the complexity inherent to a chaotic evolution can be explored to propitiate unique levels of efficiency and flexibility in several technological applications. The importance of this work can be evaluated by the increasing number of subsequent publications that are based on it. Chaos control is still a subject of intensive research, whose results provide applications in several fields of science and technology. New approaches and methodologies of chaos control are regularly proposed [Belbruno (2004)], [Bollt and Meiss (1995a)], [Bollt and Meiss (1995b)], [Macau (2000)], [Macau and Caldas (2002)], [Macau (2003)], aiming to improve the known methods and expand application possibilities.

The chaos control method proposed in 1990, and nowadays named as *OGY* after its idealizers, explores the fact that the chaotic invariant has an infinity number of different unstable periodic orbits embedded in it [Ott (1993)]. Due to transitivity, another property inherent to chaotic dynamic [Ott (1993)], the chaotic orbits visit the neighborhood of all of these unstable periodic orbits. When a chaotic trajectory becomes close enough to a chosen unstable periodic orbit, small perturbations can be applied to an accessible parameter such that system's evolution remain in the periodic orbit; consequently, the system is stabilized on a regular and periodic evolution. Without any appropriate perturbation, system's evolution could eventually become close to the periodic orbit but due to the ergodicity property it moves away from it. The method is supported by the fundamental property of chaotic dynamic, the sensitivity to initial conditions in a way that parameter perturbations required to stabilize a periodic orbit have a low amplitude, which implies a low cost energy in technical applications. Since there exist an infinite number of unstable periodic orbits embedded in the chaotic invariant set, each of them can be associated to a specific desirable operational characteristic of the system, given to it a high flexibility.

In practical terms, the main characteristic of OGY method is that it can be implemented using only data derived from time series. It means that no knowledge about system's equations is needed, but the possibility to acquire data needed. This fact has been confirmed through several experiments, such as the ones carried out by [Ditto *et al.* (1990)], [Garfinkel *et al.* (1992)], [Azevedo and Rezende (1991)], [Schiff *et al.* (1994)], [Bielawski *et al.* (1994)], [Meucci *et al.* (1994)], and many others. A special distinction can be given to work of Schiff and collaborators in 1994, that presents an in-vitro experiment with brain tissue, where the chaotic dynamic is characterized, unstable periodic orbits are identified, and OGY method is applied to stabilize the trajectory close to a periodic orbit using only data from time series to apply the method.

The present text gives a brief overview on chaos control and introduces the OGY method. It was structured to provide to readers basic concepts regarding chaotic dynamic phenomena necessary to understand OGY method. To present these concepts we make use of a simple nonlinear dynamical system. The chaotic behavior presented on it is characterized and its fundamental properties are depicted. Those who intend to plunge into this subject can find many reference, among which we can cite: [Belbruno (2004)], [Macau and Grebogi (2001)], [Boccaletti *et al.* (2000)].

2. Chaotic Dynamic

Chaotic Dynamic is part of a more extensive field of knowledge named *Dynamics*. The last one has its origins by the year 1600, when Newton introduced the differential equations, the laws of motion, and the law of universal gravitation that allowed him to explain Kepler's law for planetary motion [Newton (1687)]. With these contributions, Newton was able to analytically solve the two-body-problem, and determine the motion law of Earth trajectory around Sun, using the two-body gravitational interaction law proportional to the inverse of the squared distance between the two bodies. Generations of other mathematicians and physicians have tried in vain to use Newton's analytical methods and extend his results to the three-body-problem, comprehending for example, Earth, Sun and Moon. After years of great efforts, the researchers came to the conclusion that the *three-body-problem* has no analytical solution based on just elementary functions.

By the end of the 19th century, the famous French mathematician Henri Poincaré [Poincaré (1899)] revolutionized the methods applied for solving particularly the three-body-problem, but also differential equations in general. He proposed an analysis method emphasizing not quantitative but qualitative aspects of the solutions. In doing so, instead of looking for planetary trajectories as a function of the time, Poincaré wanted to answer the following question: "Is Solar System stable?", "Is there any Planet that can be out of Solar influence by a time interval?", "Can any planetary trajectory go to infinity?". Poincaré developed powerful geometric methods to be able to analyze these questions. These methods are responsible for a revolution in *Dynamics*, giving it the actual profile, and making it a essential tool for studying complex phenomena in Physics (flows turbulence, weather forecast, optics), Chemistry (chemical reactions), Biology (neurobiology, competition between species), Engineering (communication, cryptography), and so on. Moreover, Poincaré was the first one to conjecture the existence of *chaos*, in which a deterministic system exhibits an aperiodic motion extremely sensitive to initial conditions.

After Poincaré, the study on Dynamic focuses on the understanding of nonlinear oscillators and the application of them in physics and engineering; and the geometrical analysis methods proposed by Poincaré were improved by Birkhoff, Kolmogorov, Arnold, Moser, and others, propitiating a better comprehension of classical mechanics.

The accessibility to computers by the end of 50's decade represented a turning point in Dynamic. As so numerical experiments with differential equations could be carried on, making possible to analyze the way that such equations transform a set of initial conditions along the time. The numerical experiments done by the American meteorologist Edward Lorenz in 1963 [Lorenz (1963)], led to the discovery of *chaotic motion* over a *strange attractor*. By that time, Lorenz studied a simplified model of atmospheric convection, aiming to understand changes in meteorological conditions. In his analysis, Lorenz has experienced a type of solution that never converges to an equilibrium or periodic solution, but oscillates irregularly and aperiodically trapped in a region of the state space. Besides that, Lorenz realized the existence of *sensitivity to variations on initial conditions*, that is a *fundamental property* inherent to chaotic systems. Hence, starting the simulations from initial conditions slightly different, the temporal evolution would become totally different and uncorrelated over the time. Such discovery implies that the system Lorenz has studied is *inherently unpredictable*, that is, small error arising from measurement of atmospheric state variables are quickly amplified leading to predictions totally different from what happens in the real world. The final work of Lorenz introduces the simplest version of his system that could preserve the characteristics of chaotic evolution: It is a third-order system of nonlinear differential equations known in present days by *Lorenz Equations*,

$$\begin{cases} \dot{x} = \sigma(y - x), \\ \dot{y} = rx - y - xz, \\ \dot{z} = xy - bz, \end{cases} \quad (1)$$

whose state space evolution oscillates in a irregular form, never repeating a state, forming the Lorenz attractor with its fractal structure.

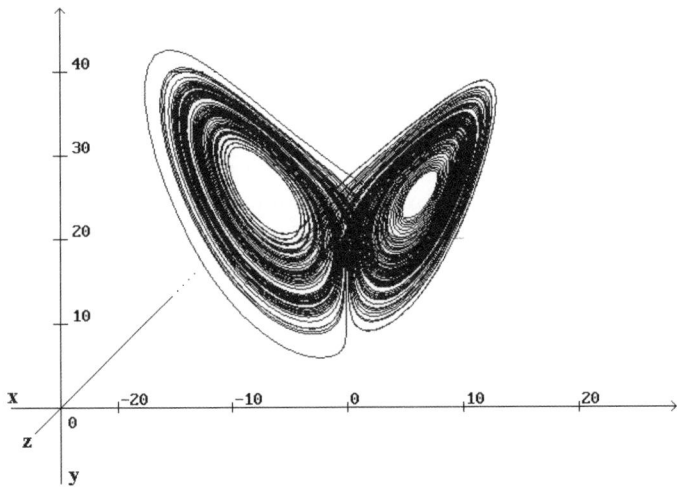

Fig. 1. **Lorenz Attractor**: Figure shows Lorenz attractor obtained when parameter values are $\sigma = 10$, $b = 8/3$, $r = 28$, and initial conditions are taken close to the origin.

Integrating the system (1) with parameter values $\sigma = 10$, $b = 8/3$, $r = 28$, starting from an initial condition close to the origin, it comes that the trajectory in the 3-dimensional space is confined into a structure that reminds a pair of butterfly wings. Such structure (set) is called *Lorenz chaotic attractor*, and has fractal geometry with fractal dimension approximately 2.05, see Fig. 1. Several other chaotic dynamical systems describe a fractal geometric figure in their state space, which is invariant, has an unique shape, and stays trapped in certain region in the state space. This invariant set is called *chaotic attractor*. We call *basin of attraction* of the chaotic attractor, the set of initial conditions whose trajectories converge into it.

The chaotic dynamic has as fundamental property the *sensitive dependence to variations in initial conditions* [Alligood *et al.* (1996)]. It means, as said previously, that trajectories starting from initial conditions slightly different, quickly deviate from each other over the time, becoming after all uncorrelated. Let $\mathbf{x}(\mathbf{t} = \mathbf{0})$ be a point of a trajectory at instant $t = 0$, and consider a neighbor point $(\mathbf{x}(\mathbf{t} = \mathbf{0}) + \delta_{t=0})$ very close to it. When the dynamic is chaotic, the distance between the trajectories, given by $\delta(t)$, increases exponentially in time, i.e., $\|\delta(t)\| \approx \|\delta_0\|e^{\lambda t}$, hence neighbor trajectories diverge exponentially in average.

In the Lorenz system, numerical analysis give a value $\lambda \approx 0.9$. This concept can be generalized as follows. Consider the time evolution of an "infinitesimal" sphere of initial conditions; The temporal evolution distorts such sphere that it becomes an ellipsoid. Let $\delta_k(t), k = 1,...n$ be the length of the main k-axis of the ellipsoid, considering a n-dimensional space. Therefore, in average, $\delta_k(t) \approx \delta_k(0)e^{\lambda t}$, where exponents λ_k are denominated *Lyapunov exponents* (for a rigorous definition, see [Alligood et al. (1996)]). Positive exponents result in increasing of ellipsoid's diameter during time evolution; In the case of Lorenz system, one of the exponents is positive with value equals to 0.9 for the parameter values considered in the example, while the other two components are negative.

3. Properties of Chaotic Dynamic

In this section, we display the key characteristics properties of a chaotic dynamical system. As a example, we consider a simple diode circuit whose nonlinearity are present in both the the diode model and the voltage source.

3.1. *The diode circuit*

Since the last two decades, electronic circuits that present chaotic behavior have been an important framework exploited in many technological applications in communication systems [Hayes et al. (1993); Baptista et al. (2000)], medical diagnostic devices [Witkowski et al. (1995)], image processing [Witkowski et al. (1995)], associative memory [Zhao and Macau (2001)] and others. Because of its handiness, electronic circuits are also a convenient stage to be used on experimental exploration regarding nonlinear and chaotic phenomena.

Let us here consider an electronic circuit composed of a sinusoidal voltage source V_{in}, a resistor R, an inductor L, and a diode D (a highly nonlinear element)[Rauch (1998); Prusha (1997)], as shown in Fig. 2.

According to Kirchhoff's voltage law, the voltage V_d across the diode is related with the input voltage generator V_{in} and the circuit current (I) by:

$$L\frac{dI}{dt} = V_0 \text{sen}(2\pi f t) - RI - V_d, \qquad (2)$$

where V_0 is the amplitude (in Volts) and f is the frequency (in Hertz) of the input voltage (V_{in}), I is the loop current.

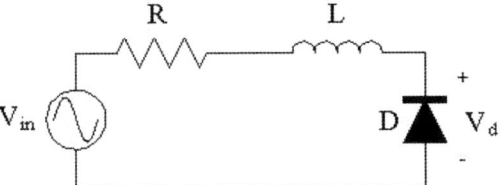

Fig. 2. **Diode Circuit**: the circuit is composed of sinusoidal voltage source V_{in}, a resistor R, an inductor L, and a diode D; where V_d is the voltage across the diode.

This equation (2) can be rewritten into the following set of equations:

$$\begin{cases} \frac{dq}{dt} = I, \\ L\frac{dI}{dt} = V_0\text{sen}(2\pi ft) - RI - V_d, \end{cases} \quad (3)$$

where q is the accumulated charge in diode junction.

The system model's equations (3) can still be converted into the following system of first-order autonomous differential equations:

$$\begin{cases} \frac{dq}{dt} = I, \\ L\frac{dI}{dt} = V_0\text{sen}(\theta) - RI - V_d, \\ \frac{d\theta}{dt} = 2\pi f. \end{cases} \quad (4)$$

Let us now take in consideration the nonlinear properties presented on diodes. In the literature [Pierce (1972); Gray and Searle (1977); Sedra and Smith (2000)], there are different diode models that capture nonlinear effects on diodes. These effects are correlated with the operational frequency regions in which the device operates. We consider here the high-frequency model that was proposed by Sedra, Pierce and others [Pierce (1972); Gray and Searle (1977); Sedra and Smith (2000)]. It is a piecewise-linear capacitances model, with a small offset voltage, as follows:

$$V_d = \frac{|q|(C_2 - C_1)}{2C_2C_1} + \frac{q(C_2 + C_1)}{2C_2C_1} + E_0, \quad (5)$$

where C_1 e C_2 are capacitances, E_0 is a constant voltage due to a minimum constant charge.

Note that for $q > 0$, $V_d = q/C_1 + E_0$, and for $q < 0$, $V_d = q/C_2 + E_0$. As so, from the model proposed by Sedra [Sedra and Smith (2000)], it comes that $C_2 = C_j$, and $C_1 = C_d/C_j \simeq C_d$ ($C_d \simeq 1000 \times C_j$), in which C_j and C_d are junction and diffusion capacitances respectively.

In doing so, we have defined a simple model for the diode, therefore the set of first-order autonomous differential equations for the system can be written as follows:

$$\begin{cases} \frac{dq}{dt} = I, \\ L\frac{dI}{dt} = V_0 \text{sen}(\theta) - RI - \left(\frac{|q|(C_j - C_d)}{2C_j C_d} + \frac{q(C_j + C_d)}{2C_j C_d} + E_0 \right), \\ \frac{d\theta}{dt} = 2\pi f. \end{cases} \quad (6)$$

3.1.1. *Diode circuit dynamics*

This simple diode circuit presented here can exhibit a very complex dynamic. For a detailed analysis we perform several numerical experiments and use appropriated tools for nonlinear analysis.

The numerical simulations were performed using a 4th-order Runge-Kutta integration method with a fixed step-size of 10^{-6} time units, setting a $0.18mH$ inductor (L), a $4.5ohm$ resistor (R), and a D1N1206C diode. The D1N1206C diode has, from the Electronics Workbench (circuit simulation software) model a $453pF$ diffusion capacitance (C_d), a $0.52V$ junction voltage (V_j), and an estimated $30nF$ junction capacitance (C_j).

We explore system's dynamic by analyzing its evolution under variation of its parameters, that can be called in this case *control parameter*. Although no analytical solution is obtained, it is possible to identify the main characteristics of the solutions and get a qualitatively understanding of the system behavior by observing its *state space* [Alligood et al. (1996); Devaney (1993, 1989); Ott (1993); Fieldler-Ferrara and Prado (1995)].

The set of values for (q, I, θ) is the system's *state space*. The state space shows the trajectory of the system in regime.

The results from numerical simulations revel the existence of orbits of different periods and even suggest "chaotic" behavior for some control parameter values. The *state space* is constructed by plotting the orbit (q, I, θ) in the space $(q \times I \times \theta)$. In the following figures we suppress the state θ by introducing a *Poincaré section* at $\theta = 2\pi$. The system under study has a periodic forcing term (V_{in}) with period 2π, therefore a section in $\theta = 2\pi$ becomes a natural choice in the $3-dimensional$ state space, i.e., the Poincaré section is obtained by sampling the trajectory in the $(q \times I)$ plane each time θ is equal to multiples of the forcing term period (T). Hence the flow analysis is reduced to a Poincaré map analysis. This map is used along this chapter to analyze the system (6).

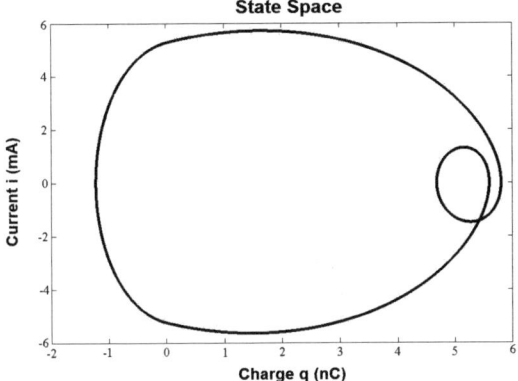

Fig. 3. **Period-2 Orbit**: State space showing the diode circuit period-2 orbit for parameter values $f = 333kHz$ and $V_0 = 0,8V$.

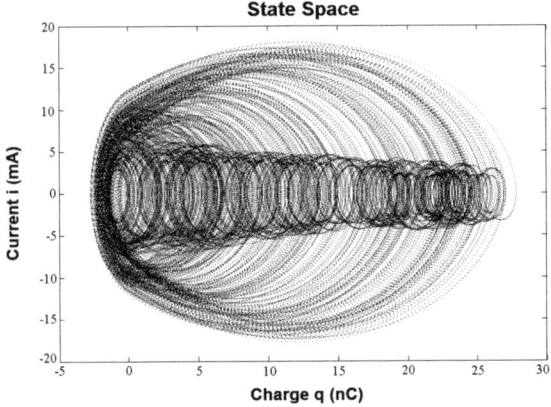

Fig. 4. **Chaotic Orbit**: State space showing the diode circuit chaotic orbit for parameter values $f = 333kHz$ and $V_0 = 2,3V$.

The curves in Figs. 3 and 4 are called *trajectories* or *orbits* [Fieldler-Ferrara and Prado (1995)].

Figures 3 and 4 show the so called *trajectories* (or *orbits*, [Fieldler-Ferrara and Prado (1995)]) for different values of the control parameter, however it does not allow to pin point properly how the transition among different behaviors occurs. The transition scenario can be better identified by using *Bifurcation Diagram*.

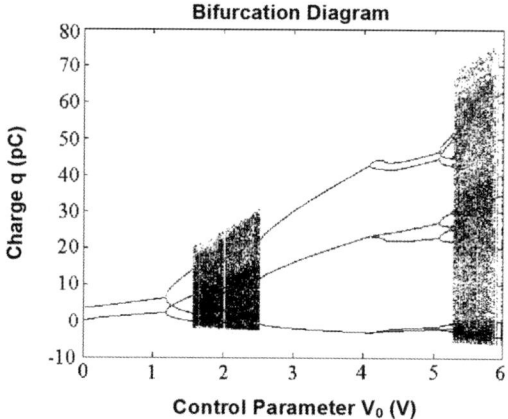

Fig. 5. **Bifurcation Diagram**: Figure shows the diode circuit bifurcation diagram obtained by stroboscopically sampling the state variable q, with a sampling period 2π (Poincaré section $\theta = 2\pi$), while control parameter V_0 is continuously varied.

Bifurcation means splitting into two parts. In the study of dynamical systems the term bifurcation is commonly used to describe any sudden change of system's behavior due to parameter variations. The term bifurcation refers to a split into two distinct regions: one above, and other below a critical parameter value for which a change of behavior occurs.

Bifurcation Diagram: Figure 5 shows the bifurcation diagram obtained by stroboscopically sampling the state variable q, with a sampling period 2π (Poincaré section $\theta = 2\pi$), while control parameter V_0 is continuously varied. The reason for choosing a sampling period of 2π is that the system's dynamic, if not chaotic, should present a periodic regime with a time period that is a multiple of the forcing term period time.

In the bifurcation diagram of Fig. 5 we can observe the existence of periodic orbits, and also dense orbits that may be chaotic ones. It is possible to identify two different *transition to chaos* in the present case. There exist a period-2 orbit that duplicates twice to give a period-8 orbit (it is not evident in the bifurcation diagram of Fig. 5). When $V_0 \sim 1,6008892V$, the period-8 orbit disappears due to a *saddle-node bifurcation*, and a chaotic orbit takes place. The second transition to chaos is through a *period-doubling cascade* in which the period-3 orbit goes through successive period-doubling till the chaotic regime. Table 1 shows the parameter values for each period-doubling.

Table 1. Period-doubling bifurcations.

Period - Doubled Period	V_0
3 - 6	4,117187
6 - 12	5,062500
12 - 24	5,227500
Chaos	5,260000

3.1.2. Chaotic dynamic: Fundamental properties

Definition 13.1. A dynamical system **F** is said to be chaotic if it presents the following properties [Devaney (1993)]:

(1) Sensitive dependence on initial conditions (impressibility);
(2) It is transitive on its invariant set;
(3) It has embedded on the chaotic invariant set infinite and numerable periodic orbits (unstable) of **F** any period.

Definition 13.2. *Sensitivity to initial conditions*: A dynamical system **F** is sensitively dependent on initial conditions if $\exists \beta > 0$ so that for $\forall \mathbf{x}$ and $\forall \epsilon > 0$, $\exists \mathbf{y}$ to a distance ϵ of **x** and a k so that distance between $\mathbf{F}^k(\mathbf{x})$ and $\mathbf{F}^k(\mathbf{y})$ is less than β [Devaney (1993)].

Figure 6 illustrates this property. The parameter values are $f = 330KHz$ and $V_0 = 5.3V$, for which the diode circuit looks like to present

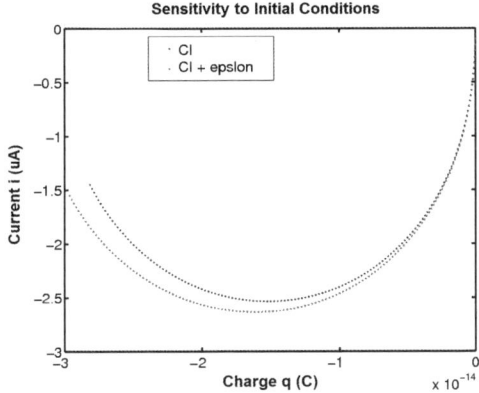

Fig. 6. **Sensitivity to Initial Conditions**: The orbits iterated from nearby initial conditions ϵ deviate exponentially in average from each other over the time iterations.

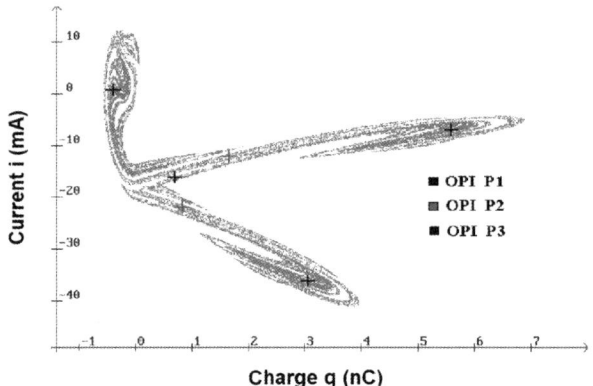

Fig. 7. **Periodic Dense Orbits**: Figure shows a chaotic attractor typical of the diode circuit for parameter values $f = 330KHz$ and $V_0 = 5, 3V$. It also illustrates the existence of three different unstable periodic orbits embedded in the chaotic attractor.

a chaotic regime. The distance between two different initial conditions is $\epsilon = 10^{-5}$. Observe that after 150 iterations these orbits that initiated in nearby points are now considerable distant from each other.

Definition 13.3. *Dense set*: Consider a set X and a subset Y of X. Y is dense in X if, and only if, $\forall x \epsilon X$, $\exists y \epsilon Y$ arbitrarily close. Equivalently, Y is dense in X if, and only if $\forall x \epsilon X$ exists a sequence $y_n \epsilon Y$ that converges to x [Devaney (1993)].

Figure 7 shows a chaotic attractor of the diode circuit, generated on the Poincaré section for parameter values $f = 330KHz$ and $V_0 = 5, 3V$. We illustrate the existence of three different unstable periodic orbits embedded in the chaotic attractor; However, besides the illustrated orbits, there are an infinite number of unstable periodic orbits, of different periods, embedded in the chaotic attractor.

Definition 13.4. *Transitivity*: a dynamical system **F** is transitive if, for any points x and y, and any $\epsilon > 0$, exist a third point z distant from x by $\epsilon' < \epsilon$, whose trajectory comes close to a distance $\epsilon'' < \epsilon$ from y [Devaney (1993)]; as shown in Fig. 8.

In other words, a transitive orbit starting from a point x inside the chaotic invariant set can come as close as we wish to any other point y of the chaotic invariant set after sufficient long time. In the case of the diode

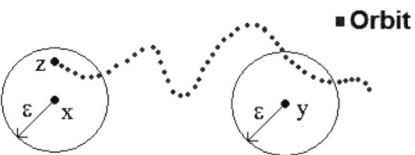

Fig. 8. **Transitivity**: The system is transitive if, for any points x and y, and any $\epsilon > 0$, exist a third point z distant from x by $\beta < \epsilon$, whose trajectory comes close to a distance $\epsilon' < \epsilon$ from y.

circuit, one can start at a point x far from the period-1 orbit by a distance β ($\beta \leq \epsilon$) and come close to the period-2 point $(0, 833936937021647, -21, 9162155422169)$ by a distance less than ϵ, as shown in Fig. 7. Setting $\beta = 10^{-7}$ and $\epsilon = 10^{-2}$, the time interval spent for leaving the $\beta - neighborhood$ of x and reach the $\epsilon - neighborhood$ of y is estimated to be 66871 interaction of the Poincaré map; for $\epsilon = 10^{-4}$, this time interval increases to 41671990 interactions.

In doing so, we can state that periodic orbits embedded in the chaotic attractor can be "reached" from any point located in the chaotic invariant set. Such characteristics makes feasible the OGY chaos control method.

4. Chaos Control

The concept of *chaos control* was coined and conceived in the last decade of the 20th century [Ott *et al.* (1990)]. Very recently, the original paper named "Controlling Chaos" [Ott *et al.* (1990)] that introduced this idea was considered to be as one of the most important and significant of the 20th century. In this article [Ott *et al.* (1990)], chaos control is presented as a method for stabilizing chaotic behavior. It wisely exploited the key properties presented on the chaotic dynamics. Basically, given a trajectory, it counts on the transitive property so that this trajectory came very close to a specific unstable periodic orbit. As the trajectory is sufficient close to this orbit, small (*low energy*) judiciously (*feedback method*) chosen perturbations are applied to keep the trajectory on the unstable periodic orbit. As so, the chaotic behavior is stabilized and a former "irregular" trajectory starts to behave "regularly" (periodically). As the chaotic invariant set presents infinite periodic orbits of any period, in theory infinite periodic motion can be achieved.

Let us in the next section focus in detail how the method works.

4.1. *OGY chaos control*

In dissipative chaotic systems, unstable periodic points (*UPO*) have associated with them directions along which trajectories converge or diverge to them, called *stable* and *unstable manifolds*, respectively [Alligood et al. (1996)]. On the chaotic attractor the UPOs have at least one *stable manifold* and one *unstable manifold*, i.e., they are saddle points. The dynamics in the neighborhood of a UPO can be approximated by a linearized local map described by its *Jacobian matrix* calculated on the UPO. The Jacobian *eigenvalues* quantify the local amplitude scale, while its *eigenvectors* determine the local directions of the manifolds.

The *OGY method* [Ott et al. (1990)] strategy takes advantage of the UPOs to stabilize chaotic evolutions. To accomplish this, two different steps are needed: learning and control. The first one involves three parts: to define a proper return map, to set of the control point, and to carry out sensitivity analysis.

The first part of the learning step consists in specifying the *return map* for the chaotic system [Baker (1995); Shimazaki (2001); Ditto et al. (1995)].

Next, the return map is used to localize the *control point*. A control point is the unstable periodic point about it the system will be stabilized so that it provides a desirable behavior for the system. Once the control point is assigned, one determines its local dynamics by estimating its Jacobian matrix.

After all, one analyzes the effect of small perturbations over the control parameter. This analysis is carrying out with the systems local dynamics in the neighborhood of the UPO.

In the control step, we calculate the small perturbations in the control parameter necessary to control the system, and apply it to the system when a trajectory comes close to the control point. To control a chaotic system, it is necessary to confine its iterations in the Poincaré section to a small neighborhood of the control point. When an iteration is close to this neighborhood, the value of control parameter (pc) is perturbed from its nominal value (pc_0) by a value δpc, "changing" the location of the control point and its manifolds. The perturbation δpc is chosen such that the next iteration of the return map lays on the stable manifold of the original control point (control point for $pc = pc_0$). The method "forces" a point in the Poincaré section to stay in the neighborhood of a unstable periodic point, differing it from other methods [Baker (1995); Shimazaki (2001); Ditto et al. (1995)].

The next sections apply the method to the system under study (diode circuit).

4.1.1. *Finding a fixed point*

Fixed points, and consequently periodic orbits, can be determined with a little modification in the algorithm to generate a Poincaré section: we calculate the distance between two consecutive points in the Poincaré section, say (q_n, I_n) and (q_{n+1}, I_{n+1}), and compare it with a predefined distance $\epsilon \ll 1$. If the calculated distance is less than ϵ, it indicates that trajectory is very close to a fixed point, say (q_F, I_F) [Baker (1995)].

This procedure can be easily modified to finding periodic orbits of any period. A period-k orbit can be determined by checking the distance between (q_n, I_n) and (q_{n+k}, I_{n+k}) [Baker (1995)], for example.

4.1.2. *Analyzing the effect of perturbations*

Next step consists in analyzing how perturbations in control parameter affect fixed point coordinates. Close to (q_F, I_F), a small perturbation δV_0 (remember V_0 is the diode circuit control parameter) of V_0 results in a "new" fixed point (q'_F, I'_F) approximately given by:

$$\begin{pmatrix} q'_F \\ I'_F \end{pmatrix} \approx \begin{pmatrix} q_F \\ I_F \end{pmatrix} + \delta V_0 \begin{pmatrix} \partial q_F / \partial V_0 \\ \partial I_F / \partial V_0 \end{pmatrix}. \tag{7}$$

The vector value $(\partial q_F/\partial V_0, \partial I_F/\partial V_0)$, can be estimated by determining the changes on the fixed points coordinates after small perturbations are applied on V_0. Despite some dispersions, linear approximation can be used to determine this vector value, so that the slopes of the curves $q_F \times V_0$ and $I_F \times V_0$, adjusted by minimum-square method, can provide a good approximation for the vector value $(\partial q_F/\partial V_0, \partial I_F/\partial V_0)$ [Baker (1995); Shimazaki (2001); Ditto et al. (1995)]. Small magnitude variations are tolerated by the control mechanism [Baker (1995)].

4.1.3. *M transformation*

Generally, *M transformation* is a nonlinear mapping of (q_n, I_n) into (q_{n+1}, I_{n+1}). However, in small regions of the state space, it can be approximated by a linear map. Close to the fixed point, the map can be written as:

$$\begin{pmatrix} q_{n+1} \\ I_{n+1} \end{pmatrix} \approx \begin{pmatrix} q_F \\ I_F \end{pmatrix} + M \begin{pmatrix} q_n - q_F \\ I_n - I_F \end{pmatrix}, \qquad (8)$$

where M is a transformation matrix 2×2 in this case.

Contracting the notation, define:

$$\begin{pmatrix} \Delta q_n \\ \Delta I_n \end{pmatrix} \equiv \begin{pmatrix} q_n - q_F \\ I_n - I_F \end{pmatrix} \text{ and } \begin{pmatrix} \Delta q_{n+1} \\ \Delta I_{n+1} \end{pmatrix} \equiv \begin{pmatrix} q_{n+1} - q_F \\ I_{n+1} - I_F \end{pmatrix}. \qquad (9)$$

Hence, Eq. (8) can be written as:

$$\begin{pmatrix} \Delta q_{n+1} \\ \Delta I_{n+1} \end{pmatrix} \equiv M \begin{pmatrix} \Delta q_n \\ \Delta I_n \end{pmatrix}. \qquad (10)$$

As said previously, there are directions associated to the fixed point along which trajectories come close to it: for forward iterations, in the case of stable manifold; For backward iterations, in the case of unstable manifold. In a neighborhood of the fixed point, vectors in the same direction of the manifolds does not change direction after M transformation, and consequently, neither eigenvalues of M.

The local map M can be determined from system's evolution starting from n initial conditions in the neighborhood of the fixed point, for one iteration period. The following equations show how to determine M:

For one initial condition $(\Delta q_n, \Delta I_n)$, one can write:

$$\begin{pmatrix} \Delta q_{n+1} \\ \Delta I_{n+1} \end{pmatrix} = M \begin{pmatrix} \Delta q_n \\ \Delta I_n \end{pmatrix}, \qquad (11)$$

equivalently,

$$\begin{aligned} \Delta q_{n+1} &= m_{11} \Delta q_n + m_{12} \Delta I_n, \\ \Delta I_{n+1} &= m_{21} \Delta q_n + m_{22} \Delta I_n. \end{aligned} \qquad (12)$$

Repeating it for n initial conditions and grouping equations, follows:

$$(\Delta q_{n+1} \; \Delta I_{n+1})_{nX2} = (\Delta q_n \; \Delta I_n)_{nX2} \, \Theta, \qquad (13)$$

$$\left[(\Delta q_n \; \Delta I_n)^T (\Delta q_n \; \Delta I_n) \right]^{-1} (\Delta q_n \; \Delta I_n)^T (\Delta q_{n+1} \; \Delta I_{n+1}) = \Theta, \qquad (14)$$

$$M = \Theta^T. \qquad (15)$$

The *eigenvalues* and *eigenvector* associated to matrix M are respectively λ_u and λ_s, e_u and e_s.

Vectors f_u and f_s perpendicular to e_s and e_u respectively, can be calculated by the relation:

$$f_u \cdot e_s = 0 \text{ and } f_s \cdot e_u = 0, \tag{16}$$

and, for normalization,

$$f_u \cdot e_u = 1 \text{ and } f_s \cdot e_s = 1. \tag{17}$$

To establish the control algorithm, we also utilize the fact that M map can be written as a combination of the quantities mentioned above as follows:

$$M = \lambda_u e_u \cdot f_u + \lambda_s e_s \cdot f_s. \tag{18}$$

4.1.4. *Control algorithm*

The control objective is to make system's trajectories convergent to a fixed point (periodic point) in the Poincaré section. It can be achieved by adjusting the control parameter in order to move the intersection point (of the trajectory in the Poincaré section) closer to the stable manifold of the fixed point (periodic point). Once close enough to the stable manifold, the intersection point is naturally attracted to it, and so is convergent to the fixed point [Ott *et al.* (1990); Boccaletti *et al.* (2000)]

The fixed point (q_F, I_F) is localized in the intersection between stable and unstable manifold. If V_0 is perturbed by δV_0, the fixed point is moved to a new position (q'_F, I'_F), such that:

$$\begin{pmatrix} q'_F \\ I'_F \end{pmatrix} = \begin{pmatrix} q_F \\ I_F \end{pmatrix} + \delta V_0 \begin{pmatrix} \partial q_F / \partial V_0 \\ \partial I_F / \partial V_0 \end{pmatrix}. \tag{19}$$

Adding the perturbation to V_0, the point becomes (q'_{n+1}, I'_{n+1}):

$$\begin{pmatrix} q'_{n+1} \\ I'_{n+1} \end{pmatrix} = \begin{pmatrix} q'_F \\ I'_F \end{pmatrix} + M \begin{pmatrix} \Delta' q_n \\ \Delta' I_n \end{pmatrix}, \text{ where } \begin{pmatrix} \Delta' q_n \\ \Delta' I_n \end{pmatrix} = \begin{pmatrix} q_n - q'_F \\ I_n - I'_F \end{pmatrix}. \tag{20}$$

The new point (q'_{n+1}, I'_{n+1}) results from the application of map M over (q_n, I_n), with control parameter being $V_0 + \delta V_0$. The vector from the original fixed point (q_F, I_F) to the new fixed point (q'_{n+1}, I'_{n+1}) is denoted by:

$$\begin{pmatrix} \Delta q'_{n+1} \\ \Delta I'_{n+1} \end{pmatrix} = \begin{pmatrix} q'_{n+1} - q_F \\ I'_{n+1} - I_F \end{pmatrix}. \tag{21}$$

Finally, including the perturbation δV_0 to the control parameter in the map M, the vector (21) can be written as:

$$\begin{pmatrix} \Delta q'_{n+1} \\ \Delta I'_{n+1} \end{pmatrix} = \begin{pmatrix} q'_F \\ I'_F \end{pmatrix} - \begin{pmatrix} q_F \\ I_F \end{pmatrix} + M \begin{pmatrix} q_n - q'_F \\ I_n - I'_F \end{pmatrix} \quad (22)$$

and finally, after appropriate substitutions, as:

$$\begin{pmatrix} \Delta q'_{n+1} \\ \Delta I'_{n+1} \end{pmatrix} = \begin{pmatrix} \partial q_F/\partial V_0 \\ \partial I_F/\partial V_0 \end{pmatrix} \delta V_0 + M \left[\begin{pmatrix} \Delta q_n \\ \Delta I_n \end{pmatrix} - \begin{pmatrix} \partial q_F/\partial V_0 \\ \partial I_F/\partial V_0 \end{pmatrix} \delta V_0 \right]. \quad (23)$$

Using M in its expanded form (18), the vector (21) can still be written as:

$$\begin{pmatrix} \Delta q'_{n+1} \\ \Delta I'_{n+1} \end{pmatrix} = \begin{pmatrix} \partial q_F/\partial V_0 \\ \partial I_F/\partial V_0 \end{pmatrix} \delta V_0 +$$

$$+ (\lambda_u e_u \cdot f_u + \lambda_s e_s \cdot f_s) \cdot \left[\begin{pmatrix} \Delta q_n \\ \Delta I_n \end{pmatrix} - \begin{pmatrix} \partial q_F/\partial V_0 \\ \partial I_F/\partial V_0 \end{pmatrix} \delta V_0 \right]. \quad (24)$$

The control objective is to move the point (q'_{n+1}, I'_{n+1}) close to the stable manifold of the fixed point, and consequently, the vector $(\Delta q_{n+1}, \Delta I_{n+1})$ has to be aligned with the stable manifold. It means δV_0 has to be such that:

$$f_u \cdot \begin{pmatrix} \Delta q'_{n+1} \\ \Delta I'_{n+1} \end{pmatrix} = 0. \quad (25)$$

When this result is combined with Eq. (24), the condition (25) leads to the final result for the OGY method:

$$\delta V_0 = \frac{\lambda_u}{\lambda_u - 1} \cdot \frac{f_u \cdot \begin{pmatrix} \Delta q_n \\ \Delta I_n \end{pmatrix}}{f_u \cdot \begin{pmatrix} \partial q/\partial V_0 \\ \partial I/\partial V_0 \end{pmatrix}}. \quad (26)$$

The expression (26) determines the accurate perturbation δV_0 necessary to alter the trajectory so that the next intersection point in Poincaré section is close enough to the stable manifold of the fixed point (q_F, I_F). The expression (26) is used to calculate the perturbation value δV_0 at each integration period (Poincaré section cycle).

Figure 9 gives a geometrical interpretation of the OGY method. Let a nonlinear map $F(X)$ whose state X_n in the instant n is illustrated by X in the figure, and a fixed point P of the map. In the next iteration of the map, at instant $(n+1)$, the state would be X_{n+1}, illustrated by a circle,

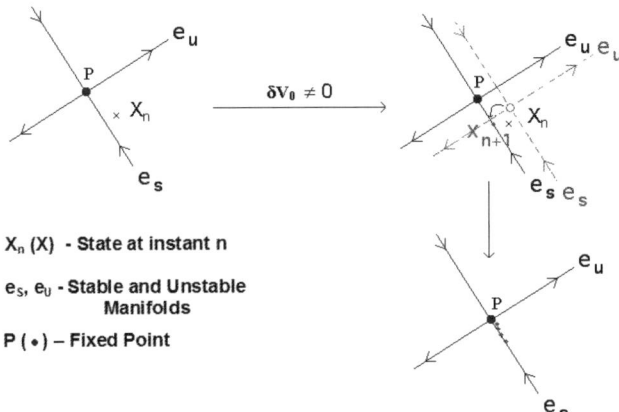

Fig. 9. **OGY Control - Geometric Interpretation**: Let a nonlinear map $F(X)$ whose state X_n in the instant n is illustrated by X in the figure, and a fixed point P of the map. In the next iteration of the map, at instant $(n+1)$, the state would be X_{n+1}, illustrated by a circle, which stable and unstable manifolds are represented by the sectioned lines. However, by applying the OGY method, with an appropriate perturbation δpc, X_{n+1} can lay on the stable manifold of the fixed point P such that, in the further iterations, the trajectory converges to the fixed point.

which stable and unstable manifolds are represented by the sectioned lines. However, by applying the OGY method, with an appropriate perturbation δpc, X_{n+1} can lay on the stable manifold of the fixed point P such that, in the further iterations, the trajectory converges to the fixed point.

5. Application

Embedded in the chaotic attractor there are an infinite number of unstable periodic orbits. It is possible to identify with some precision the one that we wish to use as control set point. The local dynamics of this UPO can be determined and the OGY method used to stabilize the system on this UPO.

Let us applied this procedure on the diode circuit. For the following $f = 333 KHz$ and $V_0 = 2.3V$ it presents a chaotic behavior. Starting from an appropriate (coherent scale and close to the attractor) initial condition, we start to integrate the system. The transitivity property guarantees that in some iteration the trajectory will become as close as we wish to a known fixed point (periodic point in the Poincaré section), at this instant, OGY control is activated.

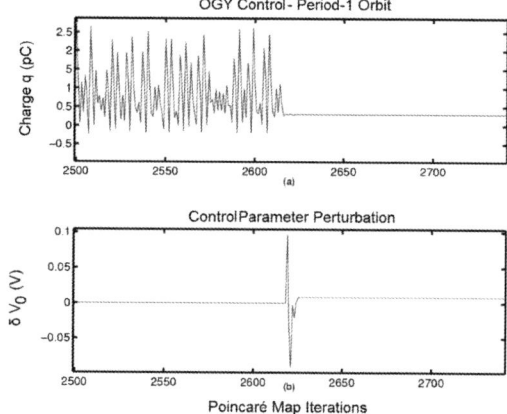

Fig. 10. **OGY Control - Period-1**: (a) shows the sequence of iterations (in the Poincaré section) of state variable q; the trajectory initially chaotic becomes periodic-1 by the action of OGY control; (b) shows the value of the control parameter V_0 after OGY control is activated; observe there is a transition time where the perturbation amplitudes are higher, followed by a controlled regime where perturbation amplitudes are constant and low.

The OGY method can be easier verified with a period-1 point. Let us consider as an example the period-1 point $P1 : (0, 302974, -7, 261744)$. Then we set a circle with ray $r = 0,01$ about the $P1$ point; once the distance between the trajectory and the $P1$ point becomes less than r, the OGY control is activated. Figure 10(a) shows the sequence of iterations (in the Poincaré section) of state variable q; the trajectory initially chaotic becomes periodic-1 by the action of OGY control. Figure 10(b) shows the value of the control parameter V_0 after OGY control is activated. Note that there is a transition time in which the perturbation amplitudes are higher, followed by a controlled regime in which the perturbation amplitudes are constant and low.

The time evolution behavior can be observed by plotting the time series of the state variables. Figures 11(a) and 11(b) show, respectively, time series for charge q and current i. The trajectories, initially chaotic, give place to an orbit very close to the period-1 orbit. It would correspond exactly to the period-1 orbit if the intersection with the Poincaré section could exactly match the period-1 point $P1$.

Also in the time domain, we can analyze the behavior of the state variable by means of state space $q \times i$ (suppressing the state relative to time evolution, θ), see Fig. 12. The dark curve represents the chaotic trajectory;

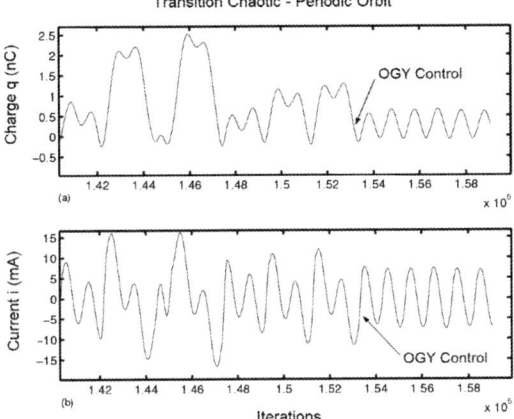

Fig. 11. **OGY Control - State Variables**: (a) and (b) show, respectively, time series of charge q and current i. The trajectories initially chaotic give place to an orbit very close to the period-1 orbit.

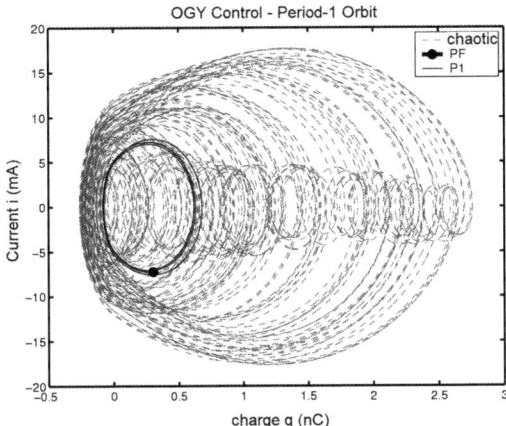

Fig. 12. **OGY Control - State Space**: dark curve represents the chaotic trajectory; once it is to a distance less than r from the point $P1$ (black circle), OGY is activated and the period-1 orbit (pale gray curve) is controlled.

at the moment the trajectory is at a distance less than a pre-set r (region represented by the black circle) from the point $P1$, the OGY control is activated and the period-1 orbit (pale gray curve) takes place. Note that the pale gray period-1 close trajectory does not remain perfectly in a period-1

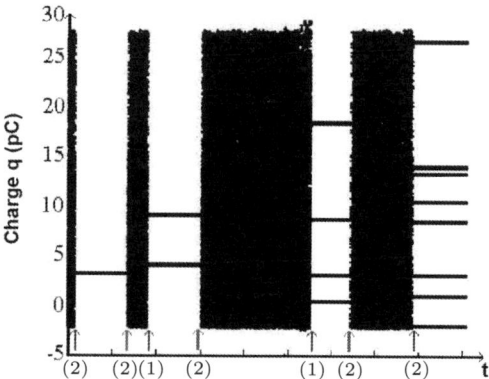

Fig. 13. **Chaos Control - OGY Method**: Figure shows system's trajectory in its chaotic transient followed by a controlled regime of different periodic orbits (period-1, 2, 4 and 8, respectively). The arrows (2) indicate the instant control is activated; the arrows (1) indicate the instant control is deactivated and trajectories are again chaotic.

cycle, it happens for two reasons: First, we have chosen $r = 0,01$ around the period-1 point $P1$, that might be considered too big in many cases. Second, the perturbation to the control parameter is adjusted just once each integration period (when the trajectory intersects Poincaré section), in some cases it might be necessary to adjust the perturbation in subperiodic instants.

In theory, the OGY method can be applied to any periodic orbit. Figure 13 shows system's trajectory in its chaotic transient followed by a controlled regime of different periodic orbits (period-1, 2, 4 and 8, respectively). The blue arrows indicate the instant distance between trajectory and fixed point (periodic point) becomes less than the pre-set value r, and the control is activated; the red arrows indicate the instant control is deactivated and trajectories are again chaotic.

OGY control method has been a subject of greatly interest in nonlinear dynamics field, mainly because it is one of the simplest but powerful existent methods, and it is applicable to a variety of dynamical systems.

References

Alligood, K. T., Sauer, T. and Yorke, J. A. (1996). *Chaos: An introduction to Dynamical Systems* (Springer-Verlag).

Arecchi, F., Meucci, R., Puccioni, G. and Tredicce, J. (1982). Experimental evidence of subharmonic bifurcations, multistability, and turbulence in a q-switched gas laser, *Phys. Rev. Lett.* **49**, pp. 1217–1220.

Azevedo, A. and Rezende, S. (1991). Controlling chaos in spin-wave instabilities, *Phys. Rev. Lett.* **66**, pp. 1342–1345.

Baker, G. (1995). Control of the chaotic driven pendulum, *Am. J. of Physics* **63**, 9, pp. 832–838.

Baptista, M., Macau, E., Grebogi, C., Lai, Y. and Rosa, E. (2000). Integrated chaotic communication scheme, *Phys. Rev. E* **62**, p. 4835.

Belbruno, E. (2004). *Capture Dynamics and Chaotic Motions In Celestial Mechanics* (Princeton).

Bielawski, S., Derozier, D. and Glorieux, P. (1994). Controlling unstable periodic orbits by a delayed continuous feedback, *Phys. Rev. E* **49**, pp. 971–974.

Boccaletti, S., Grebogi, C., Lai, Y., Mancini, H. and Maza, D. (2000). The control of chaos: Theory and applications, *Physics Reports* **329**, pp. 103–197.

Bollt, E. and Meiss, J. (1995a). Controlling chaotic transport through recurrence, *Physica D* **81**, pp. 280–294.

Bollt, E. and Meiss, J. (1995b). Targeting chaotic orbits to the moon through recurrence, *Phys. Rev. A* **204**, pp. 373–378.

Devaney, R. (1989). *An Introduction to Chaotic Dynamical Systems* (Persus Books).

Devaney, R. (1993). *A First Course in Chaotic Dynamical Systems* (Addisson-Wesley Publishing Company).

Ditto, W., Rauseo, S. and Spano, M. (1990). Experimental control of chaos, *Phys. Rev. Lett.* **65**, pp. 3211–3214.

Ditto, W., Spano, M. and Linder, J. (1995). Techniques for the control of chaos, *Physica D* **86**, pp. 198–211.

Fieldler-Ferrara, N. and Prado, C. (1995). *Caos, Uma Introduo* (Ed. Edgard Blücher).

Garfinkel, A., Spano, M., Ditto, W. and Weiss, J. (1992). Controlling cardiac chaos, *Science* **257**, pp. 1230–1233.

Gray, P. and Searle, C. (1977). *Princípios de Eletrônica* (Livros Técnicos e Científicos Ed. S.A).

Hayes, S., Grebogi, C. and Ott, E. (1993). Communicating with chaos, *Phys. Rev. Lett.* **70**, p. 3031.

Hudson, J. and Mankin, J. (1981). Chaos in belousov-zhabotinsky reaction, *J. Chem. Phys.* **74**, pp. 6171–6177.

Lorenz, E. N. (1963). Deterministic nonperiodic flow, *J. Atmos. Sci.* **20**, pp. 130–141.

Macau, E. (2000). Using chaos to guide a spacecraft to the moon, *Acta Astronautica* **47**, pp. 871–878.

Macau, E. (2003). Exploiting unstable periodic orbits of a chaotic invariant set for spacecraft control, *Celes. Mech. Dyn. Sys.* **87**, pp. 291–305.

Macau, E. and Caldas, I. (2002). Driving trajectories in chaotic scattering, *Phys. Rev. E* **65**, p. 2621.

Macau, E. and Grebogi, C. (2001). Driving trajecories in chaotic systems, *Int. J. Bifurcat. Chaos* **11**, pp. 1423–1442.

Meucci, R., Gadomski, W., Ciofini, M. and Arecchi, F. (1994). Experimental control of chaos by means of weak parametric perturbations. *Phys. Rev. E* **49**, pp. 2528–2531.

Newton, I. (1687). Principia.

Ott, E. (1993). *Chaos in Dynamical Systems* (Cambrige University Press).

Ott, E., Grebogi, C. and Yorke, J. (1990). Controlling chaos, *Phys. Rev. Letters* **64**, 11, pp. 1196–1199.

Pierce, J. (1972). *Dispositivos de Juno Semicondutores* (Ed. USP).

Poincaré, H. (1899). Les methodes nouveles de la mécanique celeste, *Gauthier-Villars*.

Prusha, B. (1997). Measuring feigenbaum's in a bifurcating eletric circuit, *The College of Wooster*.

Rauch, A. (1998). Chaos in a driven nonlinear electrical oscillator: Determining feigenbaum's delta, *The College of Wooster*.

Schiff, S., Jerger, K., Duong, D., Chang, T., Spano, M. and Ditto, W. (1994). Controlling chaos in brain, *Nature* **370**, pp. 615–620.

Sedra, A. and Smith, A. (2000). *Microeletronics* (Ed. Makron Books).

Shimazaki, H. (2001). Topics in chaos control theory.

Witkowski, F., Kavanagh, K., Penkoske, P., Plonsey, R., Spano, M., Ditto, W. and Kaplan, D. (1995). Evidence for determinism in ventricular fibrilation, *Phys. Rev. Lett.* **75**, p. 1230.

Zhao, L. and Macau, E. (2001). A network of dynamically coupled chaotic maps for scene segmentation, *IEEE T.Neural Network* **12**, p. 1375.

Chapter 14

CHAOS STABILIZATION IN THE THREE BODY PROBLEM

Arsen R. Dzhanoev[1,*], Alexander Loskutov[1],
James E. Howard[2] and Miguel A. F. Sanjuán[3]

[1] *Physics Faculty, Moscow State University,
119992 Moscow, Russia*

[2] *Laboratory for Atmospheric and Space Physics and
Center for Integrated Plasma Studies University of Colorado,
Boulder CO 80309 USA*

[3] *Departamento de Fisica, Universidad Rey Juan Carlos,
28933 Mostoles, Madrid, Spain*

*janoev@polly.phys.msu.ru

A new type of orbit in the three-body problem is constructed. It is analytically shown that along with the well known chaotic and regular orbits in the three-body problem there also exists a qualitatively different type of orbit which we call "stabilized". The stabilized orbits are a result of additional orbiting bodies that are placed close to the triangular Lagrange points. The results are well confirmed by numerical orbit calculations.

1. Introduction

The three-body problem appears in practically all fields of contemporary physics from studies on microscopic systems to macroscopic ones: quantum mechanics [1], ionic oscillations [2], protein-folding [3], planetary systems formation [4] etc. The problem considers three particles of mass m_i with positions $\mathbf{r}_i$ which are each moving under an attractive force from all the other bodies where particle index $i = 1, 2, 3$. The system is characterized by a set of differential equations

$$m_i \ddot{\mathbf{r}}_i = \sum_{j=1, i \neq j}^{3} \frac{\gamma m_i m_j (\mathbf{r}_j - \mathbf{r}_i)}{|\mathbf{r}_i - \mathbf{r}_j|^3}, \qquad (1)$$

where γ is the gravitational constant that, by appropriately choosing dimensions, could be ignored by setting it equal to unity. These equations define the phase flow in an 18 dimensional phase space. Exploiting the symmetry afforded by (1) leads to a 12 dimensional phase space with 10 integrals of motion. These are the only known integrals. Henri Poincaré suggested that highly complex behavior could occur in the the three-body problem. It is reasonable, if we cannot find a general solution, to examine special solutions and particular features (see for example [5]). A natural starting place is the restricted three-body problem. In the restricted three-body problem, m_3 is taken to be small enough so that it does not influence the motion of m_1 and m_2 (called primaries), which are assumed to be in circular orbits about their center of mass. For the restricted three-body problem it was analytically verified that the complex behavior is due to the existence of transverse heteroclinic points. A well-known example of the chaoticity of the restricted three-body problem is the Sitnikov problem.

2. The Sitnikov Problem

The Sitnikov problem consists of two equal masses M (primaries) moving in circular or elliptic orbits about their common center of mass and a third test mass μ moving along the straight line passing through the center of mass normal to the orbital plane of the primaries (see Fig. 1).

The circular problem was considered first by McMillan in 1913 [6]. He found the exact solution of the equations of motion when the eccentricity of the primaries $e = 0$ and showed that it can be expressed in terms of elliptic integrals. Detailed discussion on this case has been done by Stumpff [7]. This problem became important when Sitnikov [8] in 1960 investigated the elliptic case of $e > 0$ and proved the possibility of the existence of oscillatory motions which were earlier predicted by Chazy in 1922–32. Alekseev [9] in 1968–69 proved that in the Sitnikov problem all of the possible combinations of final motions in the sense of Chazy are realized. Later in 1973 the alternative proof of the Alexeev results was done by Moser [10]. Since then the Sitnikov problem has attracted the attention of many other authors. Here we mention some of them. An interesting work in a qualitative way was carried out by Llibre and Simó [11] in 1980 and later by C.Marchal [12] in 1990. Hagel [13] derived an approximate solution of the differential equation of motion for particle μ by using a Hamiltonian in action angle variables. An explicit numerical study of the great variety of possible structures in phase space for the Sitnikov problem has been done by Dvorak [14]. Using

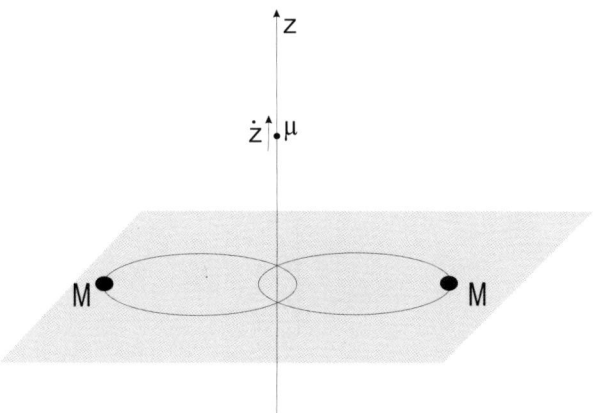

Fig. 1. Geometry of the Sitnikov problem.

Melnikov's method H.Dankowicz and Ph.Holmes [15] were able to show the existence of transverse homoclinic orbits. They proved that for any but a finite number of values of the eccentricity e the system is non-integrable, chaotic.

The main objective of this Chapter is to show through analytical and numerical methods the existence of stabilized orbits in this special restricted three-body problem and consequently, in the general three-body problem. The equation of motion can be written, in scaled coordinates and time as

$$\ddot{z} + \frac{z}{[\rho(t)^2 + z^2]^{3/2}} = 0, \qquad (2)$$

where z denotes the position of the particle μ along the z-axis and $\rho(t) = 1 + e\cos(t) + O(e^2)$ is the distance of one primary body from the center of mass. Here we see that the system (2) depends only on the eccentricity, e, which we shall assume to be small.

We first consider the circular Sitnikov problem i.e. when $e = 0$, for which

$$H = \frac{1}{2}v^2 - \frac{1}{\sqrt{1+z^2}} \qquad (3)$$

$$v = \dot{z}.$$

The level curves $H = h$, where $h \in [-2, +\infty)$, partition the phase space (v, z) into qualitatively different types of orbits as shown in Fig. 2. We are interested in solutions that correspond to the level curves $H = 0$, namely two parabolic orbits that separate elliptic and hyperbolic orbits and can be

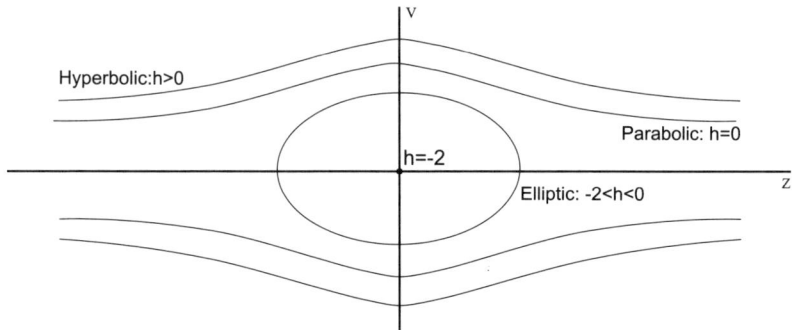

Fig. 2. Phase portrait for the unperturbed Sitnikov problem.

considered as a separatrix between these two classes of behavior. Further from Eq. (2) it is easy to show that there are just two fixed points: (0,0) — the center at the origin and $(\pm\infty, 0)$. Making use of the McGehee transformation [15] the fixed points at $(\pm\infty, 0)$ correspond to hyperbolic saddle points. Then taking into account that parabolic orbits act as connections or heteroclinic orbits between these two fixed points one may conclude that the stable and unstable manifolds of saddles correspond to the parabolic orbits.

To make clear how this problem is related to heteroclinic orbits, let us employ the non-canonical transformation [15]:

$$z = \tan u, \qquad v = \dot{z}, \qquad u \in \left[-\frac{\pi}{2}, \frac{\pi}{2}\right], v \in R. \qquad (4)$$

Then the Hamiltonian for the equation (2) in the new variables (u,v) has the form:

$$H(u,v) = \frac{1}{2}v^2 - \frac{1}{\rho(t)^2 + \tan^2 u} = \\ H_0(u,v) + eH_1(u,v,t,e), \qquad (5)$$

where $H_0(u,v) = \frac{1}{2}v^2 - \cos u$. One can see that when $e = 0$ the form of the Hamiltonian that obtained after the non-canonical transformation exhibits the pendulum character of motion.

Based on this connection between the dynamics of the nonlinear pendulum and the Sitnikov problem one can show that if $e \in (0,1)$ then for all but a possibly finite number of values of e in any bounded region, the system (2) is chaotic [15]. In this work we consider only small values of e. Hence due to the KAM-theory [16], since our system has 3/2 degrees of freedom the invariant tori bound the phase space and chaotic motion is finite and takes place in a small vicinity of a separatrix layer.

3. Stabilization of Chaotic Behavior in the Vicinity of a Separatrix

As mentioned in the introduction, our analysis is directed to the stabilization of this chaotic behavior in the elliptic Sitnikov problem. In general, this problem is related to the stabilization and control of unstable and chaotic behavior of dynamical systems by external forces. Since there are situations for which chaotic behavior might be undesirable, different methods have been developed in the past years to suppress or control chaos. The idea that chaos may be suppressed goes back to the publications [17, 18] where it has been proposed to perturb periodically the system parameters. The method of controlling chaos has been introduced in the paper [19] (the history of this question see in review [20]). A comprehensive study of chaotic systems with external controls was done in [21, 22]. Further we will give a brief review of these results. In this section we apply the Melnikov method, which gives a criterion of the chaos appearance, to the analysis of the system behavior under external perturbations. The idea is that such an approach can give us an analytical expression of the perturbations which leads to the chaos suppression phenomenon.

We explain the idea by using a general two-dimensional dynamical system subjected to a time-periodic external perturbation, and consequently possessing a three dimensional phase space.

3.1. *Melnikov function*

It is well known that in Hamiltonian systems, separatrices can split. In this case stable and unstable manifolds of a hyperbolic point do not coincide, but intersect each other in an infinite number of homoclinic points (usually the motion in the $(n+1)$–dimensional phase space $(x_1, \cdots, x_n, t)$ is considered in the projection onto a n–dimensional hypersurface $t = $ const (Poincaré section)). The presence of such points gives us a criterion for the observation of chaos. This criterion can conveniently be obtained by the *Melnikov function* (MF), which "measures" (in the first order of a small perturbation parameter) the distance between stable and unstable manifolds.

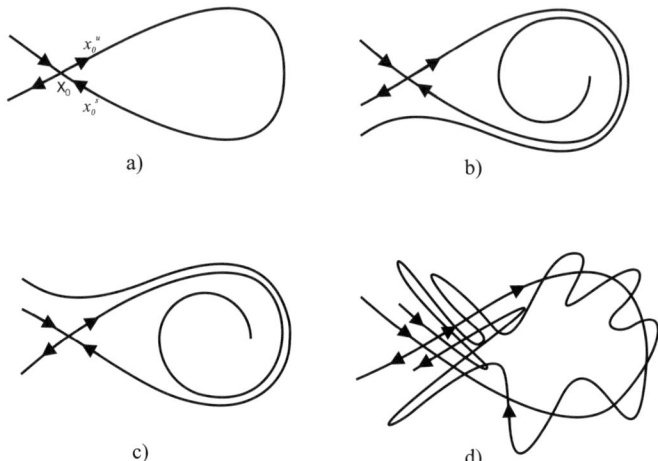

Fig. 3. Poincaré section $t = \text{const} \pmod{T}$ of the system (Eq. 6) for $\varepsilon = 0$ (Fig. 3(a)) and $\varepsilon \neq 0$ (Figs. 3(b) to 3(d)).

Melnikov analysis is based on the paper [23]. First, we consider a two-dimensional dynamical system under the action of a periodical perturbation with the property of having a unique saddle point:

$$\dot{\mathbf{x}} = \mathbf{f}_0(\mathbf{x}) + \varepsilon \mathbf{f}_1(\mathbf{x}, t), \tag{6}$$

Let furthermore $\mathbf{x}_0$ be the separatrix of the unperturbed system $\dot{\mathbf{x}} = \mathbf{f}_0(\mathbf{x})$. Then the MF at any given time t_0 is defined as follows:

$$D(t_0) = -\int_{-\infty}^{+\infty} \mathbf{f}_0 \wedge \mathbf{f}_1 \bigg|_{\mathbf{x} = \mathbf{x}_0(t - t_0)} dt,$$

where the integral is taken along the unperturbed separatrix $\mathbf{x}_0(t - t_0)$ and the integrand is $\mathbf{f}_0 \wedge \mathbf{f}_1 = f_{0x} f_{1y} - f_{0y} f_{1x}$.

In general, in dissipative systems one can observe three possibilities for the MF: either $D(t_0) < 0$ (Fig. 3(b)), $D(t_0) > 0$ (Fig. 3(c)) for any t_0 or $D(t_0)$ changes its sign for some t_0 (Fig. 3(d)). Only in the last case chaotic dynamics arises. Thus, the MF determines the character of the motion near the separatrix. Note that the Melnikov method has a perturbative (to first order) character, thus, its application is allowed only for trajectories which are sufficiently close to the unperturbed separatrix.

3.2. Function of stabilization

The Melnikov method has been applied in a lot of typical physical situations (see Refs. [24–29]) in which homoclinic bifurcations occur. Here we consider an application of the Melnikov method to the analysis of the chaos suppression phenomenon in systems with separatrix loops. Such an approach allows us to find an analytical expression of the perturbations for which the Melnikov distance $D(t_0)$ does not change sign (see also [30]) suppressing the chaotic behavior and stabilizing the orbits of the system.

We consider the problem of stabilization of chaotic behavior in systems with separatrix contours that can be described by Eq. (6)

$$\dot{\mathbf{x}} = \mathbf{f}_0(\mathbf{x}) + \varepsilon \mathbf{f}_1(\mathbf{x}, t),$$

where $\mathbf{f}_0(\mathbf{x}) = (f_{01}(\mathbf{x}), f_{02}(\mathbf{x}))$, $\mathbf{f}_1(\mathbf{x}, t) = (f_{11}(\mathbf{x}, t), f_{21}(\mathbf{x}, t))$. For this equation the Melnikov distance $D(t_0)$ is given by $D(t_0) = -\int_{-\infty}^{\infty} \mathbf{f}_0 \wedge \mathbf{f}_1 dt \equiv I[g(t_0)]$. Let us assume that $D(t_0)$ changes its sign. To suppress chaos we should get a *function of stabilization* $\mathbf{f}^*(\omega, t)$ that leads us to a situation where separatrices do not intersect:

$$\dot{\mathbf{x}} = \mathbf{f}_0(\mathbf{x}) + \varepsilon \left[\mathbf{f}_1(\mathbf{x}, t) + \mathbf{f}^*(\omega, t) \right], \qquad (7)$$

where $\mathbf{f}^*(\omega, t) = (f_1^*(\omega, t), f_2^*(\omega, t))$. Suppose $D(t_0) \in [s_1, s_2]$ and $s_1 < 0 < s_2$. After the stabilizing perturbation $\mathbf{f}^*(\omega, t)$ is applied we have two cases: $D^*(t_0) > s_2$ or $D^*(t_0) < s_1$, where $D^*(t_0)$ is the Melnikov distance for system (7). We consider the first case (analysis for the second one is similar). Then

$$I[g(t_0)] + I[g^*(\omega, t_0)] > s_2, \qquad (8)$$

where $I[g^*(\omega, t_0)] = -\int_{-\infty}^{+\infty} \mathbf{f}_0 \wedge \mathbf{f}^* dt$. Expression (8) is true for all left hand side values of inequality that is greater than s_2. It is derived that $I[g(t_0)] + I[g^*(\omega, t_0)] = s_2 + \chi = const$, where $\chi, s_2 \in \mathbb{R}^+$. Therefore $I[g^*(\omega, t_0)] = const - I[g(t_0)]$. On the other hand, $I[g^*(\omega, t_0))] = -\int_{-\infty}^{\infty} \mathbf{f}_0 \wedge \mathbf{f}^* dt$. We choose $\mathbf{f}^*(\omega, t)$ from the class of functions that are absolutely integrable on an infinite interval such that they can be represented in Fourier integral form. Then $\mathbf{f}^*(\omega, t) = \text{Re}\{\hat{A}(t)e^{-i\omega t}\}$. Here we suppose that $\hat{A}(t) = (A(t), A(t))$ i.e., assume that the regularizing perturbations applied to both components of Eq. (7) are identical. Therefore $-\int_{-\infty}^{\infty} \mathbf{f_0} \wedge \left\{ \hat{A}(t) e^{-i\omega t} \right\} dt = \text{const} - I[g(t_0)]$. The inverse Fourier transform yields: $\mathbf{f_0} \wedge \hat{A}(t) = \int_{-\infty}^{\infty} (I[g(t_0)] - \text{const}) e^{i\omega t} d\omega$. Hence,

$$A(t) = \frac{1}{f_{01}(x) - f_{02}(x)} \int_{-\infty}^{\infty} (I[g(t_0)] - \text{const}) e^{i\omega t} d\omega. \tag{9}$$

Here $A(t)$ can be interpreted as the amplitude of the "stabilizing" perturbation. Thus, for system (6) the external stabilizing perturbation has the form:

$$f^*(\omega, t) = \text{Re} \left[\frac{e^{-i\omega t}}{f_{01}(x) - f_{02}(x)} \int_{-\infty}^{\infty} (I[g(t_0)] - \text{const}) e^{i\omega t} d\omega \right]. \tag{10}$$

Here it is significant to note that in the conservative case: const=0.

Let us now consider the stabilization problem for systems of the type

$$\begin{aligned} \dot{x} &= P(x, y), \\ \dot{y} &= Q(x, y) + \varepsilon [f(\omega, t) + \alpha F(x, y)], \end{aligned} \tag{11}$$

where $f(\omega, t)$ is a time periodic perturbation, $P(x, y)$, $Q(x, y)$, $F(x, y)$ are some smooth functions and α is the dissipation.

We investigate the case which is typical for applications with a single hyperbolic point at the origin $x = y = 0$ when $P(x, y) = y$. Let $x_0(t)$ be the solution on the separatrix. In the presence of the perturbation the Melnikov distance $D(t_0)$ for the system (11) may be written as

$$D(t_0) = -\int_{-\infty}^{\infty} y_0(t - t_0) \left[f(\omega, t) + \alpha F(x_0, y_0) \right] dt \equiv I[g(\omega, \alpha)], \tag{12}$$

where $y_0(t) = \dot{x}_0(t)$. Let us suppose again that the Melnikov function (12) changes sign, i.e., the separatrices intersect. We will find an external regularizing perturbation $\mathbf{f}^*(\omega, t) = \text{Re}\{\hat{A}(t) e^{-i\omega t}\}$ that stabilizes the system dynamics:

$$\begin{aligned} \dot{x} &= y, \\ \dot{y} &= Q(x, y) + \varepsilon [f(\omega, t) + \alpha F(x, y) + f^*(\omega, t)]. \end{aligned} \tag{13}$$

It is significant to note that since the system (11) depends on parameter α then such stabilization should be made at every fixed value of this parameter and further, instead of $I[g(\omega,\alpha)]$, we will write $I[g(\omega)]$. For (13) we have $f_{01} = y$, $f_{02} = Q(x,y)$ and $\hat{A}(t) = (0, A(t))$. Consequently the value $A(t)$ has a form

$$A(t) = \frac{1}{y_0(t-t_0)} \int_{-\infty}^{\infty} (I[g(\omega)] - \text{const}) e^{i\omega t} d\omega. \quad (14)$$

So, for (13) the stabilizing function can be represented as

$$f^*(\omega,t) = \text{Re}\left[\frac{e^{-i\omega t}}{y_0(t-t_0)} \int_{-\infty}^{\infty} (I[g(\omega)] - \text{const}) e^{i\omega t} d\omega\right]. \quad (15)$$

Now, let us find a regularizing perturbation in the case when the Melnikov function $D(t,t_0)$ admits an *additive* shift from its critical values.

Again, we analyze the case when $D^*(t_0) > s_2$ is satisfied. Suppose that α_c corresponds to the critical value of the Melnikov function, $I_c = I[g(\omega, \alpha|_{\alpha=\alpha_c})]$. Then, a subcritical Melnikov distance can be expressed as $I_{out} = I_c - a$, where $a \in \mathbb{R}^+$ is constant. Assuming that the system perturbed by $f^*(\omega,t)$ exhibits regular behavior, we have

$$I' = I_{out} + I[g^*(\omega)] > s_2. \quad (16)$$

Here $I[g^*(\omega)] = -\int_{-\infty}^{+\infty} y_0(t-t_0) f^*(\omega,t) dt$. On the other hand, it is obvious that we can take any I' a fortiori greater than I_c:

$$I' = I_c + a > s_2. \quad (17)$$

Now, equating the left-hand sides of (16) and (17), we obtain $I[g^*(\omega)] = 2a$. Substituting $f^*(\omega,t) = \text{Re}\{A(t)e^{i\omega t}\}$ into the expression for $I[g^*(\omega)]$, we find $-\int_{-\infty}^{\infty} e^{i\omega t} A(t) y_0(t-t_0) dt = 2a$. The inverse Fourier transform yields $A(t) y_0(t-t_0) = -2a \int_{-\infty}^{\infty} e^{-i\omega t} d\omega$. Hence,

$$A(t) = -\frac{2a}{y_0(t-t_0)} \int_{-\infty}^{\infty} e^{-i\omega t} d\omega = -\frac{4\pi a \delta(t)}{y_0(t-t_0)}. \quad (18)$$

Thus, the dynamics of the systems that admit an additive shift from the critical value of the Melnikov function $D(t_0)$ are regularized by the perturbation:

$$f^*(\omega, t) = -\frac{4\pi a \delta(t)}{y_0(t - t_0)} \cos(\omega t), \tag{19}$$

where $\delta(t)$ is a Dirac delta–function defined as follows:

$$\delta(t) = \begin{cases} 0, & t \neq 0, \\ \infty, & t = 0. \end{cases}$$

In the general case, if $f_0 = (f_{01}(x), f_{02}(x))$, then we obviously obtain

$$f^*(\omega, t) = -\frac{4\pi a \delta(t)}{f_{01}(x) - f_{02}(x)} \cos(\omega t). \tag{20}$$

From the physical point of view the dynamics of the chaotic system are stabilized by a series of "kicks". This result could be easily extended on case when the stabilizing function $f^*(\omega, t)$ is Gaussian function. The orbit that was chaotic and became regular under the influence of the external perturbation we call the *stabilized orbit*.

4. Stabilization of Chaotic Behavior in the Extended Sitnikov Problem

In the vicinity of the orbiting primaries there exist five equilibrium points lying in the $z = 0$ plane. The points L_1, L_2, L_3 are unstable and collinear with the primaries, while each of L_4 and L_5 forms an equilateral triangle with the primaries and are stable, depending on the mass of the primaries. Let us now consider two bodies of mass m that are placed in the neighborhood of the stable triangular Lagrange points of the Sitnikov problem (see Fig. 4). Here we treat only the hierarchical case: $\mu \ll m < M$. In the new configuration that constitutes the extended Sitnikov problem a particle of mass μ experiences forces from the primaries and masses m placed close to L_4 and L_5. These forces are perpendicular to the primaries plane and therefore the particle's motion remains on the z axis. Since the bodies of mass m orbit around their common center of mass, their distance ρ' alternates between ρ'_{min} (periastron) and ρ'_{max} (apastron), consequently the forces between these bodies and the particle μ increase in a close encounter to the barycenter and vanish as the bodies move away from z axis. So we can achieve the situation where the influence of bodies that are placed in the

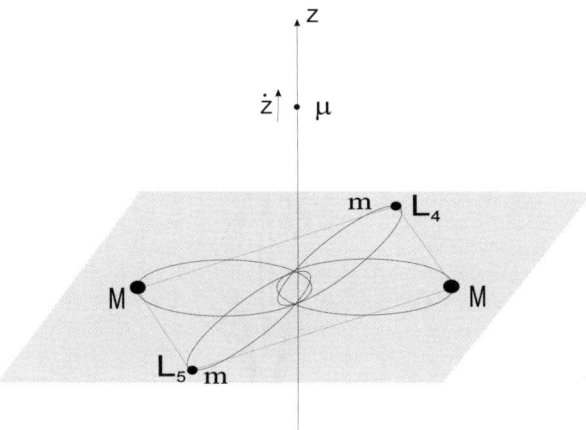

Fig. 4. Geometry of the Extended Sitnikov problem.

vicinity of the triangular Lagrange points of the particle μ can be presented as a series of periodic Gaussian function-like impulses.

Recall that earlier we showed how the elliptical Sitnikov problem deals with heteroclinic orbits that lead to non-integrability due to the existence of transverse heteroclinic points. It was done through the connection between this problem and the pendular character of motion described by (5). Therefore taking into account the new configuration of the restricted three-body problem (Fig. 4) we may say that there is a connection between the extended elliptical Sitnikov problem and the motion of the chaotic nonlinear pendulum with an external impulse-like perturbation. The Hamiltonian of such system changes to

$$H(u,v) = \\ H_0(u,v) + e\left[H_1(u,v,t,e) + H_1^*(u,v,t,e)\right], \quad (21)$$

where $H_1^*(u,v,t,e)$ - the part of Hamiltonian that is responsible for impulsive forces that the particle μ experiences from bodies in the neighborhood of L_4 and L_5.

Now taking into account the result of the previous paragraph we conclude that the forces which the particle experiences from bodies in the neighborhood of L_4 and L_5 act on the chaotic behavior of μ as an external stabilizing perturbation and the system (21) represents the system with the stabilized chaotic behavior that corresponds to the stabilized orbits in the extended Sitnikov problem. As mentioned before, in the circular Sitnikov

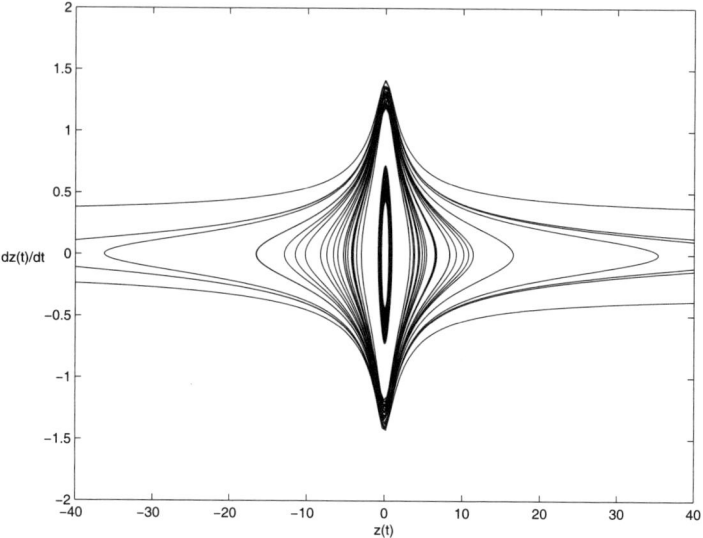

Fig. 5. Phase portrait of the particle motion in Sitnikov problem.

problem when $e = 0$ the phase space is partitioned into invariant curves corresponding to different energies. For $e > 0$ this structure is broken. This is apparent for eccentricity $e = 0.07$ in Fig. 5: just a few invariant curves survive. In this figure we also can see hyperbolic and parabolic orbits that correspond to energy $h \geq 0$. These orbits escape to infinity with positive or zero speed respectively. Now, if we consider the extended Sitnikov problem then one can see (Fig. 6) that all orbits are in a bounded region and there are no escape orbits. So one may infer that stabilized orbits of the pendulum system with Hamiltonian (21) correspond to the stabilized orbits in the extended Sitnikov problem, thus confirming the conclusion that we made before. The extension of the analysis carried out above to the corrections of higher order in ε of Eq. (2) and numerical verification of the obtained results could be found in [31].

In summary, we have performed a study of the existence of a qualitatively different orbit from those previously known in the three-body problem: the stabilized orbits. On the basis of the elliptic Sitnikov problem we constructed a configuration of five bodies which we called the extended Sitnikov problem and showed that in this configuration along with chaotic and regular orbits the stabilized orbits could be realized.

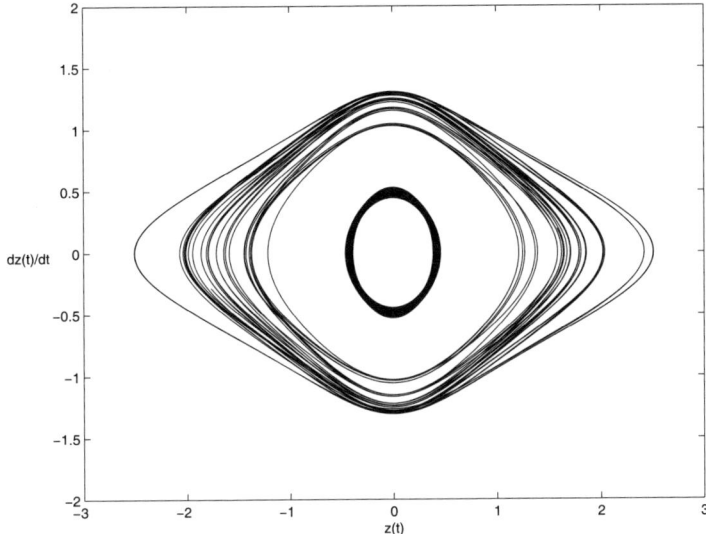

Fig. 6. Phase portrait of the stabilized particle orbits in the extended Sitnikov problem.

Acknowledgments

We thank Carles Simó and David Farrelly for valuable discussions. A. Dzhanoev acknowledges that this work is supported by the Spanish Ministry of Education and Science under the project number SB2005-0049. Financial support from project number FIS2006-08525 (MEC-Spain) is also acknowledged. This work was supported in part by the Cassini project.

References

[1] V. Aquilanti and S. Tonzani, The Journal of Chemical Physics **120**, 4066 (2004).
[2] D. Farrelly and J. E. Howard, Phys. Rev. A **49**, 1494 (1994).
[3] M. R. Ejtehadi, S. P. Avall and S. S. Plotkin, PNAS **101**, 15088 (2004).
[4] R. Dvorak, J. Kurths, *Chaos and Stability in Planetary Systems*, Lecture Notes in Physics, 683, (Springer, Berlin Heidelberg, 2005).
[5] C. Moore, Phys. Rev. Lett. **70**, 3675 (1993).
[6] W. D. MacMillan, *Astron. J.* **27**, 11 (1913).
[7] K. Stumpff, Himmelsmechanik Band II VEB, (Berlin, 1965)
[8] K. A. Sitnikov, Dokl. Akad. Nauk USSR **133**, 303 (1960).
[9] V. M. Alexeev, Math. USSR Sbornik **5**, 73; **6**, 505; **7**, 1 (1969).

[10] J. Moser, *Stable and random motions in dynamical systems*, (Prinston Univ. Press, Princeton, N.J., 1973).
[11] L. Llibre and C. Simó, Publicaciones Matemàtiques U.A.B. **18**, 49 (1980).
[12] C. Marchal, *The three body problem*, Studies in Astronautics, 4, Elsevier, Amsterdam (1990).
[13] J. Hagel, Celes. Mech. and Dyn. Sys. **53**, 267 (1992).
[14] R. Dvorak, Celes. Mech. and Dyn. Sys. **56**, 71 (1993).
[15] H. Dankovicz and Ph. Holmes, J. Differential Equations **116**, 468 (1995).
[16] V. I. Arnold, Usp. Math. Nauk **18**, 91 (1963).
[17] V. V. Alekseev and A. Loskutov, Moscow Univ. Phys. Bull. **40**, 46 (1985).
[18] V. V. Alexeev and A. Loskutov, Sov. Phys.-Dokl. **32**, 270 (1987).
[19] E. Ott, C. Grebogi, J. A. Yorke, Phys. Rev. Lett. **64**, 1196 (1990).
[20] A. Loskutov, Discr. and Continuous Dyn. Syst., Ser. B **6**, 1157 (2006).
[21] A. R. Dzhanoev, A. Loskutov et al., Discr. and Continuous Dyn. Syst., Ser. B **7**, 275 (2007).
[22] T. Schwalger, A. R. Dzhanoev,A. Loskutov, Chaos, **16**, 2, 023109 (2006).
[23] V. K. Mel'nikov, Trans. Moscow Math. Soc., **12**, 3 (1969).
[24] R. Chacón, Phys. Rev. E, **51**, 761 (1995)
[25] J.A. Almendral, J.M. Seoane, M.A.F. Sanjuán, Recent Res. Devel. Sound & Vibration **2**, 115 (2004).
[26] F. Cuadros, R. Chacón, Phys. Rev. E, **47**, 4628 (1993).
[27] R. Lima and M. Pettini, Phys. Rev. E, **47**, 4630, (1993).
[28] R.Lima, M.Pettini, Phys. Rev. A, **41**, (1990), (1990).
[29] J.L. Trueba, J.P. Baltánas, M.A.F. Sanjuán, Chaos Solitons and Fractals, **15**, 911 (2003).
[30] A. Loskutov, A.R. Dzhanoev, J. Exp. Theor. Phys., **98**, 1194 (2004).
[31] A. R. Dzhanoev, A. Loskutov, J. E. Howard and M.A.F. Sanjuán: submitted to Phys. Rev. E, (2009)

Chapter 15

CONTROLLING THE CHAOS USING FUZZY ESTIMATION IN A GYROSTAT SATELLITE

Ardeshir Guran

Institute of Structronics, 275 Slater Street, 9th Floor, Ottawa, Canada
ardeshir.guran@mcgill.ca
Faculty of Engineering, K. N. Toosi University of Technology,
Pardis Street, Tehran, Iran
guran@kntu.ac.ir

In this paper, we present a study of the dynamical behavior in a Kelvin type gyrostat satellite. We firstly obtain the Hamiltonian equations of our model by using Cardan angles as generalized coordinates. Then, we make this Hamiltonian dimensionless and calculate motion equations for this dimensionless system. The study of the Poincare's sections of this system shows us that chaotic motion regimes are present for specific parameter values. The main goal of this work is the finding of stabilizing orbits by using a control technique, the fuzzy control of Poincare map method, so that it can be applied to stabilize special periodic orbits in this system. Finally, we expect that the technique can be useful for a better understanding of control theory and their applications in gyrostat problems.

Keywords: Gyrostat, Cardan angles, Poincare section, Chaotic motion, Fuzzy control, Clustering.

1. Introduction

A Kelvin type satellite consists of two rigid parts, an axi-symmetric rotor, R inside a bigger platform, P. We assume that the center of mass of our satellite is rotating on a circular orbit around a central mass which can be the Earth and that the rotor angular velocity is very high. Also, the platform can rotate slowly in comparison to rotor's velocity. These satellites are known as gyrostat satellites [1].

Chaotic motions in nonlinear systems arise in many real problems. Investigating chaos in satellite dynamic was started in the works by Liu et al. [2, 3]. The authors have shown that chaotic motion is possible in different kinds of satellites such as satellite in circular orbits [4] and also gyrostats in a central gravitational field [5-9]. Since the pioneering work on controlling chaos due to Ott, Grebogi, and Yorke [10], named OGY, different control schemes have been proposed that allow one to obtain a desired response from a dynamical system by applying some small but accurately chosen perturbations [11, 12].

The methods stated to control chaos can be classified in feedback and non-feedback methods [13, 14], depending on how they interact with the system. Feedback methods of chaos control, as the celebrated OGY [10], stabilize one of the unstable orbits that lie in the chaotic attractor by using small state-dependent perturbations into the system. However, in experimental implementations, the fast response that these methods require cannot usually be provided. For these situations, non-feedback methods are more useful. Non-feedback methods have been mainly used to suppress chaos in periodically driven dynamical systems.

Also in the period of time of OGY work, the Pyragas method, based on delayed feedback control was presented [12, 15]. In recent years, some chaos controller based on fuzzy systems, have been proposed [16-18]. In [16] the idea of chaos control by fuzzy systems is introduced and the Chua's circuit was controlled via fuzzy systems. The fuzzy estimation of OGY and Pyragas controllers are also used for chaos control and is applied to a Bonhoeffer-Van der Pol oscillators as shown in Ref. [18]. In Ref. [19] the author considered the fuzzy control of Poincaré maps, and two algorithms for chaos control based on fuzzy systems are proposed for stabilizing the fixed points or unstable periodic orbits. The first algorithm provides a fuzzy system for the controller using the clustering technique and the second one design the controller by fuzzy table look up method. The advantage of the proposed algorithms is that only the state variables of the system on a Poincaré section are used for chaos control and there is no need to know the mathematical model of the system and its Poincaré map. Because these controllers are constructed on the Poincaré sections, the method of this paper can be used for both discrete and continuous systems.

In this paper, nonlinear governing equations of a gyrostat without any restriction to small angles or perturbations are adapted from the work described in Ref. [1]. Later, attitude dynamic of this system is investigated in Poincare map. The fixed points in these Poincare maps are found using a recursive method and stabilized using the fuzzy control method presented in [19].

This paper is organized as follows. In Sec. 2 we present a complete description of our model, the Kelvin type satellite. Section 3 presents a complete estimation of the parameters of our model. Section 4 provides a full description of the control method used for the stabilization of our system, namely Fuzzy Control. Numerical evidence of the robustness of the Fuzzy Control technique is given in Sec. 5. Conclusions and discussions of the main results of this paper are presented in Sec. 6.

2. Model Description

We now introduce our prototype model, the Kelvin type Gyrostat Satellite (see Fig. 1). In their equations, we assume that both gyrostat and rotors are rotating about axis z. A Cxyz coordinate system is fixed to the gyrostat. The center of mass, C, rotates around the Earth in a circular orbit. θ_2, θ_1 and θ_3 are three Cardan angles about axis y_1, x_0 and z, respectively. Kinematic and potential energy of this system can be calculated as

$$T = \frac{A}{2}[\dot{\theta}_1^2 \cos^2\theta_2 + \dot{\theta}_2^2 + \Omega^2(\sin^2\theta_1 + \cos^2\theta_1 \sin^2\theta_2) - 2\dot{\theta}_1\Omega\cos\theta_1\sin\theta_2\cos\theta_2$$
$$+ 2\dot{\theta}_2\Omega\sin\theta_1] + \frac{C^P}{2}(\omega_z^P)^2 + \frac{C^R}{2}(\omega_z^R)^2 \quad (1)$$

and

$$V = \frac{3}{2}(C-A)\Omega^2 \sin^2\theta_2. \quad (2)$$

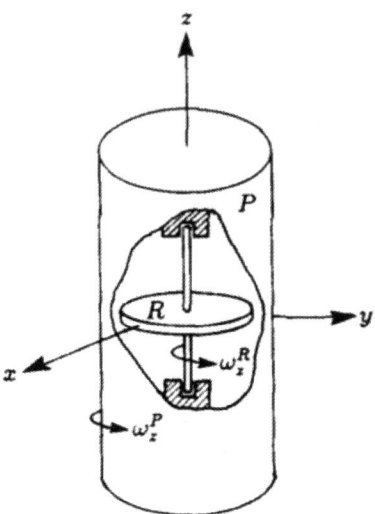

Fig. 1. A Kelvin-type gyrostat consisting of a platform and a single rotor [1].

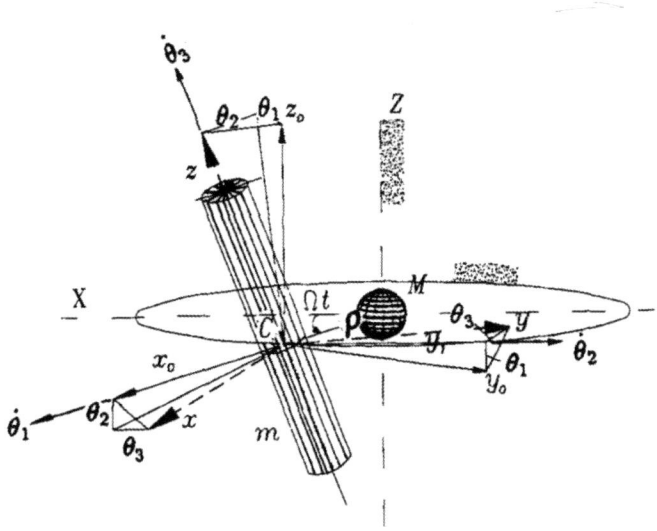

Fig. 2. Schematic representation of gyrostat motion in a circular orbit (01, 02, 03 are Cardan angles of the first kind) [1].

In these relations, $A = A^P + A^R$, $C = C^P + C^R$ are moments of inertia of satellite and ω_z^R, ω_z^P are angular velocities of rotor and platform about z axis, respectively. $\Omega = \sqrt{\dfrac{\mu}{Q^3}}$ is the orbital angular velocity of the whole system, where $\mu = GM$ and Q is the radius of the satellite orbit around the Earth, and θ_3 is a cyclic coordinate (see Fig. 2). So, we can conclude that P_3 is constant during the motion. Now, if we assume that the angular momentum of the rotor is constant too, we can write

$$\omega_z^P = \dot{\theta}_1 \sin\theta_2 + \dot{\theta}_3 + \Omega\cos\theta_1 \sin\theta_2 \qquad (3)$$
$$= \omega_0 = const.$$

Using this relation, the Hamiltonian of this system can be obtained as

$$H = P_1\dot{\theta}_1 + P_2\dot{\theta}_2 + P_3\dot{\theta}_3 - L = \frac{1}{2A}\left(\frac{P_1^2}{\cos^2\theta_2} + P_2^2\right) + P_1\Omega\cos\theta_1 \tan\theta_2$$
$$- \frac{P_1 P_3 \sin\theta_2}{A\cos^2\theta_2} - P_2\Omega\sin\theta_1 - P_3\Omega\frac{\cos\theta_1}{\cos\theta_2} + \frac{P_3^2}{2A}\tan^2\theta_2 + \frac{3}{2}(C - A)\Omega^2\sin^2\theta_2, \qquad (4)$$

where

$$P_1 = A(\dot{\theta}_1 \cos^2\theta_2 - \Omega\cos\theta_1 \sin\theta_2 \cos\theta_2) + P_3 \sin\theta_2,$$
$$P_2 = A(\dot{\theta}_2 + \Omega\sin\theta_1), \qquad (5)$$
$$P_3 = C^P\omega_z^P + C^R\omega_z^R = const.$$

Now, by introducing dimensionless variables $\tau = t\sqrt{H/A}$ and $p_i = \dfrac{P_i}{\sqrt{HA}}$ we will have

$$h = \frac{1}{2}\left(\frac{p_1^2}{\cos^2\theta_2} + p_2^2\right) + \sqrt{\gamma}\, p_1\Omega\cos\theta_1 \tan\theta_2 - \sqrt{\gamma}\lambda_2\lambda_3 p_1 \frac{\sin\theta_2}{\cos^2\theta_2}$$
$$- \sqrt{\gamma}\, p_2 \sin\theta_1 - \gamma\lambda_2\lambda_3 \frac{\cos\theta_1}{\cos\theta_2} + \frac{\gamma}{2}\lambda_2^2\lambda_3^2 \tan^2\theta_2 + \frac{3}{2}\gamma(\lambda_1 - 1)\sin^2\theta_2, \qquad (6)$$

where h is the dimensionless Hamiltonian equation. In this equation, $\lambda_1 = C/A$, $\lambda_2 = \omega_0/\Omega$ and $\lambda_3 = C^*/A = \dfrac{C}{A} + \left(\dfrac{\omega_z^R}{\omega_z^P} - 1\right)\dfrac{C^R}{A}$ are control parameters of the system. Also, $\gamma = A\Omega^2/H$ is the ratio of gravitational energy of the rotational energy of the satellite.

By letting h = 1, the equations of motion will be obtained as

$$\dot{\theta}_1 = \frac{p_1}{\cos^2 \theta_2} + \sqrt{\gamma} \cos \theta_1 \tan \theta_2 - \sqrt{\gamma} \lambda_2 \lambda_3 \frac{\sin \theta_2}{\cos^2 \theta_2},$$

$$\dot{p}_1 = \sqrt{\gamma} p_1 \sin \theta_1 \tan \theta_2 + \sqrt{\gamma} p_2 \cos \theta_1 - \gamma \lambda_2 \lambda_3 \frac{\sin \theta_1}{\cos \theta_2},$$

$$\dot{\theta}_2 = p_2 - \sqrt{\gamma} \sin \theta_1, \qquad (7)$$

$$\dot{p}_2 = -p_1^2 \frac{\sin \theta_2}{\cos^3 \theta_2} - \sqrt{\gamma} p_1 \frac{\cos \theta_1}{\cos^2 \theta_2} + \sqrt{\gamma} \lambda_2 \lambda_3 p_1 \frac{1 + \sin^2 \theta_2}{\cos^3 \theta_2}$$

$$- \gamma \lambda_2^2 \lambda_3^2 \frac{\sin \theta_2}{\cos^3 \theta_2} + \gamma \lambda_2 \lambda_3 \cos \theta_1 \frac{\sin \theta_2}{\cos^2 \theta_2} - 3\gamma(\lambda_1 - 1)\sin \theta_2 \cos \theta_2.$$

Studying the behaviour of chaotic systems is much simpler when discretized. The idea of reducing the study of continuous time systems to the study of an associated discrete time system is due to Poincaré (1899). As a matter of fact, associated to an ordinary differential equation we can construct a discrete time dynamical system which is called a Poincaré map [20]

$$\vec{X}(n+1) = P(\vec{X}(n), u(n)), \qquad (8)$$

where $P(.,.)$ is the Poincaré map, $\vec{X}(n)$ is the state vector on the Poincaré section in which the Poincaré map is defined, and $u(n)$ is the controlling action. The fixed point for a chaotic system is defined as the state which maps into itself through the Poincaré map. In other words, this specific trajectory of the system, beginning from a fixed point, returns to this point after a specific time named period. In Figs. 3 and 4 Poincaré maps of this equations for different values of parameters are shown. As it is easily observed, in each of them there are different types of orbits (periodic, chaotic, etc.) that can be stabilized by implementing our fuzzy control scheme we describe in the next section. We also observe KAM islands which are typical in Hamiltonian systems.

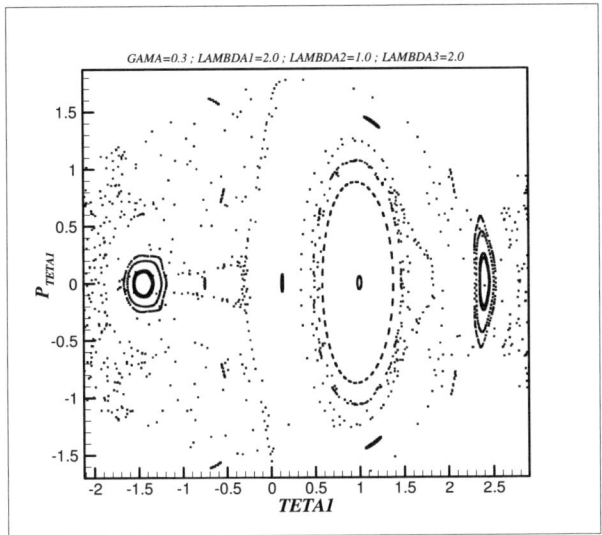

Fig. 3. Poincaré map of our Kelvin type Gyrostat satellite for parameter values as follows: $\gamma = 0.3$ and $\lambda_1 = 2.0$, $\lambda_2 = 1.0$, $\lambda_3 = 2.0$. Different dynamical behaviors are observed as periodic, chaotic, etc.

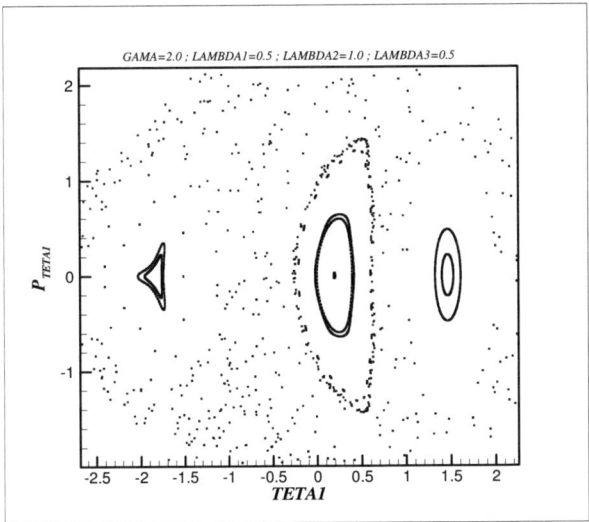

Fig. 4. Poincaré map of our Kelvin type Gyrostat satellite for parameter values as follows $\gamma = 2.0$ and $\lambda_1 = 0.5$, $\lambda_2 = 1.0$, $\lambda_3 = 0.5$. As in Fig. 3 different dynamical behaviors can be observed.

The chaotic system starting from an arbitrary point exhibits erratic behavior in the Poincaré map which fills specific areas in this surface. With an appropriate control of $u(n)$, the system can be forced towards its fixed point which is the desired state in most cases. In this case the chaotic behaviour of the system transforms into a periodic behaviour. It is assumed that the dynamic equation of the system is unknown, but the state vector, $\vec{X}(n)$, is obtainable, then the main goal is to design an identifier/controller scheme to stabilize the unstable fixed points of the system as the authors stated in Refs. [19]–[22].

3. Parameter Estimation

The parameter $\lambda_1 = C/A$ defined previously is the degree of oblateness of the satellite. For typical geometrical shapes we can assume that this degree is of order 1. As an example, in a cube this degree is exactly equal to 1 and in a cylinder $\lambda_1 = \dfrac{6r^2}{3r^2 + h^2}$, where r is the radius of the cylinder, and h is its height. We can simply assume that this parameter is constant and for example, equal to 2. We are not going to use this parameter as action control in our fuzzy system. The change of moment of inertia has some other effects on the equations of motions and also other parameters. In fact, we cannot look at this parameter as a free variable and change it without considering these effects on the whole system.

For a satellite rotating around the Earth it is well known that $\mu = 3.986005 \times 10^{14}$ m^3/s^2 and $6800 \leq Q \leq 7500$km. Therefore, angular velocity of the satellite around the Earth is small, and of order $\Omega \propto O(e^{-3})$ and the angular velocity of the platform can have different values. In especial flight missions, it is necessary for spacecraft to keep its orientation to the Earth. Hence, the angular velocity of the satellite, around its major axis, should be equal to its orbital angular velocity and therefore, $\lambda_2 = \omega_0/\Omega \approx 1$. Control of this value and keeping $\lambda_2 = 1$ is a goal of control systems in these satellites. In some other types which the orientation of the satellite is not a control goal of the system, we can assume it as an action control for obtaining especial maneuvers and orientations. The range of variation of λ_2 in these spacecrafts could be very wide. If we assume angular velocity of satellite around its major axis, ω_0 equal to unity we will have $\lambda_2 = \omega_0/\Omega \approx 1000$. Using the relation stated in Eq. (5c) and by changing the angular velocity of the rotor we can change λ_2 in the range of $-1000 \leq \lambda_2 \leq 1000$. Variation in this parameter is a result of variations in the angular velocity of rotor. On the other hand, the angular velocity of the rotor could not exceed especial values depending on the rotor characteristics. So, this range of variation for λ_2 cannot be obtained in real cases.

The last control parameter in this system is λ_3. If the rotor, R, contains only 5% of the whole satellite weight we will have, $C^R/A = 0.1$. Also, we know that $\omega_z^R \gg \omega_z^P$. Now, if we let $-50 \leq \dfrac{\omega_z^R}{\omega_z^P} \leq 50$, the range of variation of λ_3 will be $-4 \leq \lambda_3 \leq 6$.

By changing ω_z^R, both λ_2 and λ_3 will be changed. Also, these two parameters always appear in equations of motions together. So, we can let $\lambda = \lambda_2 \lambda_3$ and choose it as the control input of control system. Range of practical control inputs in this system is so that the acceptable ranges for λ_2 and λ_3 are satisfied.

Now, by using the fuzzy control method developed in Ref. [19] we can design a fuzzy controller to stabilize periodic orbits in the Poincaré map.

4. Fuzzy Control Method

One of the methods available for constructing a fuzzy model from input-output data pairs is the fuzzy clustering method. This method is especially useful when the number of input-output pairs is limited. The basic idea is to group the input-output pairs into clusters and use a specific rule for each cluster, in the form of

$$\text{IF } x \text{ in } A[x_c^l], \text{ THEN } y \text{ in } B[y_c^l], \tag{9}$$

where $A[x_c^l]$ and $B[y_c^l]$ are input and output fuzzy sets with centers at x_c^l and y_c^l, respectively and l is the number of cluster. There are several algorithms to make a fuzzy system based on the clustering. One of the simplest methods is the nearest neighborhood algorithm. This method is explained extensively in many fuzzy control books such as shown in Ref. [22]. The designed fuzzy system using singleton fuzzifier, products inference engine and center average deffuzifier based on k input-output pair clustered in this method can be written as follows:

$$f_k(x) = \frac{\sum_{l=1}^{M} a^l(k) \exp\left(-\frac{|x-x_c^l|^2}{\sigma^2}\right)}{\sum_{l=1}^{M} b^l(k) \exp\left(-\frac{|x-x_c^l|^2}{\sigma^2}\right)}, \qquad (10)$$

where M is the number of clusters constructed, x_c^l denotes the center of the lth cluster, $a^l(k)$ is the summation of all output data gathered in cluster l and $b^l(k)$ is the number of data points gathered in cluster l after examining k data points. In the above equation σ is a smoothing parameter. The smaller the σ the smaller the matching error becomes, but the less smooth the $f_k(x)$. The matching error is the difference between the actual output and the one obtained from fuzzy model. It should be noted that the number of clusters depends on the distribution of input points and the radius r.

The following algorithm is adapted from Ref. [19] to construct a suitable controller for this system. For further studies you can return to the original paper. Now, we provide a complete description of the algorithm implemented in order to stabilize the orbits in our model.

5. Algorithm I:

The Fuzzy algorithm can be built according to the following steps:

STEP-1: Let $\vec{X}(1)$ be an arbitrary point in the domain of Poincaré map, then choose a random value for $u(1)$ in its prescribed domain, and measure $\vec{X}(2)$ due to problem assumptions.

STEP-2: Repeat step 1, by setting $\vec{X}(2)$ as an starting point to generate $\vec{X}(3)$ by using a random value for $u(2)$.

STEP-3: By iterating step 2, a set of $\Gamma = \{(\vec{X}(k+1), \vec{X}(k), u(k)),\ k = 1, 2, \ldots, N\}$ for a large N, is generated.

STEP-4: The input-output data pairs for the fuzzy clustering algorithm are obtained according to:

STEP-4.1: Let $j = 0$ and $k = 1$.

STEP-4.2: Consider $(\vec{X}(k+1), \vec{X}(k), u(k))$, and then examine the following condition

$$\left| \vec{X}(k+1) - \vec{X}_F \right| \le m \left| \vec{X}(k) - \vec{X}_F \right|, \tag{11}$$

where $0 < m < 1$ is selected arbitrarily and called the approaching factor. If the above condition is satisfied then let $j = j+1$, $\vec{X}_0^j = \vec{X}(k)$, $u_0^j = u(k)$. The jth input-output data pair is let to be $(\vec{X}_0^j; u_0^j)$.

STEP-4.3: Iterate step 4.2. By $k = k+1$.

STEP-5: Now the clustering algorithm is applied on $(\vec{X}_0^j; u_0^j)$ to obtain $U(.)$.

6. Numerical Results

The periodic orbits of the Poincaré map in Figs. 3 and 4 can be found by using a recursive method. The central periodic orbit shown in Fig. 3 takes place for:

$$\begin{aligned}
\theta_1 &= 1.00001, \\
p_{\theta_1} &= 0, \\
\theta_2 &= 0, \\
p_{\theta_2} &= 2.152278.
\end{aligned} \quad (12)$$

Results which are obtained from fuzzy control are shown in Figs. 5 and 6. Figures 5 and 6 show convergence to the fixed point of the Fig. 3 once our fuzzy control is implemented. After approximately 100 iterations in our control scheme, our orbit trend to a fixed point.

Finally, Fig. 7 shows the phase space of our model after applying our control method in which we easily observe the convergence to a fixed point for which the stabilization is completely achieved.

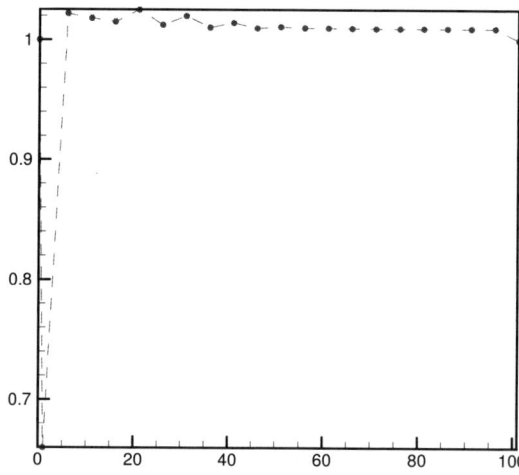

Fig. 5. Trend of θ_1 convergence to the fixed point on Poincaré map in which horizontal axis shows number of iterations.

Controlling the Chaos using Fuzzy Estimation 421

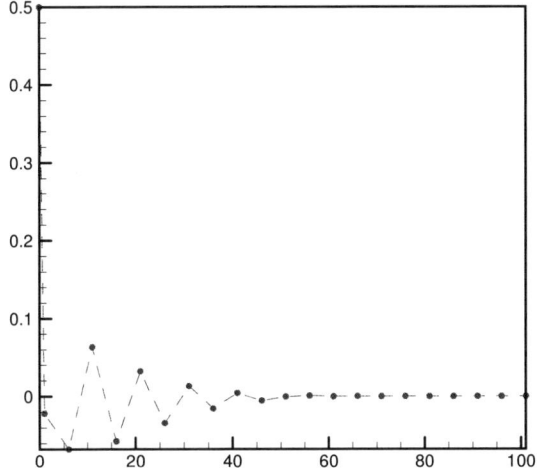

Fig. 6. Trend of p_{θ_1} convergence to the fixed point on Poincaré map, in which horizontal axis shows number of iterations.

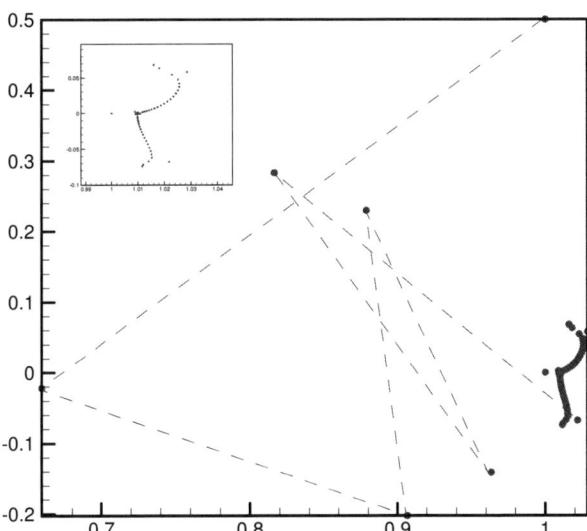

Fig. 7. Convergence of points on $\theta_1 - p_{\theta_1}$ in the Poincaré map. The stabilization in a periodic orbit is obtained.

7. Concluding Remarks

Summarizing, we have implemented a control technique to stabilize orbits in chaotic or periodic regimes in a Kelvin type gyrostat satellite. The Hamiltonian equations of our model are obtained by using Cardan angles as generalized coordinates. Our control scheme, named Fuzzy control, is fully described and applied to stabilize periodic or chaotic orbits. This technique is successfully applied for special orbits found in phase space. Finally, we expect our technique to be applied in other gyrostat models, e.g. Refs. [23]–[50] and different physical situations where control techniques are required.

Acknowledgments

The author would like to thank Dr. Jesus Seoance and Prof. Miguel A. F. Sanjuan from Nonlinear Dynamics, Chaos and Complex Systems group, Department of Physics, Rey Juan Carlos University, Madrid, Spain, for reviewing this work and making many constructive comments. I am also indebted to my colleagues in the Mechatronics group of K. N. Toosi University of Technology, in particular Prof. Ali Ghaffari, Dr. Ali Nahvi, Dr. A. K. Mousavian and Dr. Alireza Fatehi for their friendship and kind hospitality during the preparation of this book chapter.

References

1. A. Guran, 1993, Chaotic motion of a Kelvin type gyrostat in a circular orbit, Acta Mechanica, 98 (1993) 51-61.
2. Y. Z. Liu, L. Q. Chen, Nonlinear problems in spacecraft attitude dynamics, in: Weizang Chien (Ed.), Proceedings of the Third International Conference on Nonlinear Mech. Shanghai University Press, 1998, pp. 80–86.
3. Y. Z. Liu, L. Q. Chen, G. Cheng, X. S. Ge, Stability, bifurcation and chaos in spacecraft attitude dynamics, Adv. Mech. 30(3) (2000) 351–357 (in Chinese).
4. A. Guran, X. Tong, F.P.J. Rimrott, Instabilities in a spinning axi-symmetric rigid satellite, Mech. Res. Commun. 18 (5) (1991) 287–291.
5. A. Guran, On the stability of a spinning satellite in a central force Field, In: Bifurcation and chaos: analysis, algorithms, applications (Seydel, R., Schneider, E., Kupper, T., Troger, H. , eds), Birkhauser, 1991, pp. 149-153.
6. X. Tong, B. Tabarrok, F.P.J. Rimrott, Chaotic motion of an asymmetric gyrostat in the gravitational field, Int. J. Non-linear Mech. 30 (3) (1995) 191–203.

7. A.C. Or, Chaotic motions of a dual-spin body, J. Appl. Mech. 65 (1) (1998) 150–156.
8. J. H. Peng, Y. Z, Liu, Chaotic motion of a gyrostat with asymmetric rotor, Int. J. Non-linear Mech. 35 (3) (2000) 431–437.
9. G. Cheng, Y. Z. Liu, Chaotic motion of an asymmetric gyrostat in the magnetic Field of the earth, Acta Mech. 141 (2000) 125–134.
10. Ott E, Grebogi C, Yorke JA. Controlling chaos. Phys Rev Lett 1990; 64:1196–9.
11. Ott E. Chaos in dynamical systems. 2nd ed. Cambridge University Press; 2002. p. 61–3.
12. Chen G. Controlling chaos and bifurcation in engineering systems. CRC Press; 2000. p. 46–65.
13. Boccaletti S, Grebogi C, Lai Y. C., Mancini H, and Maza D. The control of chaos: Theory and applications. Phys. Rep. 2000. 329, 103-197.
14. Lima R, and Pettini M. Suppression of chaos by resonant parametric perturbations. Phys. Rev. A. 1990. 41, 726-733.
15. Z. M. Ge, C.-I. Lee, H. H. Chen, S. C. Lee, Nonlinear dynamics and chaos control of a damped satellite with partially-filled liquid, J. Sound Vibration 217 (1998) 807–825.
16. Pyragas K. Continuous control of chaos by self-controlling feedback, Phys Lett A., 1992; 170:421-428.
17. Calvo O, Cartwright JHE. Fuzzy control of chaos. Int J Bifurcat Chaos 1998; 8:1743–7.
18. Chen L, Guanrong Chen G., Lee YW. Fuzzy modeling and adaptive control of uncertain chaotic systems, Information Sciences 1999;121:27-37.
19. Guran, A. Stabilizing periodic orbits of chaotic regimes in a Kelvin type gyrostat satellite, In: Engineering Mechanics (Jiri Naprstek and Cyril Fischer, eds), Academy of Sciences of the Czech Republics, 2009, pp. 70-71.
20. Moser J. Stable and Random Motions in Dynamical Systems, Princeton University Press, Princeton, NJ, 1973.
21. Stephen Wiggins, Introduction to Nonlinear Dynamical Systems and Chaos, Springer-Verlag, NY, 1990.
22. Li-Xin Wang, A Course in Fuzzy Systems and Control, Prentice Hall, NJ, 1997.
23. S.V. Kovalevskaya, Sur le probleme de la rotation d'un corps solide autor d'un point fixe. Acta. Math. 12 (1889).
24. V. Volterra, Sur la theories des variations des latitudes, Acta Math. 22 (1899).
25. N. Ye. Zhukovsky, The motion of a rigid body having cavities filled with a homogeneous capillary liquid. In Collected Papers, Vol. 2. Gostekhizdat, Moscow (in Russian) (1949) 152-309.
26. A. Wangerin, Über die bewegung miteinander verbundener körper, Universitäts-Schrift Halle, 1889.
27. P.W. Likins, Spacecraft Attitude Dynamics and Control - A Personal Perspective on Early Developments, J. Guidance Control Dyn. Vol. 9, No. 2 (1986) 129-134.
28. G.J. Cloutier, Stable Rotation States of Dual-Spin Spacecraft, Journal of Spacecraft and Rockets, Vol. 5, No. 4 (1968) 490-492.
29. P.M. Bainum, P.G. Fuechsel, D.L. Mackison, Motion and Stability of a Dual-Spin Satellite With Nutation Damping, Journal of Spacecraft and Rockets, Vol. 7, No. 6 (1970) 690-696.

30. V. Schlegel, A. Guran, F. P. J. Rimrott, An elastic oblate axisymmetric reference gyro for attitude studies,Proceedings of the Twelfth Canadian Congress of Applied Mechanics (1989) 396-397.
31. A. Guran, F.P.J. Rimrott and F. Sharifi, On the stability of viscoelastic spinning columns, J. Structures and Machines, vol. 19, no. 4 (1991) 437–455.
32. A. Guran, Classification of singularities of a torque-free gyrostat satellite, Mechanics Research Communications, vol. 19, no5 (1992)465-470.
33. K.J. Kinsey, D.L. Mingori, R.H. Rand, Non-linear control of dual-spin spacecraft during despin through precession phase lock, J. Guidance Control Dyn. 19 (1996) 60-67.
34. C.D. Hall, Escape from gyrostat trap states, J. Guidance Control Dyn. 21 (1998) 421-426.
35. C.D. Hall, Momentum Transfer Dynamics of a Gyrostat with a Discrete Damper, J. Guidance Control Dyn., Vol. 20, No. 6 (1997) 1072-1075.
36. A.E. Chinnery, C.D. Hall, Motion of a rigid body with an attached spring-mass damper, J. Guidance Control Dyn. Vol. 18, No. 6 (1995) 1404-1409.
37. A.I. Neishtadt, M.L. Pivovarov, Separatrix crossing in the dynamics of a dual-spin satellite, J. Appl. Math. Mech., V. 64 (5) (2000) 709-714.
38. M. Inarrea, V. Lanchares, Chaotic pitch motion of an asymmetric non-rigid spacecraft with viscous drag in circular orbit, Int. J. Non-Linear Mech. 41 (2006) 86-100.
39. V. Lanchares, M. Icarrea, J.P. Salas, Spin rotor stabilization of a dual-spin spacecraft with time dependent moments of inertia, Int. J. Bifurcation Chaos 8 (1998) 609-617.
40. A. Guran (Editor), Nonlinear Dynamics, Richard Rand 50th Anniversary volume, World Scientific, NY, 1997.
41. A. Guran, A. Bajaj, Y. Ishida, N. Perkins, G. D'Eluterio, C. Pierre, Stability of Gyroscopic Systems, World Scientific, NY, 1999.
42. J. Kuang, S. Tan, K. Arichandran, A.Y.T. Leung, Chaotic dynamics of an asymmetrical gyrostat, Int. J. Non-Linear Mech. 36 (2001) 1213-1233.
43. F.R. Gantmaher, L.M. Levin, The theory of rocket uncontrolled flight, Fizmatlit, Moscow, 1959 (in Russian).
44. V.S. Aslanov, A.V. Doroshin, About two cases of motion of unbalanced gyrostat. Izv. Ross. Akad. Nauk. MTT. V.4 (2006) 42-55. (Transactions of the Russian Academy of Sciences: Mechanics of solids. In Russian).
45. V.S. Aslanov, A.V. Doroshin, Stabilization of the descent apparatus by partial twisting when carrying out uncontrolled descent. Cosmic Research, Vol. 40, No. 2 (2002)193-200.
46. V.S. Aslanov, A.V. Doroshin, G.E. Kruglov, The motion of coaxial bodies of varying composition on the Active Leg of descent, Cosmic Research, Vol. 43, No. 3 (2005) 213-221.
47. V. S. Aslanov, A. V. Doroshin, The motion of a system of coaxial bodies of variable mass, J. Appl. Math. Mech. Vol. 68 (2004) 899-908.
48. A.V. Doroshin, Phase space research of one non-autonomous dynamic system, Proceedings of the 3rd WSEAS/IASME International Conf. on Dynamical System and Control. Arcachon, France (2007) 161-165.
49. A. Guran, Control of chaos in a spinning kelvin type gyrostate satellite under the influence of gravitational force field, The 3rd International IEEE Scientific

Conference on Physics and Control (PhysCon 2007), University of Potsdam, Germany(2007) 118.
50. A.V. Doroshin, Synthesis of attitude motion of variable mass coaxial bodies, WSEAS Transactions on Systems and Control archive, Volume 3 , Issue 1 (2008) 50-61.

Index

adaptive feedback adjustment, 45
Aihara's chaotic neural network (CNN) model, 23
anti-control, 197, 198, 225, 226
anti-control of chaos, 73, 198, 224
anti-controlled, 226
associative memory, 22, 26
associative memory dynamics, 24–26, 35
attractors, 190
autocorrelation function, 33, 34

basin of attraction, 134, 375
beam, 194, 199–201
best shape, 227
bifurcation, 380
bifurcation diagram, 271, 275, 379
bifurcation structure, 45
bouncing ball map, 317
brain, 292
breather, 153
BT-maps, 62
buckled beam, 189, 201, 202, 227
bursting frequency, 296
bursting neurons, 292, 295
bursting phases, 300
BZ reaction model, 55

Cardan angles, 409
central manifold, 350
chaos, 374
chaos (anti-control), 191
chaos control, 190, 191, 193–196, 199, 202, 339, 372, 383
chaos controller, 196

chaotic, 199
chaotic attractor, 192, 202, 207, 217, 219, 375
chaotic behavior, 190–192, 196, 202
chaotic dynamics, 190, 197, 212, 371
chaotic evolution, 372
chaotic fluctuations, 198
chaotic in the sense of Devaney, 76, 95
chaotic in the sense of Li-Yorke, 76
chaotic in the sense of Wiggins, 77
chaotic invariant, 372
chaotic motion, 196, 198, 201, 202, 409
chaotic neural network model, 22
chaotic neuron map, 50
chaotic phenomena, 194, 203
chaotic properties, 190, 192
chaotic regimes, 195
chaotic region, 193, 228
chaotic response, 194, 195
chaotic saddle, 315
chaotic sets, 202
chaotic trajectories, 199
chaotic transients, 190
chaotic zones, 192
chaotification, 73
Chile, 277
Chua's circuit, 259
city, 269
clipping, 258
clustering, 409
CMOS based VLSI, 265
CMT technique, 351
complex behavior, 272
Constant Feedback Method, 47

427

control of chaos, 73, 103, 189, 190, 193, 195–198, 200, 201, 291
control strategy, 284
control term, 3
control through time-delayed feedback, 308
controlling chaos, 25, 190
coupled-expanding map, 78
crisis, 272
crisis-free, 50
crisis-induced intermittency, 159
critical behavior, 275
criticality, 276
cross-well attractors, 216
cross-well chaos, 206, 217, 219

database, 248
deep-brain stimulation, 293
degree of oblateness, 416
delay feedback method, 31
delayed feedback control algorithm, 141
dense set, 382
deterministic system, 374
diffusion of test particles, 16
discontinuities, 58
double-inverted pendulum, 198
double-well oscillator, 130
drag, 270
Duffing equation, 200, 215, 217
Duffing model, 228
Duffing oscillator, 193, 194, 200, 215
Duffing-like oscillators, 199
Duffing-Ueda, 200
dynamic range, 239
dynamics, 199

$\mathbf{E} \times \mathbf{B}$ drift, 1
eigenvalues, 384
eigenvectors, 384
electrical activity, 293
elliptic orbits, 396
encoding information, 247
equilibrium, 374
escape, 201, 202, 211, 212, 215
escape time, 174

escapes, 174
excitation shape, 201, 205, 222
excitations, 210
expanding fixed point, 77
exponential map, 50
extended delayed feedback control (EDFC), 107
extended Sitnikov problem, 404

feedback and nonfeedback methods, 147
feedback control, 74, 357
feedback gain, 106
Feingenbaum number, 367
fixed point, 75, 95, 416
Floquet exponents (FEs), 111
Floquet multiplier (FM), 106
frequency locking, 302
frequency of control, 255
fuel, 270
function of stabilization, 401
fuzzy control, 409

game, 285
global, 207, 228
global anti-control, 225
global control, 208, 213, 215, 217, 218, 220–222, 346
green wave, 275
gyrostat, 409

Hamming distance, 28, 29, 38, 39
Helmholtz equation, 203
Helmholtz oscillator, 199, 211, 213, 215, 223
Helmholtz-Duffing, 199, 219
Helmholtz-Duffing oscillator, 220
heteroclinic, 224
heteroclinic bifurcation, 189, 193, 202, 205, 225, 229
heteroclinic chaos, 196
heteroclinic intersections, 202
heteroclinic orbits, 398
higher-order control terms, 7
hilltop saddle, 204, 205, 212, 213, 215, 216, 219, 220, 227

homoclinic, 189
homoclinic bifurcation, 201, 202, 207, 208, 211–216, 218, 219, 221, 222, 224, 227, 228
homoclinic intersection, 206, 212
homoclinic intersections, 333
homoclinic loop, 213
homoclinic solutions, 228
homoclinic tangencies, 221
horseshoe map, 327
hyperbolic object, 343
hyperbolic orbits, 397
hyperchaotic circuit, 257

impatient, 280
impulsive control, 55
induced map, 95
induced system, 95
information processing, 22, 26, 30, 40, 41
information storage, 246
invariant set, 75
invariant tori, 399
inverted pendulum, 198, 204, 205, 209, 211–213, 215, 219
invertible linear map, 80

jams, 267
Jerk equations, 257

KAM-theory, 3, 399
Kelvin-type gyrostat, 412

largest curves, 50
largest Lyapunov exponent, 25, 32
laser, 255
linear feedback, 357
local control, 340
locally stable attractors, 53
logistic map, 49, 136, 316
look-up table, 245
Lorenz attractor, 250
Lorenz chaotic attractor, 375
Lorenz equations, 374
Lorenz system, 111
Lyapunov exponent, 272, 376

map, 316
mathematical pendulum, 196, 219
maximum line, 49
mean-field, 300
Melnikov chaos, 201
Melnikov function, 212
Melnikov method, 204, 205, 213, 217, 225, 227, 399
Melnikov predictions, 215
Melnikov technique, 196, 197
microbeam, 222, 223
mod-operation, 82, 96
multi-dimensional systems, 250

Nejmark-Sacker (discrete Hopf) bifurcation, 112
neuronal spikes, 250
noise, 316
non-hilltop saddle, 211
non-resonant operator, 5
nonlinear dynamical system, 373
nonlinear pendulum, 398
nonrobustness, 41

odd number limitation, 111
OGY method, 192, 193, 197, 200, 291, 372
OGY Pyragas controllers, 410
OGY strategy, 199
OGY technique, 196, 339
OGY-based methods, 196
one-side, 215, 225, 228
one-side anti-control, 226
one-side control, 207, 216, 217, 220
one-side optimal control, 210
one-side optimal excitation, 208, 216
optimal, 228
optimal chaos control, 350
optimal control, 189, 201, 208, 213, 223, 224
optimal excitation, 206, 208–210, 213–216, 219
optimal impulses, 210
optimal problem, 213
optimal solution, 206, 213, 222
optimal value, 223

optimal way, 189
optimal-harmonic, 210
optimal-optimal, 210
optimality, 190
optimally, 200
optimization, 228
optimization problem, 189, 206, 209, 213, 215, 220–222, 225, 228, 229
optimized, 229
optimizing, 228
orbit, 75
order parameter, 303
overturning, 198, 225, 226

parabolic orbits, 397
parameter estimation, 416
parameter modulated method, 29
Parkinson's disease, 293
Parrondo paradox, 285
partial control, 316
pathological rhythms, 293
pattern recognition, 26
pendulum, 194, 196, 197, 210
period doubling, 272
period locking, 51
period-doubling bifurcation, 65, 111
period-doubling cascade, 380
periodic boundary condition, 91
periodic orbits, 371
periodic point, 75, 95
periodic windows, 47
phase control, 148
phase control scheme, 161
phase portrait, 398
phase shifts, 271
phase space constraint control, 27
phase synchronization, 300
phase transition, 280
Pinning method, 26, 27
Poincaré section, 378, 409
pole placement technique, 340
potential well, 191, 202, 210, 220
power-law coupling, 299
pre-images of periodic solutions, 63
problem of optimization, 205, 219, 221

processing information, 248
proportional feedback control (PFC), 109
pulsive feedback, 356

quarter car, 357

Rössler model, 55
Rössler system, 111
refractoriness, 23, 27, 28, 32
resonance, 271, 275
resonant operator, 5
restricted three-body problem, 396
return map, 384
robustness against noise, 59
rolling friction, 270
Rulkov, 294, 295

saddle points, 398
saddle-node (tangent) bifurcations, 65
saddle-node bifurcation, 380
safe sets, 316
sawtooth functions, 84, 98
scale-free networks, 292, 297
scaling laws, 277
scaling region, 273
scanning interval, 53
scrambled set, 76
search, 248
sensitive dependence on initial conditions, 75
sensitivity to initial conditions, 372
separatrix, 398
separatrix layer, 399
shape of the excitation, 189, 192, 198, 200, 206, 220, 227–229
single-humped maps, 49
single-well, 206, 217
single-well Duffing oscillator, 201
Sitnikov problem, 396
small-world, 294
smart matter, 253
snap-back repeller, 77
stabilized orbits, 397
stable manifold, 384
state space, 371

strange attractor, 374
subcritical Hopf bifurcation, 111
sup-norm, 81
superstable, 61, 242
suppression of chaos, 190
synaptic weight, 24
synchronization, 271, 293

2 orbit, 242
targeting techniques, 347
temporal shape, 228
tent, 316
the largest value, 48
three-body problem, 395
threshold coupling method, 28
threshold mechanism, 239
time delay feedback, 26
time-periodic signal, 305
topological conjugate, 76
topologically transitive, 75
tracking, 56

traffic light, 269
transient chaos, 315
transient dynamics, 281
transient processes, 46
transients, 284
transition to chaotic phase
    synchronization of bursting, 305
transverse heteroclinic points, 396
triangular Lagrange points, 404
turbulent map, 78
two-well, 220
two-well Duffing oscillator, 201
two-well potential, 207

universal, 277
unstable manifold, 384
unstable periodic orbits, 45, 103, 338, 372

van der Pol oscillator, 120
visualization, 46